GROUP AND CROWD BEHAVIOR FOR COMPUTER VISION

Computer Vision and Pattern Recognition Series

Series Editors

Horst Bischof   Institute for Computer Graphics and Vision, Graz University of Technology, Austria

Kyoung Mu   Department of Electrical and Computer Engineering, Seoul National University, Republic of Korea

Sudeep Sarkar   Department of Computer Science and Engineering, University of South Florida, Tampa, United States

Also in the Series:

Lin and Zhang, Low-Rank Models in Visual Analysis: Theories, Algorithms and Applications, 2017, ISBN: 9780128127315

Zheng et al., Statistical Shape and Deformation Analysis: Methods, Implementation and Applications, 2017, ISBN: 9780128104934

De Marsico et al., Human Recognition in Unconstrained Environments: Using Computer Vision, Pattern Recognition and Machine Learning Methods for Biometrics, 2017, ISBN: 9780081007051

Saha et al., Skeletonization: Theory, Methods and Applications, 2017, ISBN: 9780081012918

GROUP AND CROWD BEHAVIOR FOR COMPUTER VISION

Edited by

VITTORIO MURINO

MARCO CRISTANI

SHISHIR SHAH

SILVIO SAVARESE

ACADEMIC PRESS

An imprint of Elsevier

Academic Press is an imprint of Elsevier
125 London Wall, London EC2Y 5AS, United Kingdom
525 B Street, Suite 1800, San Diego, CA 92101-4495, United States
50 Hampshire Street, 5th Floor, Cambridge, MA 02139, United States
The Boulevard, Langford Lane, Kidlington, Oxford OX5 1GB, United Kingdom

Notices

Knowledge and best practice in this field are constantly changing. As new research and experience broaden our understanding, changes in research methods, professional practices, or medical treatment may become necessary.

Practitioners and researchers must always rely on their own experience and knowledge in evaluating and using any information, methods, compounds, or experiments described herein. In using such information or methods they should be mindful of their own safety and the safety of others, including parties for whom they have a professional responsibility.

To the fullest extent of the law, neither the Publisher nor the authors, contributors, or editors, assume any liability for any injury and/or damage to persons or property as a matter of products liability, negligence or otherwise, or from any use or operation of any methods, products, instructions, or ideas contained in the material herein.

Library of Congress Cataloging-in-Publication Data
A catalog record for this book is available from the Library of Congress

British Library Cataloguing-in-Publication Data
A catalogue record for this book is available from the British Library

ISBN: 978-0-12-809276-7

For information on all Academic Press publications visit our website at
https://www.elsevier.com/books-and-journals

Working together
to grow libraries in
developing countries

www.elsevier.com • www.bookaid.org

Publisher: Joe Hayton
Acquisition Editor: Tim Pitts
Editorial Project Manager: Naomi Robertson
Production Project Manager: Kiruthika Govindaraju
Cover Designer: Christian Bilbow

Typeset by VTeX

CONTENTS

4.   Exploring Multitask and Transfer Learning Algorithms for Head Pose Estimation in Dynamic Multiview Scenarios   67

Elisa Ricci, Yan Yan, Anoop K. Rajagopal, Ramanathan Subramanian, Radu L. Vieriu, Oswald Lanz, Nicu Sebe

5.   The Analysis of High Density Crowds in Videos   89

Mikel Rodriguez, Josef Sivic, Ivan Laptev

Part 3.   Metrics, Benchmarks and Systems

13. Integrating Computer Vision Algorithms and Ontologies for Spectator Crowd Behavior Analysis   297
Davide Conigliaro, Roberta Ferrario, Céline Hudelot, Daniele Porello

14. SALSA: A Multimodal Dataset for the Automated Analysis of Free-Standing Social Interactions   321
Xavier Alameda-Pineda, Ramanathan Subramanian, Elisa Ricci, Oswald Lanz, Nicu Sebe

ABOUT THE EDITORS

Vittorio Murino is a Full Professor at the University of Verona, Italy, and Director of the PAVIS (Pattern Analysis and Computer Vision) department at the Istituto Italiano di Tecnologia. He received the Laurea degree in Electronic Engineering in 1989 and the PhD in Electronic Engineering and Computer Science in 1993 from the University of Genova, Italy.

His main research interests include computer vision, pattern recognition, and machine learning, and more specifically, statistical and probabilistic techniques for image and video processing, with applications in (human) behavior analysis and related applications such as video surveillance, biomedical imaging, and bioinformatics.

Prof. Murino is a coauthor of more than 400 papers published in refereed journals and international conferences, and a member of the editorial board of Computer Vision and Image Understanding, Machine Vision & Applications, and Pattern Analysis and Applications journals.

Finally, he is a Senior Member of the IEEE and Fellow of the IAPR.

Marco Cristani is an Associate Professor at the University of Verona, External Collaborator at the Istituto Italiano di Tecnologia, Genova, Italy and an Associate Member at the National Research Council. After receiving his Master's Degree (having developed his thesis partially at the Instituto Superior Tecnico, Lisbon under the supervision of Prof. Mario Figueiredo), he obtained his PhD in Computer Science at the  Department of Computer Science at the University of Verona in 2006. During the period of 2004–2005 he was a Research Scholar at the University of Southern California, under the guidance of Prof. Ram Nevatia. He has been and is currently, a scientific collaborator of a number of national and European projects. His research interests include statistical pattern recognition, and especially generative modeling, applied to video surveillance, social signaling, and multimedia in general.

Shishir Shah is a Professor and Associate Chair in the Department of Computer Science at the University of Houston, USA. He studied Mechanical Engineering as an undergraduate at The University of Texas at Austin, where he received his BS degree in 1994. He received his MS and PhD degrees in Electrical and Computer Engineering from The University of Texas at Austin. His main research interests include

computer vision, pattern recognition, and machine learning with specific focus on human behavior modeling and analysis, scene understanding, video analytics, biometrics, and microscopy image analysis. Over the past 20 years, his research has contributed to knowledge in the areas of computer vision for navigation, surveillance, object recognition, biomedical image analysis, and pattern recognition. He has authored numerous papers on object recognition, sensor fusion, statistical pattern analysis, biometrics, and video analytics. Prof. Shah currently serves as an Associate Editor for Image and Vision Computing and the IEEE Journal of Translational Engineering in Health and Medicine. He is a Senior Member of the IEEE.

Silvio Savarese is an Assistant Professor of Computer Science at Stanford University. He earned his PhD in Electrical Engineering from the California Institute of Technology in 2005 and was a Beckman Institute Fellow at the University of Illinois at Urbana–Champaign during 2005–2008. He joined Stanford in 2013 after being an Assistant and then Associate Professor (with tenure) of Electrical and Computer Engineering  at the University of Michigan, Ann Arbor, from 2008 to 2013. His research interests include computer vision, object recognition and scene understanding, shape representation and reconstruction, human activity recognition and visual psychophysics. He is a recipient of several awards including a Best Student Paper Award at CVPR 2016, the James R. Croes Medal in 2013, a TRW Automotive Endowed Research Award in 2012, an NSF Career Award in 2011, and Google Research Award in 2010. In 2002 he was awarded the Walker von Brimer Award for outstanding research initiative.

CHAPTER 1

The Group and Crowd Analysis Interdisciplinary Challenge

Vittorio Murino*, Marco Cristani†, Shishir Shah‡, Silvio Savarese§
*Pattern Analysis and Computer Vision (PAVIS), Istituto Italiano di Tecnologia, Genova, Italy
†University of Verona, Verona, Italy
‡University of Houston, Houston, TX, USA
§Stanford University, Stanford, CA, USA

Contents

1.1 THE STUDY OF GROUPS AND CROWDS

Understanding activities and human behavior from images and videos is an active research area in computer vision and has a large impact to many real-world applications. These include surveillance, assistive robotics, autonomous driving, data analytics, to cite a few. The research community has put significant focus on analyzing the behavior of individuals and proposed methods that can understand and predict behavior of humans as they are considered in isolation. More recently, however, the attention has shifted to the new issues of analyzing and modeling gatherings of people, commonly referred as groups or crowds, depending on the number of people involved. The research done on these two topics has brought about many diverse ad-hoc methodologies and algorithms, and has led to a growing interest in this topic. This has been supported by multiple factors. Firstly, the advancement of the detection and filtering strategies running on powerful hardware has encouraged the development of algorithms able to deal with hundreds of different individuals, providing results that were unthinkable just few years ago. Concomitantly, there has been a broader availability of new types of sensors, and the possibility of mounting these sensors on cutting edge devices, from glasses to drones. Such sensory devices have made it possible to observe people from radically different points of view, in a genuine ecologic, noninvasive manner, and for long durations, namely, from ego-vision settings to bird-eye views of people. Moreover, the advancement of social signal processing [1,2] has brought in the computer vision and pattern recognition community new models imported from the social sciences, able to read between-the-lines of simple locations and velocities assumed by the individuals,

Group and Crowd Behavior for Computer Vision
DOI: 10.1016/B978-0-12-809276-7.00001-1

using advanced notions of proxemics and kinesics [3,1]. Finally, the industry, governments, and small companies are asking our community for methods to understand and model groups and crowd, for public order and safety, social robotics, advanced profiling and many other applications.

The study of groups and crowds has generally been considered as having its roots in sociology and psychology. Human behavior, in general, has been extensively studied by sociologists to understand social interactions and crowd dynamics. It has been argued that characteristics that dictate human motion constitute a complex interplay between human physical, environmental, and psychosocial characteristics. It is a common observation that people, whenever free to move about in an environment, tend to respect certain patterns of movement. More often, these patterns of movement are dominated by social mechanisms [3]. The study of groups and crowds from the computer vision perspective has typically been modeled as a three-level approach. At the low level, given a video, humans are detected [4,5], then tracked [6,7], and then tracklets are grouped to form trajectories [8]. At the mid-level, machine learning techniques are used to identify groups by clustering trajectories [9]. At the higher level, a semantic understanding of the group behavior is obtained, like classifying actions such as "walking in groups", "protesting", "group vandalism", etc.

The low level algorithms have been widely studied in computer vision [10–12] with promising results. However, algorithms at the middle and high level have only been explored in recent times.

Algorithms at the higher level can either explicitly model human behavior and their interactions in the group and with the environment, or a model can be created through observations by assuming that the human behavior is encapsulated in the learning process. Khan and Shah [13] observed and learned a group's rigid formation structure to classify the activity and successfully applied it to parades. Ryoo and Aggarwal [14,15] represented and learned various types of complex group activities with a programming language-like representation, and then recognized the displayed activities based on the recognition of activities of individual group members. On the other hand, human behavioral models can be used to predict the human interactions with each other. Helbing and Molnar [16] proposed the social force model, which assumes humans as particles and models the influence of other humans and the environment as forces. Furthermore, Pellegrini et al. [17] and Choi et al. [18], as well as [19,20], proposed models that anticipate and avoid collisions of a human with other humans and the actual scene physical structure. These models assume that the humans partaking in the group follow the existing social norms and hence can be used to model specific categories of people, and even crowds.

Typically, in crowded scenes, people are engaged in multiple activities resulting from inter- and intra-group interactions. This poses a rather challenging problem in analyzing group events due to variations in the number of people involved, and more specifically the different human actions and social interactions exhibited within people and groups [21–24]. Understanding groups and their activities is not limited to only analyzing movements of individuals in group. The environment in which these groups exist provides important contextual information that can be invaluable in recognizing activities in crowded scenes [25,26]. Perspectives from sociology and psychology embedded into computer vision algorithms show that human activities can be effectively understood by considering implicit cognitive processes as latent variables that drive positioning, proximity to other people, movement, gesturing, etc. [27,16,28–30]. For example, exploring the spatial and directional relationships between people can facilitate the detection of social interactions in a group. Thus, activity analysis in low-density crowded scenes can often be considered a multistep process, one that involves individual person activity, individuals forming meaningful groups, interaction between individuals, and interactions between groups [28]. In general, the approaches to group activity analysis can be classified into two categories: bottom-up and top-down. The bottom-up (BU) approaches rely on recognizing activity of each individual in a group. Vice versa, top-down (TD) methods recognize group activity by analyzing at the group level rather than at the individual level. Since BU algorithms address the understanding of activities at the individual level, they are limited in recognizing activities at the group level. Conversely, TD approaches show better contextual understanding of activities of a group as a whole, but they are not robust enough to recognize activities at the individual level.

On the other side, when the density of people becomes too high, also in dependence of the camera perspective view, individuals and even groups cannot be distinguished anymore, and a more holistic analysis should be performed to figure out the behavior of a *crowd*. Analyzing crowd scenes can be categorized into three main topics, i.e., (i) crowd density estimation and people counting, (ii) tracking in crowd, and (iii) modeling crowd behaviors [31]. Recently, some works on pedestrian path predictions in crowded scenes were also proposed [32]. The goal of these methods is to predict the pedestrian pathway in advance, given the past walking history and the surrounding environment (obstacles, scene geometry, etc.). This is yet another interesting application in crowd scenarios having the aim of, for instance, estimating entry/exit points in a specific area or to find the main people walking pathways or standing areas, so that this information can be possibly used to set up open spaces.

Estimating the number of people in a crowd is a cardinal stage for several real-world applications such as safety control, monitoring public transportation, crowd simulation and rendering for animation and urban planning. Many interesting works are present in the literature addressing this target [33–35], however, automated crowd density estimation still remains an open problem in computer vision due to extreme occlusion conditions and visual ambiguities of the human appearances in such scenarios [36].

Tracking individuals (or objects) in crowd scenes is another challenging task [37,38] which involves, other than severe occlusions, cluttered background and pattern deformations, which are common complexities in visual tracking. In practice, the efficiency and effectiveness of crowd trackers is largely dependent on crowd density and dynamics, people social interactions as well as the crowd's psychological characteristics [39,40, 36].

Typically, the primary goal of modeling crowd behaviors is to allow the identification of abnormal events such as, for instance, riots, panic, and violence acts [41]. Despite recent works in this direction, detecting crowd abnormalities still remains an open and challenging problem mainly because of the "loose" definition of abnormality which is strongly context dependent [42,43]. For example, riding a bike in a street is a normal action, whereas it may be considered abnormal in another scene with a different context such as a park or a sidewalk. Similarly, people gathering for a social event is typically a normal situation, while a similar gathering to "protest against something" can be an abnormal event, which may deserve attention and needs to be detected. Several methods have been devised to analyze crowd behavior. One of the most influential works still derived from the Social Force Model (SFM) [16], and was proposed by Mehran et al. [44]. It adopted the SFM and particle advection scheme for detecting and localizing abnormal behavior in crowd videos. To this end, it considered the entire crowd as a set of moving particles whose interaction force was computed using SFM. The interaction force mapped onto the video frames identifies the force flow of each particle and is used as the basis for extracting features which, along with a bag-of-words strategy, is used to classify each frame as either normal or abnormal. Another interesting work can be found in [45], where a general framework able to measure the degree of collectiveness for different crowd systems was proposed. It was able to efficiently estimate the structural properties of the collective manifold in various types of crowds. Collectiveness was defined as the degree of individuals acting as a union in collective motion, and this was used to estimate and categorize crowd dynamics. Other works can be quoted from the relatively few labs in the world addressing such issues, indeed presenting their works in this book, but this is out of the scope of this chapter. A more accurate and up-to-date analysis of the state of the art can be found in some of the following chapters related to crowd behavior.

In conclusion, it is worth to note that the main challenge of this domain stems from the lack of adequate training samples, both synthetic and in-the-wild, to allow a proper learning of a crowd model. This drastically limits the generalization capability of current crowd models, since they remain not capable of capturing the large intra–class variations of crowd behaviors [31], or conversely it is difficult to generalize such models to different domains or scenarios [46].

1.2 SCOPE OF THE BOOK

The aim of this book is to communicate to a wide range of researchers in computer vision, machine learning on the one side, and applied social sciences on the other, innovative ideas and solutions for exploiting the potential synergies emerging from the study of groups and crowds, for a range of different applications and areas. Some of the contributing ideas for this book emerged from discussions that took place at the *Workshop on Group and Crowd Behavior Analysis and Understanding* (GROW 2015), held in conjunction with the 2015 Computer Vision and Pattern Recognition Conference.

Contributions and discussion in the book address many unresolved issues in the computer vision community, like: What is a group and what is a crowd? Is it true that the difference among them is a matter of number of people involved? Are there cues other than the spatial proximity and oriented velocity that could be used to detect them? Are there different types of groups and crowds? For example, in the sociological literature, social studies defined different kinds of crowd (spectator crowd, casual crowd, protest crowds, and others), and it could be interesting to understand whether these definitions could have some computational counterpart [47].

The chapters in this book are grouped by specific computational stages that address specific challenges related to the analysis and modeling of groups and crowds.

The first part of the book presents relevant features and representations, with the aim of recognizing the presence of groups and crowds in the input data (images, video). This part is oriented to individuate and discuss low-level processing methods applicable for detection of when and where a group or crowd is placed in the scene, and spans from the use of people detectors toward more ad-hoc strategies to individuate group and crowd formations. In particular, Chapter 2 focuses on sociological notions that can be embedded into computer vision methods. It offers, first, an overview on the sociological understandings of social interaction, thereby proposing a distinction between different kinds of interaction, namely *unfocused*, *common-focused*, and *jointly-focused*. Second, drawing also on such a typology, it presents a taxonomy of small to large social gatherings – that is, groups and crowds – and provides some clarifying examples. Chapter 3 discusses the most common features and definitions from the sociological science used to detect and track groups of people that are interacting. In particular, two types of cues are discussed: the low-level cues used to determine the spatial properties of each person in a scene (person position and head/body orientation), and the high-level features that agglomerate or use the low-level cues to implement sociological and biological definitions (frustum of visual attention). These cues are then exploited to detect and track groups of people, offering a comprehensive comparative analysis. Chapter 4 reviews recent works addressing human pose estimation under target motion. In particular, the chapter examines the use of transfer learning and multitask learning for head pose estimation. Transfer learning is particularly useful when the training and the test data have different attributes (e.g., training data contains pose annotations for static targets, but test data involves moving targets), while multitask learning can be explicitly designed to address facial appearance variations under motion. Chapter 5 first provides a broad overview of a host of recent computer vision studies that have made tremendous progress in the ability to effectively understand and predict the behavior of individuals in high-density crowded scenes. Many of these studies incorporate various contextual constraints that can be used to tackle the challenges of high-density crowd behavior understanding. Chapter 6 scales up the analysis to more than hundred million individuals in crowded urban spaces. Thanks to one of the largest networks of cameras (more than hundred cameras per site) that capture the trajectories of pedestrians in crowded train terminals, up to a thousand pedestrians has been tracked simultaneously. The chapter gives insight on how this has been possible in terms of features and tracking methodologies. Chapter 7 presents a subject centric group feature, explicitly suited for person reidentification. The feature has been inspired by the observation that people often tend to walk alongside others or in a group. Co-travelers' information, including geometry and visual cues, can reduce the reidentification ambiguity and lead to better accuracy.

The second part of the book discusses methods for analyzing the behavior of groups and crowd. Building on methods for detecting groups and crowds, these methods focus on how to extract semantic information from them. Predicting and tracking the movement of a group, the formation or disaggregation of a group/crowd, together with the identification of different kinds of groups/crowd depending on their behavior, are some of the specific questions being addressed. In particular, Chapter 8 discusses the role of leaders in group and crowd scenarios, starting from sociological findings. The aim is to provide a learning framework that can be useful for the visual recognition of both groups and associated leaders, specifically casting it into a structured learning framework, which can be solved efficiently using Structural SVMs. To this end, the roles of leaders in forming and structuring a group and the mutual influence groups and leaders have for their automatic visual detection has been demonstrated through empirical evidence and original experiments. Chapter 9 presents two families of methods to forecast human trajectories in crowded environments. The first one is based on the popular Social Force Model, where the causalities behind human navigation are hand-designed by a set of functions properly defined given the knowledge of the underlying physics-based social behavior. The second method is a fully data-driven approach based on Recurrent Neural Networks not imposing any hand-designed functions or explicit mobility based constraints. For the latter, Long–Short Term Memory (LSTM) networks have been properly customized to implement the capability to capture the dependencies between multiple correlated sequences (trajectories), specifically introducing a "social" pooling layer which allows the LSTMs of spatially proximal sequences to share their hidden states with each other. This architecture, named as "Social-LSTM", can automatically learn typical interactions that take place among trajectories which coincide in time. Chapter 10 addresses a typical issue of many proposed algorithms, that is, the specificity of the domain for which the algorithms are designed for (a.k.a. dataset bias) [46]. In other words, many existing works on crowd scene understanding are scene-specific, i.e., models learned from a particular scene cannot be well applied to other scenes. This is clearly a serious drawback since it limits the application of these technologies, and obliges the collection of extra training samples from a new scene, which is often unfeasible or unpractical. In this context, Chapter 10 introduces scene-independent crowd analysis using deep learning. Once generic deep models are learned from large-scale training sets, they can be applied to various crowd scenes without the need to be trained again. The tasks addressed by this approach are diverse and cover crowd density estimation, crowd counting, and crowd attribute recognition. Also, several large-scale datasets introduced recently are discussed together with issues related to annotation and scene diversity/variability. Finally, multiple architectures of deep neural networks and training strategies to learn the feature representations for crowd analysis are also illustrated. Chapter 11 presents a comprehensive study on crowd behavior understanding, specifically addressing the detection of crowd behavior anomalies. The focus is on providing a broad overview of most recent physics-based approaches along with an extensive experimentation to show the strengths and weaknesses of each proposed method. Social Force based, flow fields, energy based and substantial derivative models are introduced and reviewed to show their characteristics, performances, and comparative analyses in the task of detection of violence and panic events in video sequences, considering all the public available benchmark datasets. Chapter 12 focuses on human activity forecasting using a decision-theoretic approach. Casting the process of visual perception in a 2-layer model, recognition and prediction, Kitani et al. first describe the underlying theory that defines a (classic AI-based) decision-theoretic model in terms of human activity forecasting. Markov decision process, optimal control and inverse optimal control, and their maximum entropy variants are the topics covered, followed by three concrete examples of how the decision-theoretic approach can be applied to the task of visual activity forecasting. The first approach is addressing the inference of future trajectories of people taking into account the effects of the physical environment on the choice of the human actions. The second method also considers the influence of other human actions on the environment and again its reflex to the human activity to predict human interactions. Finally, a dual-agent interaction model is also introduced, casted as an optimal control problem.

The third part introduces the most relevant datasets and also proposes some figures of merits and metrics to evaluate the crowd behavior analysis algorithms proposed in the literature. While there have been some interesting datasets that have been proposed for various tasks related to groups and crowds, such as detection, tracking, and understanding activity, datasets for modeling and analyses of crowds in specific situations have been limited. Chapter 13 discusses the S-HOCK dataset [47] that provides annotated video acquired during a hockey competition, specifically to study spectator crowds. More-over, specific ontology is described to study spectator crowds and example applications are presented. Chapter 14 introduces *SALSA*, a multimodal dataset to free-standing conversational groups. The dataset includes video recording of 18 subjects over 60 min along with data acquired from additional sensors that were worn by each subject to cap-ture their motion, auditory content of their conversations, as well as their interactions with others in the scene. The discussion and analyses presented in the chapter highlight some of the challenges and research gaps in analyzing social scenes. While datasets are clearly important for benchmarking multiple algorithms and development of new mod-els to understand human behavior, it will never be possible to have a dataset for every possible behavior. Chapter 15 discusses this particular challenge and presents a general-izable model developed on well annotated common crowd attributes that can then be deployed to recognize novel and previously unseen crowd behaviors or attributes. The lack of metrics to specifically evaluate algorithms for the study of groups and crowds is an evident weakness of this research area. While several algorithms have been proposed in the past few years to detect social groups in a scene, also associated to several datasets, there is a lack of a well defined metric to evaluate the true existence of social groups. In this context, Chapter 16 presents *GRODE*, group detection metrics, with the intent of capturing the behavior of group detection approaches with respect to specific group cardinalities. Further, the authors present the use of the metrics to benchmark several known algorithms. Finally, Chapter 17 highlights one of the most challenging issues in video-based pedestrian detection and path prediction, i.e., the development of real-time or online methods, also capable of dealing with medium or high densities of pedestri-ans. In this perspective, this chapter first gives an overview of real-time algorithms for extracting the trajectory of each pedestrian in a crowd video using a combination of nonlinear motion models and learning methods designed on the basis of new collision-avoidance and local navigation algorithms. Moreover, the prediction of future pedestrian position over a long horizon is also obtained by combining these motion models with global movement patterns while applying Bayesian inference. The resulting tracking and prediction algorithm proves to be able to provide improved accuracy in dense settings, tens of pedestrians, at real-time rates (25–30 fps).

1.3 SUMMARY OF IMPORTANT POINTS

This book highlights the study of groups of people as the primary focus of research in conjunction with crowds. Crowds are formed primarily by groups, and not only by single individuals, so the focus on groups is beneficial to understanding crowds, and vice versa. The subject matter covered in this book aims to address a highly focused problem with a strong multidisciplinary appeal to practitioners in both fundamental research and applications. This book is dedicated to solving the problem of group and crowd analysis and modeling in computer vision, pattern recognition and social sciences, and highlighting the open issues and challenges. Despite aiming to address a highly focused problem, the techniques covered in this book, e.g., techniques of segmentation and grouping, tracking and reasoning, are highly applicable to other more general problems in computer vision and machine learning.

REFERENCES

[1] Vinciarelli Alessandro, Pantic Maja, Bourlard Herve. Social signal processing: survey of an emerging domain. Image Vis Comput 2009;27(12):1743–59. http://dx.doi.org/10.1016/j.imavis.2008.11.007.

[2] Vinciarelli Alessandro, Pantic Maja, Heylen Dirk, Pelachaud Catherine, Poggi Isabella, D'Errico Francesca, Schroder Marc. Bridging the gap between social animal and unsocial machine: a survey of social signal processing. IEEE Trans Affect Comput 2012;3(1):69–87. http://dx.doi.org/10.1109/T-AFFC.2011.27.

[3] Hall Edward T. The hidden dimension. Doubleday; 1966. 240 p.

[4] Dalal N, Triggs B. Histograms of oriented gradients for human detection. In: IEEE Computer Society conference on computer vision and pattern recognition, vol. 1. 2005. p. 886–93.

[5] Tuzel O, Porikli F, Meer P. Human detection via classification on Riemannian manifolds. In: IEEE conference on computer vision and pattern recognition. 2007. p. 1–8.

[6] Blackman SS. Multiple hypothesis tracking for multiple target tracking. IEEE Aerosp Electron Syst Mag 2004;19(1):5–18. http://dx.doi.org/10.1109/MAES.2004.1263228.

[7] Zhao T, Nevatia R. Tracking multiple humans in crowded environment. In: Proceedings of the 2004 IEEE Computer Society conference on computer vision and pattern recognition, vol. 2. 2004. p. II-406–13.

[8] Collins RT, Ge W, Ruback RB. Vision-based analysis of small groups in pedestrian crowds. IEEE Trans Pattern Anal Mach Intell 2012;34(5):1003–16. http://dx.doi.org/10.1109/TPAMI.2011.176.

[9] Choi Wongun, Shahid Khuram, Savares Silvio. Learning context for collective activity recognition. In: Proceedings of the IEEE international conference on computer vision and pattern recognition. 2011.

[10] Santhanam T, Sumathi CP, Gomathi S. A survey of techniques for human detection in static images. In: Proceedings of the second international conference on computational science, engineering and information technology. New York (NY): ACM; 2012. p. 328–36.

[11] Watada Junzo, Musa Zalili, Jain Lakhmi C, Fulcher John. Human tracking: a state-of-art survey. In: Setchi Rossitza, Jordanov Ivan, Howlett Robert J, Jain Lakhmi C, editors. Knowledge-based and intelligent information and engineering systems. Lect Notes Comput Sci, vol. 6277. Berlin, Heidelberg: Springer; 2010. p. 454–63.

[12] Morris BT, Trivedi MM. A survey of vision-based trajectory learning and analysis for surveillance. IEEE Trans Circuits Syst Video Technol 2008;18(8):1114–27. http://dx.doi.org/10.1109/TCSVT.2008.927109.

[13] Khan Saad M, Shah Mubarak. Detecting group activities using rigidity of formation. In: Proceedings of the 13th annual ACM international conference on multimedia. New York (NY): ACM; 2005. p. 403–6.

[14] Ryoo MS, Aggarwal JK. Recognition of high-level group activities based on activities of individual members. In: IEEE workshop on motion and video computing. 2008. p. 1–8.

[15] Ryoo MS, Aggarwal JK. Stochastic representation and recognition of high-level group activities: describing structural uncertainties in human activities. In: IEEE Computer Society conference on computer vision and pattern recognition workshops. 2009. p. 11.

[16] Helbing D, Molnar P. Social force model for pedestrian dynamics. Phys Rev E 1995;51(5):4282–6.

[17] Pellegrini S, Ess A, Schindler K, Van Gool L. You'll never walk alone: modeling social behavior for multi-target tracking. In: IEEE 12th international conference on computer vision. 2009. p. 261–8.

[18] Choi W, Savarese S. Understanding collective activities of people from videos. IEEE Trans Pattern Anal Mach Intell 2014;36(6):1242–57.

[19] Leal-Taixé Laura, Pons-Moll Gerard, Rosenhahn Bodo. Everybody needs somebody: modeling social and grouping behavior on a linear programming multiple people tracker. In: 2011 IEEE international conference on computer vision workshops. IEEE; 2011. p. 120–7.

[20] Yamaguchi Kota, Berg Alexander C, Ortiz Luis E, Berg Tamara L. Who are you with and where are you going? In: 2011 IEEE conference on computer vision and pattern recognition. IEEE; 2011. p. 1345–52.

[21] Ryoo M, Aggarwal J. Stochastic representation and recognition of high-level group activities. Int J Comput Vis 2011:183–200.

[22] Tran KN, Gala A, Kakadiaris IA, Shah SK. Activity analysis in crowded environments using social cues for group discovery and human interaction modeling. Pattern Recognit Lett 2014;44:49–57. http://dx.doi.org/10.1016/j.patrec.2013.09.015.

[23] Choi Wongun, Chao Yu-Wei, Pantofaru Caroline, Savarese Silvio. Discovering groups of people in images. In: European conference on computer vision. Springer; 2014. p. 417–33.

[24] Vascon Sebastiano, Mequanint Eyasu Zemene, Cristani Marco, Hung Hayley, Pelillo Marcello, Murino Vittorio. Detecting conversational groups in images and sequences: a robust game-theoretic approach. Comput Vis Image Underst 2016;143:11–24. http://dx.doi.org/10.1016/j.cviu.2015.09.012.

[25] Zhang Hongyi, Geiger Andreas, Urtasun Raquel. Understanding high-level semantics by modeling traffic patterns. In: Proceedings of the IEEE international conference on computer vision. 2013. p. 3056–63.

[26] Ramanathan Vignesh, Yao Bangpeng, Fei-Fei Li. Social role discovery in human events. In: Proceedings of the IEEE conference on computer vision and pattern recognition. 2013. p. 2475–82.

[27] Smith K, Ba SO, Odobez J-M, Gatica-Perez D. Tracking the visual focus of attention for a varying number of wandering people. IEEE Trans Pattern Anal Mach Intell 2008;30:1212–29.

[28] Cristani Marco, Bazzani Loris, Paggetti Giulia, Fossati Andrea, Tosato Diego, Del Bue Alessio, Menegaz Gloria, Murino Vittorio. Social interaction discovery by statistical analysis of F-formations. In: Proceedings of the British machine vision conference. 2011. p. 23.1–12.

[29] Setti Francesco, Russell Chris, Bassetti Chiara, Cristani Marco. F-formation detection: individuating free-standing conversational groups in images. PLoS ONE 2015;10(5):e0123783.

[30] Sedda Anna, Manfredi Valentina, Bottini Gabriella, Cristani Marco, Murino Vittorio. Automatic human interaction understanding: lessons from a multidisciplinary approach. Front Human Neurosci 2012;6. http://dx.doi.org/10.3389/fnhum.2012.00057.

[31] Gong Shaogang, Loy Chen Change, Xiang Tao. In: Security and surveillance. Springer; 2011. p. 455–72.

[32] Yi Shuai, Li Hongsheng, Wang Xiaogang. Pedestrian behavior understanding and prediction with deep neural networks. In: Computer vision – ECCV 2016 – 14th European conference, proceedings, part I. 2016. p. 263–79.

[33] Chan Antoni B, Liang Zhang-Sheng John, Vasconcelos Nuno. Privacy preserving crowd monitoring: counting people without people models or tracking. In: IEEE conference on computer vision and pattern recognition. IEEE; 2008. p. 1–7.

[34] Conte Donatello, Foggia Pasquale, Percannella Gennaro, Tufano Francesco, Vento Mario. A method for counting people in crowded scenes. In: 2010 seventh IEEE international conference on advanced video and signal based surveillance. IEEE; 2010. p. 225–32.

[35] Zhang Cong, Li Hongsheng, Wang Xiaogang, Yang Xiaokang. Cross-scene crowd counting via deep convolutional neural networks. In: IEEE conference on computer vision and pattern recognition. 2015. p. 833–41.

[36] Saleh Sami Abdulla Mohsen, Suandi Shahrel Azmin, Ibrahim Haidi. Recent survey on crowd density estimation and counting for visual surveillance. Eng Appl Artif Intell 2015;41:103–14.

[37] Rodriguez Mikel, Ali Saad, Kanade Takeo. Tracking in unstructured crowded scenes. In: 2009 IEEE 12th international conference on computer vision. IEEE; 2009. p. 1389–96.

[38] Tang Siyu, Andriluka Mykhaylo, Milan Anton, Schindler Konrad, Roth Stefan, Schiele Bernt. Learning people detectors for tracking in crowded scenes. In: Proceedings of the IEEE international conference on computer vision. 2013. p. 1049–56.

[39] Ali Saad, Shah Mubarak. Floor fields for tracking in high density crowd scenes. In: Computer vision – ECCV 2008. Springer; 2008. p. 1–14.

[40] Teng Li, Chang Huan, Wang Meng, Ni Bingbing, Hong Richang, Yan Shuicheng. Crowded scene analysis: a survey. IEEE Trans Circuits Syst Video Technol 2015;25(3):367–86.

[41] Kratz Louis, Nishino Ko. Anomaly detection in extremely crowded scenes using spatio-temporal motion pattern models. In: IEEE conference on computer vision and pattern recognition. IEEE; 2009. p. 1446–53.

[42] Jiang Fan, Wu Ying, Katsaggelos Aggelos K. Detecting contextual anomalies of crowd motion in surveillance video. In: 2009 16th IEEE international conference on image processing. IEEE; 2009. p. 1117–20.

[43] Mohammadi Sadegh, Kiani Hamed, Perina Alessandro, Murino Vittorio. A comparison of crowd commotion measures from generative models. In: Proceedings of the IEEE conference on computer vision and pattern recognition workshops. 2015. p. 49–55.

[44] Mehran Ramin, Oyama Akira, Shah Mubarak. Abnormal crowd behavior detection using social force model. In: IEEE conference on computer vision and pattern recognition. IEEE; 2009. p. 935–42.

[45] Zhou Bolei, Tang Xiaoou, Zhang Hepeng, Wang Xiaogang. Measuring crowd collectiveness. IEEE Trans Pattern Anal Mach Intell 2014;36(8):1586–99. http://dx.doi.org/10.1109/TPAMI.2014.2300484.

[46] Patel Vishal M, Gopalan Raghuraman, Li Ruonan, Chellappa Rama. Visual domain adaptation: a survey of recent advances. IEEE Signal Process Mag 2015;32(3):53–69. http://dx.doi.org/10.1109/MSP.2014.2347059.

[47] Conigliaro Davide, Rota Paolo, Setti Francesco, Bassetti Chiara, Conci Nicola, Sebe Nicu, Cristani Marco. The S-HOCK dataset: analyzing crowds at the stadium. In: 2015 IEEE conference on computer vision and pattern recognition (CVPR). 2015. p. 2039–47.

PART 1

Features and Representations

CHAPTER 2

Social Interaction in Temporary Gatherings
A Sociological Taxonomy of Groups and Crowds for Computer Vision Practitioners

Chiara Bassetti
Department of Information Engineering and Computer Science, University of Trento, Trento, Italy
Department of Sociology and Social Research, University of Trento, Trento, Italy

Contents

2.1 INTRODUCTION: GROUP AND CROWD BEHAVIOR IN CONTEXT

Groups and crowds – better defined as *temporary gatherings* (cf. [17,18,35]), as we shall see – are almost ever-present in our societies: from urban spaces to airports, from ERs to malls, from theaters to arenas, we are everyday immersed in groups and crowds of different size, that gather people holding different mutual relationships, engaged in different activities and different kinds of interpersonal interaction. Understanding and analyzing group and crowd behavior is thus crucial. More specifically, modeling gatherings dynamics (e.g., assembling/disassembling) is fundamental to enhance our ability to prevent and manage possible critical situations. This is where semi/automated technologies can make a difference, and that is necessarily a multidisciplinary effort.

Research on group and crowd modeling in computer science, indeed, is highly multidisciplinary, and encompasses the social and the cognitive sciences when it comes to the analysis of human action and interaction. However, computer scholars are often confronted with surpassed conceptions of social interaction, collective action, and crowd behavior especially (cf. [35]; cf. also [3,22]). As we shall better see (Section 2.3.3), "contagion theory" represents a clear example (cf., e.g., [27,33]). Furthermore, many psychological interpretations of collective behavior are based on motivational factors that, even when properly measured and accounted for, are better predictors of long-term socio-psychological dynamics than situated behavioral ones (cf. also [4]). Yet the latter are the only ones accessible by the visual means of computer vision and, according to the epistemological foundations of Ethnomethodology [13], by any human analyst alike, that is why such an approach is particularly well-suited to be integrated into computer vision techniques.

The sociological literature upon which I draw in this chapter can be broadly defined as microsociology (cf., e.g., [7]). Whereas the term has been relatively recently introduced, microsociological foundations are to be found in classical sociological work, starting, for instance, from Simmel's [45,46] attention to the "quantitative dimension" of social formations, and Durkheim's [10] focus on the "elementary forms" of social life (e.g., in religion and rituals).[1] Durkheimian roots, indeed, lay underneath the two microsociological perspectives on which the chapter more specifically focuses – namely, Erving Goffman's (e.g., [16–21]) interactionist approach,[2] and the above-mentioned Harold Garfinkel's (e.g., [13–15]) Ethnomethodology (EM), from which Conversation Analysis (CA) (e.g., [41,39,26]) has then emerged, thereby generating the so-called EM/CA approach.[3]

From an ethnomethodological perspective, human behavior in public – that is, social inter/action or, in EM/CA terms, *action-in-interaction* – is to be regarded as an ordered activity. Such an order is based on three, interconnected key features of social inter/action, namely "accountability", "reflexivity" and "indexicality", and represents a collective, situated and processual achievement. Our actions in public, Garfinkel [13,15] states, are designed in such a way to result immediately understandable, explainable, and easily reportable if necessary – in one word, *accountable* – to the eyes of any other copresent member of (a particular) society. Further, it is not that we accompany our actions-in-interaction with explanatory accounts[4]; rather, our actions contain in themselves their own explanations, they are (designed to be) self-evident to our fellow social members – in short, our actions are *reflexively* accountable. Yet this holds true only in relation to the particular context in which interaction takes place. The reflexive accountability of our actions in public is fully entangled with the context at hand – in brief, it is *indexical*. All this helps members to know "what's going on" and hence "what to do", and to dis/confirm to themselves that "everything's normal, nothing to signal."

---

[1] On the classical foundations of microsociology, see also [11].

[2] See also [6].

[3] Actually, EM/CA can be regarded as a discipline on its own, rather than a stream within sociology.

[4] We feel this necessity only in some situations, which we usually then define as awkward, embarrassing, weird, etc.

In other words, it helps to maintain the *orderliness* of social interaction. This is not, however, a given, predetermined outcome. As I mentioned, it is something members achieve together, in always different particular ways in different particular circumstances, precisely through their both sequential and cooccurrent actions-in-interaction.

The derived ethnomethodological analytical principles, therefore, entail a particular attention

- to the *sequentiality* of interaction, which is seen as an unfolding process (who did what when);
- to the *perspective of the participants* at any given point of such a process (vs. the perspective of the analyst who knows what happens next), since interactants are those for whom actions are designed, and those whose subsequent actions constitute, among the rest, interpretations of the former ones;
- to the context of the interaction, in a perspective that sees *contexts-as-constitutive-of-their-actions and actions-as-constitutive-of-their-contexts*.

The local context consists of the situated circumstances in which the interaction takes place and the socio-cultural contours of the occasion alike. Goffman [17] captures this duality in talking of "social occasion" and "social situation." The former consists of the set of general social characteristics of the circumstances; it basically amounts to answering the question "What are we doing?" – think of a party, a conference dinner, a picnic, a walk in the city, an evening at the theater, a night in the club, a day at the office, an afternoon at the mall, a university lecture, shopping for groceries, movie night at a friend's place, etc. Social occasions provide the general scheme or *frame* (structural social context) within which temporary gatherings develop and dissolve in each specific social situation (situational social context), that is, that specific party, dinner, picnic, etc. (cf. [18,20]).

All in all, context is crucial to analyze behavior in public – and this is particularly so when visual means are the only available ones. In what follows I delineate some of the basic dimensions along which we can categorize social contexts, in the attempt to offer to computer vision practitioners some tools for the semi/automated analysis of groups and crowds. I start from the fundamental dimension consisting of the *kind of social interaction* that prevails within the context, and I present a threefold typology (Section 2.2). Then I draw on such a typology and provide a taxonomy by considering it together with the *number of gathered persons*, the *private vs. public* nature of the setting, and the *static vs. dynamic* character of its proxemic arrangement (Section 2.3). I close with a couple of examples of integration of sociological theory and/or empirical analysis into computer vision methods and techniques.

2.2 SOCIAL INTERACTION: A TYPOLOGY AND SOME DEFINITIONS

(Nonmediated) Social inter/action takes place during and through (nonmediated) temporary gatherings, which can be defined as any set of two or more individuals in the immediate presence of each other at a given time. That is, when people are copresent, some kind of *behavior in public*, in Goffmanian terms, or *action-in-interaction*, in EM/CA jargon, takes place. The latter may or may not involve common or concerted – i.e., collective – action. Therefore, following Goffman [18,21], a first distinction lies between

1. *unfocused interaction*, that is, individual action in public, and
2. *focused interaction*, that is, collective action in public.

The latter occurs whenever two or more individuals willingly agree – although such an agreement is rarely verbalized – to sustain for a time a single focus of attention [17]. This may happen in "free conversation" [17], or within one or the other situated system of activity. Systems of activity are similar to social occasions, yet they entail a more *organized* coordination and/given a *specific goal* to reach (e.g., building a roof, assembling an engine, transplanting an organ).

Furthermore, the focus of attention that people sustain together for a certain time may be mutual or not, and situated systems of activity can be just cooperative or fully collaborative. Indeed, following Kendon's elaboration of Goffman's approach to face-to-face interaction [30], we can further distinguish between

a. *common-focused interaction*, involving a common focus of attention and action (e.g., watching a movie or a match together), and
b. *jointly-focused interaction*, entailing a joint focus of attention and concerted actions (e.g., cheering, conversing).

Notice, however, that social interaction is a multiscalar entity: individual, common, and concerted actions-in-interaction are often cooccurrent and intertwined in ordinary life situations (e.g., conversing with a partner while attending a theater show). That is, as we shall better see in Section 2.3, temporary gatherings can be *nested* into each other.

2.2.1 Unfocused Interaction

Unfocused interaction occurs whenever people find themselves by chance in the immediate presence of others, such as when waiting in line at the post office or crossing the street at a traffic light junction. On such occasions, simply by virtue of the reciprocal presence, some form of interpersonal interaction takes place, regardless of individual intent. Goffman [17] distinguishes between communication, as the intentional production of social signals, and expression, as the inevitable production of social signals which someone's mere bodily and embodied presence in the public space generates, and which generates in turn others' impressions about the former. One can try to control those signals, a situated endeavor that Goffman [16] calls "impressions management", a fundamental brick of "self-presentation." Whereas, whatever the in/success of such a management, expression cannot be avoided, communication can – and, as far as unfocused interaction is concerned, it should. That is to say, in EM/CA phrasing, that *interpersonal communication is dispreferred* in unfocused interaction.

What is then preferred? A particular behavioral display, labeled "civil inattention" [18], which consists of overtly disattending others and their actions while simultaneously acknowledging their presence. Think of a hotel or hospital elevator, of a shopping boulevard, of an art gallery hall. Everybody is busy in displaying disengagement with others' activities and engagement in his/her own. Everybody, in EM/CA terms, is *doing* [*minding one's own business*]. Everybody, however, is also constantly monitoring others and their actions both for coordination purposes (e.g., not pressing together the elevator floor-button thus touching each other, not bumping into each other on the street or in front of a painting) and for checking that everything is proceeding "normally", as expected, in an ordered way (e.g., nobody is pushing the alarm button in the elevator for no reason, nobody is walking backward, spiraling or in a zigzag fashion along the boulevard, nobody is moving paintings from their exhibit position). Such an ordered way also entails, at least in modern urban life, the fact that no stranger is staring at, or trying to engage in a conversation with, another (excessive attention), nor s/he is completely disregarding the latter's (rightful) presence in the public space (absence of acknowledgment).

2.2.2 Common-Focused Interaction

When the focus of attention is common but not reciprocal, we talk of common-focused interaction. Watching a movie at the theater or the drive-in, attending a university lecture or a press conference are some examples. When engaged in this kind of interaction, people very often find themselves in the role of the spectator, and together they build up to the audience. As we all know, conversation among audience members is highly discouraged: indeed, there is something else going on, the reason why we all are here, and most if not all the people around are strangers to the other participants. Therefore, despite not being as much dispreferred as in the case of unfocused interaction (after all, we are here for the *same* reason), *interpersonal communication* does not represent the preferred behavior, and remains therefore *interstitial and peripheral* when common-focused interaction is the prevailing/official activity, that is, conversation is relegated to particular space-times (e.g., the concert hall during the interval) and/or particular degrees of concealment (e.g., murmuring in the ears of the person aside).

It should also be noticed that the spatial and proxemic setting is more likely to be static than both in unfocused and jointly-focused interaction, which also present a larger spectrum of variation as for proxemic arrangements. In Section 2.3 we shall see how such a feature of common-focused interaction affects gatherings of any size.

2.2.3 Jointly-Focused Interaction

Jointly-focused encounters, or "face engagements" [18], encompass a sense of reciprocity and mutual activity. Further, they entail a *preferential openness to interpersonal communication*, a special conversational license. Whereas for speaking during a lecture you either have to raise your hand (at a determined time) or to murmur in your colleague's ears, during a free conversation, a board game session, a meeting, a job-interview – in short, when jointly-focused interaction is taking place – interpersonal communication is not only legitimate but expected. Its absence would appear anomalous. More generally, participation is not at all peripheral but engaged; people are – and display to be – mutually involved in some joint activity [18].

Examples cover a wide range of *collaborative situated systems of activity* that entail a more or less static vs. dynamic spatial and proxemic organization: playing cards, having dinner, doing a puzzle, pitching a tent, or just conversing, whether sitting at the living-room table or standing in the theater hall or the coffee break lounge. The latter two examples constitute also instances of so-called "Free-standing Conversational Groups" (FCGs) [31], that is, small gatherings formed by people engaged in free conversation – i.e., jointly-focused interaction – and arranged in a mainly-static proxemic layout within a generally more dynamic proxemic scenario.

2.3 TEMPORARY GATHERINGS: A TAXONOMY AND SOME EXAMPLES

As I mentioned at the very beginning of this chapter, group and crowds are better defined as temporary gatherings of various sizes. This has to do, on the one hand, with the notion of group, and, on the other hand, with the debunking of what Clark McPhail [35] refers to as "the myth of the madding crowd." Whereas we shall see the latter further on (Section 2.3.3), it is important to distinguish since the beginning between groups and gatherings.

Groups are social units entailing some explicit or implicit membership and organization [17]. They can be small – such as the family, the intimate circle of close friends, or the project team – or larger, thus generally more organized and/or institutionalized – think of a class, a company, an editorial board, a jury, etc. – but they anyway entail some sense of belongingness, membership, and often identity. And they do so thanks to the fact that group members are generally involved in continued rather than occasional interaction (cf., e.g., [12]). Such a state of affairs, however, does not characterize the whole spectrum of human action-in-interaction. Temporary gatherings, on the contrary, can be defined as any set of two or more copresent individuals having some form of social interaction [18]. Therefore, they cover a wider spectrum, and they point to a situated, context-grounded social unit as opposed to the more stable one consisting of groups.

Following such a Goffmanian take on individual and collective behavior in public, we can identify different types of gatherings, depending on several, mutually interconnected dimensions (Table 2.1). The first and most important dimensions are:

Table 2.1 Temporary gatherings taxonomy

Temporary gatherings (2 to N) Two or more persons in copresence in a given space-time		
Small gatherings (2 to 6 participants) *Happen in private, semipublic and public places*	**Medium gatherings (7 to 12/30 participants)** *Happen in private but mostly in semi/public places*	**Large gatherings (13/31 to N participants)** *Happen in semipublic but mostly in public places*
• <u>Unfocused</u>, e.g., line at the shop register, watching timetables at the bus stop • <u>Common-focused</u>, e.g., television-watching during family meal • <u>Jointly-focused</u>, e.g., free-standing conversational groups, game board playing	• <u>Unfocused</u>, e.g., line at the post office • <u>Common-focused</u>, e.g., classroom, touring group at the museum • <u>Jointly-focused</u>, e.g., meeting, extended family dinner	• <u>Unfocused</u>, e.g., line at the airport check-in (Prosaic or *Casual* crowd) • <u>Common-focused</u>, e.g., Mass, sport/theater/cinema audience (*Spectator* crowd) • <u>Jointly-focused</u>, e.g., flash-mob, sport supporters (*Expressive* crowd); mob/riot/sit-in (*Protest*/Acting crowd)

Private places, e.g., home, private garden, car; *semipublic places*, e.g., classroom, office, club, party area; *public places*, e.g., open plaza, public transportation, station, walkway, park, street.

1. the prevailing *kind of social interaction*;
2. the *amount of people* being present, that is, the quantitative dimension of social life of Simmelian memory.[5]

Furthermore, two properties of the setting are particularly relevant:

3. the *privateness vs. publicness of the place* – ranging from private homes to semipublic offices and shops, to public streets and stations;
4. the *static vs. dynamic proxemic arrangement*, that is, the degree of freedom and flexibility of the spatial, positional, and orientational organization of gatherings as situated systems of activity (cf. also [5]).

As I mentioned, indeed, sometimes people maintain approximately their positions for an extended period of time within fixed physical boundaries (e.g., family meal or classroom activities); sometimes they move within a delimited area (e.g., home-garden party or

---

[5] As for quantitative thresholds, their partial arbitrariness should be mentioned, although there is some evidence ([45,25], [35, p. 314]) supporting the numerical boundaries reported in Table 2.1. However, it is also important to bear in mind that given characteristics of the context – such as the dimensions of the physical space, and the social occasion of Goffmanian memory – bestow different meanings and nuances, so to speak, onto the same number of people.

conference coffee break); and sometimes they do within a more or less unconstrained – and usually semi/public – space (e.g., conversing while walking on the street or along the beach). Whereas the first three dimensions are considered in Table 2.1, the fourth shall be only discursively taken into consideration in what follows (Sections 2.3.1–2.3.3); also, I refer to previous work [44, Table 1 and Fig. 3 especially] for more examples at the intertwinement of the first and fourth dimension.

Finally, before looking into gatherings of different sizes, it is worth mentioning that the smaller the size, the more shared the interaction focus, and the stronger the underpinning social relationships. On the other hand, the bigger the size, the less shared the interaction focus, and the harder to model people behavior.

2.3.1 Small Gatherings – Semi/Private Encounters and Group Life

Small temporary gatherings present a number of participants that range between two and (partially arbitrarily, cf. footnote 5) six persons. They take place very often in private places, with largely static arrangements, yet the range of variation is huge, as we are about to see.

People in line at the shop register, or watching timetables at the bus stop participate in small gatherings where *unfocused* interaction prevails. Unfocused ones are those among small gatherings that are most likely to take place in public or semipublic spaces (e.g., a street corner or the local grocery store, respectively), whereas focused small gatherings more often happen in private places.

Watching television together on the couch at a friend's place or during the family meal constitutes a perfect example of a *common-focused* small gathering. Window-shopping together, however, represents another example, as dynamic and public as the context may be. Like watching the sport match on the television with some friends at the local bar (rather than at somebody's home), window-shopping in the city or at the mall can be regarded as a common-focused small encounter that takes place within a larger context, and potentially a larger gathering (e.g., all the match viewers in the bar) – this is the phenomenon to which I refer as *nested gatherings*.

Similarly, yet differently, *jointly-focused* small gatherings are often coupled with private spaces and mostly static arrangements. Instances run from a group of friends playing cards or a board game in a living-room, to colleagues coworking at the blackboard in a studio; from relatives fixing a car in the garage, to young siblings drawing together on their bedroom wall. An equally good example, though, is constituted by a couple picking a perfect gift for a common friend at the book shop. Another case of nested gatherings: one (or more) jointly-focused small encounter(s) nested into a medium-sized unfocused one.

It is worth mentioning that FCGs, and *focused small gatherings* more generally, represent the foundational basis of sociality. A pupil of Goffman and an ethnographer, Gary Alan Fine made clear how face-to-face interaction within small groups represents the small-scale domain "in which broader social forces, properties and processes are enacted in practice" [11, p. 4].

2.3.2 Medium Gatherings – Semi/Public Occasions and Community Life

Medium gatherings may count roughly from half a dozen to two and a half dozen participants. Generally, these social encounters have as their preferred terrain semipublic settings, and are characterized by largely static, more or less preestablished arrangements: a classroom, meeting room, conference room, restaurant table, deck-chairs of a hotel terrace, and so on.

Waiting in line at a post office, at a touristic info point, at a student services university office, etc., are good examples of medium-sized temporary gatherings where interaction is mostly *unfocused*. Another instance consists in a group of strangers and/or acquaintances gathered at a barber shop at lunch time.

Classroom activities, in traditional teacher-centered lessons, can be considered cases of *common-focused* interaction in medium gatherings. The same holds for the more dynamic situation of a touring group at a museum. Further examples are constituted by a varied set of "inner-circle ceremonies" – i.e., ceremonies designed for small social circles, with members generally holding close relationships with each other – ranging from (intimate) weddings and funerals, to awards ceremonies and other events at the local level of community/associative life. A local press conference, with its "public-occasion flavor", can be regarded too as a common-focused medium-sized encounter (at least until Q&A time).

Work meetings and negotiation "tables" constitute instead examples of *jointly-focused* medium gatherings. An equally static yet more private setting is that of an extended family dinner. For a dynamic example consider, for instance, the *milonga* context, where tango dancing couples interact with other couples (to avoid collisions) and, simultaneously, members of each couple interact with their respective partners.

Whereas focused small gatherings are the loci of group life, *focused medium gatherings* and connected social occasions constitute the terrain onto which community life unfolds. Whether with your partner/close friend/relative or alone, you are anyway participating in a gathering of less close friends, acquaintances, and even strangers.

2.3.3 Large Gatherings – Public Events and Collective Life

Large gatherings, generally called crowds, group together a number of persons ranging from more than a dozen to hundreds and even thousands of dozens. They usually take place in public spaces, with an often explicitly public/collective purpose. Such a purpose, indeed, constitutes the dimension along which crowds have been categorized in social research. We can distinguish four principal types of crowd, variously related to the three kinds of social interaction we already identified (Section 2.2):

1. *Casual crowd.* "Prosaic" [35] or "casual" [3,22] crowds amount to large collections of people having little in common except their spatio-temporal colocation, that is, they are copresent by circumstance, generally in semi/public places. The prevailing kind of action-in-interaction, therefore, is *unfocused*. People in line at a check-in counter or waiting in an airport lounge constitute a couple of examples; people in the streets during a music festival provide another one, characterized by a more dynamic setting.

2. *Spectator crowd.* "Spectator" [35] (cf. also [2]) or "conventional" [3,22] crowds are collections of individuals and groups who gather for specific social events, such as theatrical performances, sport matches, and some huge ceremonies, whether religious or civil. Large spectator gatherings are basically audiences, and therefore point to a generally static arrangement, to a particular kind of social interaction, i.e., *common-focused*, and to a particular kind of collective action, i.e., common. Yet the context often involves also unfocused and especially jointly-focused interaction (e.g., with the strangers around and with the friend on your left in the stalls, respectively).

3. *Expressive crowd.* "Expressive" crowds [3,22] are collections of people and groups gathered for a specific social event *and* intending to act as full members of the crowd, to participate in "crowd action." Examples run, in a descending degree of dynamism of the setting, from flash-mob dancers, to Mass participants, to sport supporters (not just attendees), gathered together to dance, ritually pray, or cheer. Action therefore is concerted, and the focus of attention shared among participants. *Jointly-focused* interaction is necessarily at the center stage of expressive crowds.

4. *Protest crowd.* "Demonstration", "protest" [35] or "acting" [3,22] crowds are collections of individuals and groups who gather for specific protest events, such as mob, riot, sit-in, or march participants. Like expressive crowd members, those of a protest crowd intend to participate in collective action, whereby such an action is usually oriented toward markedly political (meaning public and collective) concerns. As a consequence, *jointly-focused* interaction is the prevailing kind of action-in-interaction in this case, too. The static vs. dynamic nature of the spatial and proxemic arrangements highly depends on the nature of the selected form of protest (e.g., sit-in vs. march).

Since the large amount of people involved, and hence the required organization and coordination, *focused large gatherings* – i.e., spectator, expressive, and protest crowd – often qualify as *social events*, rather than "mere" social occasions. That is to say, in Goffmanian terms, they often are contemporary rituals of an extraordinary vs. ordinary/everyday kind (e.g., the world championship final match vs. the local championship weekly football match). It is worth noticing that rituals can be regarded as the archetypal form of organization of large gatherings of people, that is, the archetypal form of "crowd management" (codified movement and talk, traditional dances and songs, given roles and connected spatial positions and paths, prearranged action turns, etc.), and it is probably not by chance that crowd behavior rose as an interesting topic of political and scientific debate once large gatherings witnessed a progressive increase together with the loosening of the ritual means by which they were organized.

Furthermore, it is important to underline that different types of crowd can be cooccurrent, e.g., a flash-mob at an airport while others are in line (expressive plus casual crowds), and even intertwined, e.g., cheering sport supporters, compactly packed on the stalls, within the broader, copresent audience of mere attendees (expressive plus spectator crowds). As I mentioned, indeed, gatherings of diverse sizes and kinds can be nested into each other. This means that not only common-focused large gatherings (spectator crowds), but also *jointly-focused large gatherings* (expressive and protest crowds) entail individual, common and concerted action in various and varying proportions at any given time.

The notion of crowd does not encompass a membership layer as much stable as the notion of group does – members of groups such as families or sport teams are involved in continued mutual interaction, and this is also the reason why group assembling, disassembling and reassembling dynamics markedly differ from crowd ones (think for instance to greeting sequences). However, as I mentioned in the Introduction, certain theories of collective behavior tried to assign to crowds a sort of uniform temporary membership with both cognitive and emotional nuances. Heavily relying on Goffman's interactionist approach, Clark McPhail [35,37] and colleagues empirically documented (through team ethnography, cf. [43]) and analyzed in detail (in their "elementary forms") many large gatherings of various kinds, ranging from political demonstrations and strikes to sport events. McPhail ended up with a strong critique of the above mentioned theories and of "the crowd" concept. The latter, indeed, suggests homogeneity (all members are the same), unanimity (all have the same motive/s), mutual inclusivity (all behave the same), and continuity (their mutually inclusive behavior is continuous), whereas in fact gatherings are characterized by alternating individual and collective actions, with both varying in quality, and the latter presenting varied proportions of copresent people engaged in any particular action. McPhail refers to this as *Alternating and Varied Individual and Collective Actions* (AVICA). That is, not only crowds are not anonymous, since most individuals arrive at the gathering with one or more companions (i.e., in groups), but also they do not increase "suggestibility" or alterate cognitive functions, and they are not uniquely emotional – indeed, displayed emotions vary in type, in amount, and are only seldom uniform. Therefore, the "crowd mind" is a myth, McPhail concludes, and "large gatherings" is a label that better describes the considered phenomenon.

2.4 CONCLUSION: MICROSOCIOLOGY APPLIED TO COMPUTER VISION

Collective behavior is a multiscalar entity, where mixed interactions and nested gatherings are the norm. Understanding such dynamics is fundamental for both managing and modeling them appropriately, hence for implementing technologies able to detect and reason about neither only physical, nor only cognitive, but also social aspects of a scene. This is where microsociology can be useful. On the one hand, Goffman's (and his scientific followers') theory of interaction allows characterizing and distinguishing several fundamental features of social encounters, such as focused vs. unfocused interaction, or social occasion and social situation. On the other hand, ethnomethodology provides a theory of the fundamental functioning of action-in-interaction that comes with a set of analytical principles which make such an approach perfectly suited when it comes to interdisciplinary endeavors with computer vision (and computer science at large; cf., e.g., [9,38]).

I close this chapter with a couple of examples of microsociological theory and analysis applied to computer vision. The first one concerns a specific category of jointly-focused small gatherings, namely, free-standing conversational groups. Adam Kendon's elaboration of the Goffmanian framework for the analysis of face-to-face interaction in terms of nonverbal behavior (e.g., [28,29,32]; cf. also [42]) and of proxemics especially (e.g., [31,5]) has been at the basis of a novel method for automatically identifying FCGs in images through mere positional and orientational information. Tested under five different datasets against state-of-the-art methods, the microsociology-grounded method was found to significantly outperform the others (cf. [44]). The second example regards sport spectator crowds, that is, a particular kind of common-focused large gatherings. In this case, researchers have been able to integrate not only sociological theory, but also detailed empirical analysis into novel computer vision methods for automated crowd segmentation and crowd excitement level calculation (cf. [8,1]). Novel techniques were again found to outperform extant ones.

2.5 FURTHER READING

- On the fundamental functioning of action-in-interaction: [13,15,34,40].
- On proxemics in social interaction: [5,23,24,31].
- On kinds of interaction: [18,21,30,31].
- On unfocused interaction, or behavior in public: [18].
- On small groups and face-to-face interaction: [7,12,16,17,19].
- On large gatherings: [22,35,36,43].

REFERENCES

[1] Bassetti C. A novel interdisciplinary approach to socio-technical complexity. Sociologically-driven, computable methods for sport spectator crowds' semi-supervised analysis. In: Cecconi F, editor. New frontiers in the study of social phenomena. New York: Springer; 2016. p. 117–43.

[2] Berlonghi A. Understanding and planning for different spectator crowds. Saf Sci 1995;18:239–47.

[3] Blumer H. Collective behavior. In: McClung Lee A, editor. Principles of sociology. New York: Barnes & Noble; 1951. p. 167–222.

[4] Choi YS, Martin JJ, Park M, Yoh T. Motivational factors influencing sport spectator involvement at NCAA division II basketball games. J Study Sports Athletes Educ 2009;3(3):265–84.

[5] Ciolek TM, Kendon A. Environment and the spatial arrangement of conversational encounters. Sociol Inq 1980;50:237–71.

[6] Collins R. Theoretical continuities in Goffman's work. In: Drew P, Wootton A, editors. Erving Goffman: exploring the interaction order. Oxford: Polity Press; 1988. p. 41–63.

[7] Collins R. Interaction ritual chains. Princeton: Princeton University Press; 2004.

[8] Conigliaro D, Rota P, Setti F, Bassetti C, Conci N, Sebe N, Cristani M. The S-HOCK dataset: analyzing crowds at the stadium. In: IEEE conference on computer vision and pattern recognition (CVPR). 2015. p. 2039–47. http://ieeexplore.ieee.org/xpl/articleDetails.jsp?arnumber=7298815&queryText=cvpr%202015%20s-hock&newsearch=true.

[9] Dourish P, Button G. On "technomethodology": foundational relationships between ethnomethodology and system design. Hum-Comput Interact 1998;13(4):395–432.

[10] Durkheim E. The elementary forms of the religious life. New York: Free Press; 1965.

[11] Fine GA, Harrington B, Segre S. Politics in the public sphere. The power of tiny publics in classical sociology. Sociologica 2008;1. http://www.sociologica.mulino.it/doi/10.2383/26566.

[12] Fine GA. Tiny publics. A theory of group action and culture. New York: Russell Sage Foundation; 2012.

[13] Garfinkel H. Studies in ethnomethodology. Englewood Cliffs: Prentice–Hall; 1967.

[14] Garfinkel H. Ethnomethodology's program: working out Durkheim's aphorism. Lanham: Rowman & Littlefield Publishers; 2002.

[15] Garfinkel H. Seeing sociologically: the routine grounds of social action. Boulder: Paradigm Publishers; 2006.

[16] Goffman E. The presentation of self in everyday life. New York: Anchor Books; 1959.

[17] Goffman E. Encounters: two studies in the sociology of interaction. Indianapolis: Bobbs–Merrill; 1961.

[18] Goffman E. Behavior in public places: notes on the social organization of gatherings. New York: Free Press of Glencoe; 1963.

[19] Goffman E. Interaction ritual: essays on face-to-face behavior. Chicago: Aldine Publishing; 1967.

[20] Goffman E. Frame analysis: an essay on the organization of experience. New York: Harper & Row; 1974.

[21] Goffman E. Forms of talk. Philadelphia: University of Pennsylvania Press; 1981.

[22] Goode E. Collective behavior. Fort Worth: Saunders College Publ.; 1992.

[23] Hall ET. A system for the notation of proxemic behavior. Am Anthropol 1963;65(5):1003–26.

[24] Hall ET. The hidden dimension. Garden City: Anchor Books; 1966.

[25] Hare AP. Group size. Am Behav Sci 1981;24(5):695–708.

[26] Heritage J. Conversation analysis at century's end: practices of talk-in-interaction, their distributions, and their outcomes. Res Lang Soc Interact 1999;32(1–2):69–76.

[27] Hocking JE. Sports and spectators: intra-audience effects. J Commun 1982;32(1):100–8.

[28] Kendon A. Some functions of gaze-direction in social interaction. Acta Psychol 1967;26:22–63.

[29] Kendon A. Movement coordination in social interaction: some examples described. Acta Psychol 1970;32:101–25.

[30] Kendon A. Goffman's approach to face-to-face interaction. In: Drew P, Wootton A, editors. Erving Goffman: exploring the interaction order. Cambridge: Polity Press; 1988. p. 14–40.

[31] Kendon A. Conducting interaction: patterns of behavior in focused encounters. Cambridge: Cambridge University Press; 1990.

[32] Kendon A. Gesture: visible action as utterance. Cambridge: Cambridge University Press; 2004.

[33] Levy L. A study of sports crowd behavior: the case of the Great Pumpkin incident. J Sport Soc Issues 1989;13(2):69–91.

[34] Liberman K. More studies in ethnomethodology. New York: SUNY Press; 2013.

[35] McPhail C. The myth of the madding crowd. New York: De Gruyter; 1991.

[36] McPhail C. From clusters to arcs and rings. Res Commun Sociol 1994;1:35–57.

[37] McPhail C. Keynote talk at the CVPR 2015 workshop "Group and crowd behavior analysis and understanding". In: Computer vision and pattern recognition conference. 2015.

[38] Moore RJ. Ethnomethodology and conversation analysis: empirical approaches to the study of digital technology in action. In: Price S, Jewitt C, Brown B, editors. The Sage handbook of digital technology research. London: Sage; 2013. p. 217–35.

[39] Psathas G, editor. Conversation analysis: the study of talk-in-interaction. USA: Sage; 1995.

[40] Rawls AW. Wittgenstein, Durkheim, Garfinkel and Winch: constitutive orders of sense-making. J Theory Soc Behav 2011;41(4):396–418.

[41] Sacks H. Lectures on conversation. Oxford: Blackwell; 1992.

[42] Scheflen AE. Body language and the social order. Englewood Cliffs: Prentice Hall; 1972.

[43] Schweingruber D, McPhail C. A method for systematically observing and recording collective action. Sociol Methods Res 1999;27(4):451–98.

[44] Setti F, Russel C, Bassetti C, Cristani M. F-formation detection: individuating free-standing conversational groups in images. PLoS ONE 2015;10(5):e0123783.

[45] Simmel G. The number of members as determining the sociological form of the group: I. Am J Sociol 1902;8:1–46.

[46] Simmel G. Soziologie. Leipzig: Duncker und Umbolt; 1908.

CHAPTER 3

Group Detection and Tracking Using Sociological Features

Sebastiano Vascon*,†, Loris Bazzani‡

*Department of Environmental Sciences, Informatics and Statistics, University Ca' Foscari of Venice, Mestre, Italy
†Pattern Analysis and Computer Vision, Istituto Italiano di Tecnologia, Genova, Italy
‡Department of Computer Science, Dartmouth College, Hanover, NH, USA

Contents

Group and Crowd Behavior for Computer Vision
DOI: 10.1016/B978-0-12-809276-7.00004-7

3.1 INTRODUCTION

Person detection and tracking are two important topics in computer vision [14,19,109, 70,51], representing one of the core modules of automatic video analysis. Despite still being open problems, detection and tracking of individuals have been recently reformulated to tackle new challenges, such as group analysis.

Analyzing groups means being able to extract some sociological information of the individuals from videos in an automatic way. After decades of research on the automated detection and tracking of people, the interest is now moving from encoding simple actions performed by individuals to capturing collective behaviors and social interactions [44,60,35,65,112,47,22,55,110,94]. Automatic group analysis is a topic of interest in many contexts, such as video surveillance [22], social signal processing [60,47,65,44, 12], multimedia [35], social robotics [48], and activity recognition [20].

What is a group? A group is defined in [33] as "a social unit comprising several members who stand in status and relationships with one another." There are many types of group that differ in how long a group stays unchanged (ad hoc or stable groups), informality of organization, and level of physical dispersion [42]. In this chapter, we focus on *self-organizing* groups, defined as individuals that gradually cooperate and engage with each other around some task of interest [2]. Some examples of self-organizing groups are: a team of supporters going to watch their preferred team, a family exiting from a mall and going to their car, a couple visiting a museum. Goffman [43] observed that group interactions can be categorized into those that are "focused" and those that are "unfocused." Focused interactions concern the gatherings of people to participate in an activity where there is a common focus, such as playing and watching a football match, conversing, or marching in a band. Unfocused encounters involve light interactions such as avoiding people on a busy street, briefly greeting a colleague while passing him/her in the corridor, or letting someone pass when boarding a train. This taxonomy has been exploited recently in [86].

Within the class of focused encounters, the F-formation (Facing formations) is a specific type of group interaction which requires more attention from our senses. Specifically, an F-formation arises "whenever two or more individuals in close proximity orient their bodies in such a way that each of them has an easy, direct and equal access to every other participant's transactional segment, and when they maintain such an arrangement" [21, p. 243]. Different F-formations are possible as shown in Fig. 3.2A–D, some examples of F-formations in real-world situations are illustrated in Fig. 3.1. In the case of two participants, typical F-formation arrangements are vis-a-vis, L-shape, and side-by-side. Three social spaces emerge from an F-formation: the o-space, the p-space, and the r-space (see Fig. 3.2A). The most important part is the o-space, a convex empty space surrounded by the people involved in a social interaction, in which every participant looks inward, and no external people are allowed. The p-space is a narrow strip

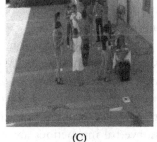

(A)   (B)   (C)

Figure 3.1 Standing conversational groups: (A) grouping at a vending machine, (B) circular F-formation, and (C) a coffee-break event.

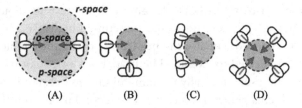

(A)   (B)   (C)   (D)

Figure 3.2 F-formations: (A) components of an F-formation, namely, o-space, p-space, and r-space; in this case, a face-to-face F-formation is sketched; (B) L-shape F-formation; (C) side-by-side F-formation; and (D) circular F-formation.

that surrounds the o–space and contains the bodies of the conversing people, while the r-space is the area outward the p–space.

In order to capture the nature of groups as defined above, the first important step is to define and engineer features or visual cues which are suitable for the task at hand. Many studies have been carried out by social psychologists to understand how people behave in public [43,21,45]. These theories can be exploited to define *sociological features* for modeling human behavior and therefore perform group detection and tracking.

The most popular features can be categorized in the followings two categories: the *low*-level features are used to determine the spatial properties of each person in a scene (person position and head/body orientation), and the *high*-level ones that are built on top of the low–level features to implement sociological definitions (e.g., the frustum of visual attention and the transactional segment).

The low–level features are cues that can be directly measured from the image or the video related to the persons of interest. Position, velocity, and head and body orientation of the individuals are examples of low–level features. They are usually the result of a computer vision algorithm which processes the image or the video in order to detect the location and/or the pose of each person. These features usually do not contain or are not based on any sociological aspect. The sociological modeling is deferred to the group detection and tracking methods. In this chapter, we will briefly analyze the main methods that can be used to extract the low-level features useful for group detection and tracking.

High-level features combine the information provided by the low-level (person position, orientation, velocity, etc.) with the notions from the sociological community in order to make the task of group detection easier. In particular, a definition from the sociologists which is recurrent in many approaches for group detection is the one of *transactional segment* [52]. A transactional segment is an area in front of a person in which nonverbal interactions are possible [91,63,87]. It is defined in terms of the biological field of view of a person and oriented as his/her gaze. In this chapter we will present two ways to characterize the frustum of visual attention used to model the transactional segment of individuals.

Group detection is a particularly challenging task because its input, i.e., low-level features (see Section 3.3.1), is very noisy and sometime unreliable. Moreover, not only the person position is needed, but also the head/body orientation which is another cue that is even more complicated to extract. The head/body orientation is of fundamental importance for detecting sociologically meaningful groups because it defines the direction of the area in which a person is focusing. The second problem is that the detection of groups is defined, in most cases, as an unsupervised clustering problem, which is by nature NP-hard to solve exactly and thus for efficiency reasons only approximated. This task is more challenging than standard clustering problems because the points to cluster (persons in a scene) are usually few, leading to a partitioning in a low density space.

In this chapter we describe in details two methods for group detection, the Robust Game-Theoretic Conversational Group detector (R-GTCG) [98] and the Dirichlet Process Mixture Model (DPMM) [111]. The first framework is a deterministic method and casts the grouping problem as an evolutionary game. The aim is to find the equilibrium points between players (a kind of Nash equilibrium). Such points correspond to the groups in a scene. In this game the persons correspond to the available strategies, and the payoff (the gain obtained by playing a certain strategy i vs. another one j) is directly related to the likelihood that the corresponding persons i and j have to interact. To quantify this likelihood, the method uses entropic measures on sociological features (Section 3.3.2). The latter framework is a stochastic method in which the grouping problem is cast as an inference in a Dirichlet process mixture model. Each person in a scene is seen as an observation from a Dirichlet process with infinite components. The groups in this context are probabilistically represented by the components. The game-theoretic approach has the advantage of being deterministic, and therefore it is easy to use in practice. However, its drawback is that it is less flexible than the probabilistic approach such as the DPMM.

Group tracking is defined as an extension of group detection over the time domain. In other words, the aim is to connect the detected groups through the video by

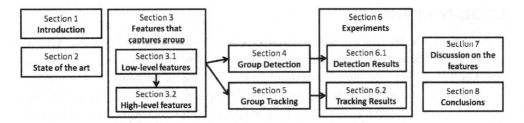

Figure 3.3 The organization of the chapter.

maintaining their identities. Group tracking usually involves the implicit task of tracking individuals, which is why the problem is usually named individual-group tracking.
The tracking of groups has a different level of complexity, compared to the tracking
of individuals. In particular, individuals are atomic entities that cannot be separated in
parts, while groups are very dynamic entities which can modify their configuration over
time. In particular, merging and splitting phenomena, absent in the individual tracking,
play a crucial role in group tracking. From a sociological perspective, whenever an individual leaves a group (an example of split) or joins a group (merge), all the social
relationships between the remaining people are revised so that the individuals produce
new entities [52]. One of the most challenging problems in individual-group tracking is
to handle those merging and splitting events, since they are responsible for the creation
or deletion of groups.

In this chapter we describe the model for individual-group tracking presented in [9]
(Section 3.5) to track both groups and individuals in a *unified* framework. The presented
framework exploits directly the dependence between the individual and the group statistics in a similar spirit of [53,20]. In particular, the support of the individual statistics
feeds the reasoning module at a group level, which itself gives feedback to the individual tracking. This allows obtaining a trade-off between individual- and group-tracking
performance. Technically, this is made possible since the group- and individual-tracking
instances combine their estimates in a single, joint state space.

Fig. 3.3 shows the graphical table of contents of the chapter. Section 3.2 describes
the state-of-the-art for group detection and tracking. Section 3.3 reports the sociological features that can be used for group detection and tracking, which are divided into
low-level features, e.g., person detection, motion properties, and body/head pose estimation, and high-level features that are usually built on top of the low-level ones. In
Section 3.4 we describe two models to detect groups, which can be used in combination of the low-level and high-level features. Section 3.5 presents a model for tracking
groups. The chapter is concluded with Sections 3.6, 3.7, and 3.8 which contain the
performance evaluation of the discussed methods, a discussion on the features, and the
conclusions, respectively.

3.2 STATE-OF-THE-ART

In this section we describe the recent literature of group detection and tracking. For the specific task of detecting groups, different approaches have been proposed in the literature. Groh et al. [44] used the relative shoulder orientations and distances (using markers attached to the shoulders) between each pair of people as a feature vector for training a binary classifier. Later Cristani et al. [22] proposed a Hough voting strategy to solve the task, the accumulated density estimates the location of the o-space. At the same time Hung and Kröse [47] consider an F-formation as a dominant set [77], the problem was cast as a maximal clique search in an edge-weighted graph. Each node in the graph is a person, and the edges between them measure the likelihood of belonging to a group. Later these two approaches have been compared by Setti et al. [84] to analyze the strengths and weaknesses for the F-formation task. They found that while the method of Cristani et al. [22] was more stable using head orientation information in the presence of noise, the method of Hung and Kröse [47] performed better when only position (and not orientation) information was available. Furthermore, they found that a substantial improvement in the performance of F-formation detection can be achieved combining a probabilistic approach (as in [22]) and graph-based clustering methods [47]. Further improvement over [22] has been proposed by Setti et al. [85] to handle the physical effect that different sizes of F-formations would have on the spatial layout of each member of the group. A similar density-based approach has also been proposed by Gan et al. [35] where the final goal of their task was to dynamically select the camera angles for automatic recording events. Tran et al. analyzed the temporal patterns of groups in order to classify the group activities [94]. Choi et al. [18] modeled different forms of group behavior trying to distinguish differing group types. In Vascon et al. [99,98] the problem of detecting conversational groups is modeled as a clustering-game combining a probabilistic model of interaction and a deterministic grouping from evolutionary game-theory. Following the success of the graph-based methods, Setti et al. [86] presented a graph-cut based minimization method for detecting F-formations using proxemic data. In the same trend, a more recent work from Zhang et al. [113] extended the work of [47] by creating a model to detect the so-called *associates* of F-formations, defined as persons that are close to a group but have not taken an active part in a conversation.

Following the taxonomy introduced by [9], the literature of group tracking can be partitioned into three categories:

1. the class of *group-based* techniques, where groups are treated as atomic entities without the support of individual tracks statistics [102,37,62,30,57];
2. the class of *individual-based* methods, where group descriptions are built by associating individuals' tracklets that have been calculated beforehand, typically, with a time lag of a few seconds [78,108,36,81,16,59,69,23,66,90]; and
3. the class of *joint individual-group* approaches, where group tracking and individual tracking are performed simultaneously [81,7,76,68,6].

In this section we describe in detail only methods belonging to the last category, but the reader can have a look at [9] for further details about the other classes of methods.

Joint individual-group-based methods deal with individuals and groups modeling and tracking at the same time. Many of the approaches maintain the structure of a graph in which connected components correspond to groups of individuals [76,41]. A two-level structure for tracking using a mass–spring model is proposed in [68]: the first level deals with individual tracking, and the second level tracks individuals that are spatially coherent as a whole. Similarly, two processes are involved in [6]: the group process considers groups as atomic entities, the individual process captures how individuals move, and revises the group posterior distribution. The drawback of these methods is that they do not consider splitting and merging events, limiting their applicability.

The model [9] presented in this chapter falls in this last category. One of its advantages with respect to the literature is that it can deal with a large and varying number of individuals and groups, unlike [76,41,38]. Moreover, the proposed formulation manages splitting and merging events, whereas other approaches (e.g., [68,6]) are more rigid. Finally, the probabilistic modeling allows associating individuals either to a single group as a hard membership or to multiple groups as a soft assignment, unlike [108,68].

3.3 SOCIOLOGICAL FEATURES

In this section we describe the existing and most popular social features of the following two categories: the *low*-level, used to determine the spatial properties of each person in a scene (person position and head/body orientation), and the *high*-level that agglomerates or uses the low-level features to implement sociological and biological definitions (subjective view frustum/transactional segments).

3.3.1 Low-Level Features

The low-level features used to eventually detect groups are cues that can be directly measured from the image or the video related to the persons of interest. Position, velocity, and head and body orientation of the individuals are examples of low-level features. These low-level features are usually the result of a computer vision algorithm which processes an image or a video in order to detect the location and/or the pose of each person.

3.3.1.1 *Person Detection*

In order to detect the position of a person in each frame of a video, there are many approaches in the literature [82,31,1,26]. One of the most famous and successful methods is the Deformable Part Model (DPM) proposed by [31] which can be used for detecting any object, not only a person. The DPM is a star-structured part-based model with a root which represent the full object and a set of parts that are free to move. Each part and the root are represented with HOG features. Training of the model without having the positions of the parts in the training set is formulated as a latent SVM, where the latent variables are the positions of the parts.

Another and more recent approach for general object detection that can be easily applied for person detection is R-CNN [40,39]. The idea is to combine the region proposal methods with the representational power of convolutional network to perform detection of objects. In practice, a set of region proposals are generated for each image, which are evaluated by a pretrained convolutional network in order to find the bounding boxes that more likely contain the objects of interest.

3.3.1.2 Person Velocity and Direction

Once the positions of the individuals are available for each frame, the velocity can be estimated by performing tracking. Tracking can be seen basically as a way to connect and smooth detections over time even in the case that some detection are missing. Many tracking methods have been proposed in the literature, e.g., particle filtering [73, 3], tracking by detection [1,10], and others [58] to name just a few.

The approaches discussed above usually estimate the position and velocity of the persons in the image space. But often this is not enough to perform group detection because groups are formed in the physical environment, and therefore it is very hard to detect them in the image space. In particular, most of the time the position in the 3D space is required in order to create a valid sociological model of interactions. For this reason, it is usually suggested to project the detected positions to the meter space. In order to do so, full or partial camera calibration has to be available. Methods such as [95,114] can be used to estimate the calibration of the camera using calibration patterns.

3.3.1.3 Head & Body Orientation

Another low-level cue for group detection is related to head and body orientation estimation. In order to perform orientation estimation, the head or body of the person should be available. One can use the same methods mentioned above to perform head or body detection. If the scene is very crowded, head detection might be preferred since it is easier to perform than full body detection because of the occlusion nuisances.

Once the position of the head is given, there are many methods for head orientation estimation in the literature which can be used. A review can be found in [72]. Most papers define head estimation as a classification problem and boosting seems to be one of the most popular methods in this context [61,101,107,97,105,74]. Among the different features used for this task (see [106] for an updated list), covariance features [96] have been exploited as powerful descriptors of pedestrians [97,105], and their effectiveness has been explicitly investigated in a comparative study [74]. When injected in boosting systems [97,105,74,93], covariances provide strong detection performance, encapsulating possible high intra-class variances (due to pose and view changes of an object of interest).

In video-surveillance applications, sometimes the resolution of the head to detect is very low. Therefore, detecting the orientation of the head can be almost impossible. In these cases, it is suggested to detect the orientation of the body as an approximation of the orientation of the head. This approximation is valid for most of the dynamic situations, where the individuals move toward a certain direction. However, it is a bad approximation on the static scenarios, for example, when people engage in F-formations [22]. In complete absence of orientational information, Hung [47] proposed a socially-motivated focus of attention (SMEFO) which approximates the head/body orientation using the relative distance between the participants in the scene.

Since every technique has its own advantages and disadvantages, the choice of method to use has to depend on the application and the data.

3.3.2 High-Level Features

High level features combine the information provided by the low-level features (person position, orientation, velocity, etc.) with the notions from the sociological community in order to make the task of group detection easier. In particular, a definition from the sociologists which is recurrent in many approaches for group detection is the one of *transactional segment* [52]. In terms of nonverbal interactions and activities, the transactional segment is an area (typically a cone) in front of a person in which such interactions are possible [91,63,87]. It is defined in terms of the biological field of view of a person and oriented as his/her gaze. Since objects and people are foveated for visual acuity, gaze direction generally provides more precise information than other bodily cues regarding the spatial localization of the attentional focus. A detailed overview of gaze-based focus of attention detection in meeting scenarios is presented in [5]. However, measuring it by using eye gaze is often difficult or impossible: either the movement of the subject is constrained, or high-resolution images of the eyes are required, which may not be practical [67,89], and several approximations are considered in many cases. For example, in [91], it is claimed that the focus of attention can be reasonably inferred by head pose, and this is the choice made in many works. Following the same hypothesis, in [87] pan and tilt parameters of the head are estimated, and the focus of attention is represented as a vector normal to the person's face, and it is employed to infer whether or not a walking person is focused on an advertisement located on a vertical glass. Since the situation is very constrained, this proposed model works pretty well, but a more complex model, considering camera position, person's position, and scene structure, is required in more general situations. The same considerations hold for the work presented in [63], where active appearance models are fitted on the face of the person in order to discover which portion of a mall-shelf is observed. In [54], the visual field is modeled as a tetrahedron associated with a head pose detector. However, their model fixes the depth of the visual field, and this is quite unrealistic.

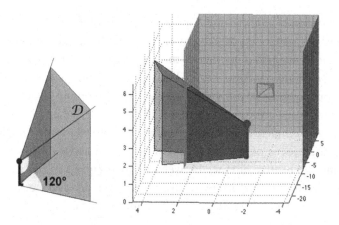

Figure 3.4 (Left) SVF model. (Center) An example of SVF inside a 3D "box" scene. In red, the surveillance camera position: the SVF orientation is estimated with respect to the principal axes of the camera. (Right) The same SVF delimited by the scene constraints (in solid blue). (For interpretation of the references to color in this figure legend, the reader is referred to the web version of this chapter.)

In the following, we present two ways to characterize the frustum of visual attention used to model the transactional segment of individuals.

3.3.2.1 3D Subjective View Frustum

The visual focus of attention is approximated by the *Subjective View Frustum* (SVF), first proposed in [29]. This feature represents the 3D visual field of a human subject in a scene. According to biological evidence [75], the SVF can be modeled as a 3D polyhedron $\mathcal{D}$ delimiting the portion of the scene that the subject is looking at as shown in Fig. 3.4 (left). The SVF is composed of three planes that delimit the view angles on the left, right, and top sides, in such a way that the angle span is 120° in both directions. The 3D coordinates of the points corresponding to the head and feet of a subject are obtained from a person tracker, while the SVF orientation is obtained by a head pose detector, which we discussed in Sections 3.3.1 and 3.3.2, respectively.

The SVF $\mathcal{D}$ is computed as the intersection of three negative half-spaces defined by their supporting planes of the left, right, and top sides of the subject. In practice, the SVF is limited by the scene when the environment is closed, like a room (see Fig. 3.4 (right)). The scene volume is similarly modeled as the intersection of negative half-spaces; consequently, the exact SVF inside the scene can be computed solving a simple *vertex enumeration* problem, for which very efficient algorithms exist in the literature [80].

This information can then be gathered in an *InterRelation Pattern Matrix* [8,28] (IRPM) that encodes the social exchanges occurred among all the persons in a scene.

The SVF of each person can be used to discover the dynamics of the interactions among two or more people. Such analysis relies on a few assumptions with respect to social cues, e.g., the entities involved in the *social interaction* stand closer than 2 m (covering thus the *socio-consultive zone* between 1 and 2 m, the *casual-personal zone* between 0.5 and 1.2 m, and the *intimate zone* around 0.4–0.5 m) [100]. The second condition for the interaction to take part is that the SVFs of two people overlap. The third condition is that their heads are positioned inside the reciprocal SVFs. In this case, it is very likely that a face-to-face conversation is happening [104,56,50]. The InterRelation Pattern Matrix (*IRPM*) records when a possible social interaction occurs, and it can be formalized as a three-dimensional matrix [34], where each entry $IRPM(i, j, t) = IRPM(j, i, t)$ is set to 1 if subjects i and j satisfy the three conditions above, during the tth time instant.

The IRPM matrix is used to analyze time intervals in which we look for social interactions. Suppose we focus on the time interval $[t - T + 1, t]$. In this case we take into account all the IRPM slices that fall in $[t - T + 1, t]$, summing them along the t direction and obtaining the *condensed* IRPM (*cIRPM*). Intuitively, the higher the entry $cIRPM_t(i, j)$, the stronger the probability that subjects i and j are interacting during the interval $[t - T + 1, t]$. Therefore, in order to detect a relation between a pair of individuals i, j in the interval $[t - T + 1, t]$, we check if $cIRPM_t(i, j) > Th$, where Th is a threshold defined a priori. This threshold filters out noisy group detection: actually, due to the errors in the tracking and in the head pose estimation, the lower the threshold, the higher the possibility of false positives detection. In the experiments, we show how the choice of the parameters T and Th impacts the results in terms of social interaction detection rates.

In order to detect groups, we treat *cIRPM* as the adjacency matrix of an undirected graph, with a vertex v_i for each person in the scene, and an edge e_{ij} if $cIRPM_t(i, j) > Th$. The *groups* present in the scene are detected by computing the connected components of the graph.

3.3.2.2 Transactional Segment-Based Frustum

The Transactional Segment-based Frustum (TSbF) is inspired by Kendon's definition of a transactional segment [52] and defines an area in front of a person where interactions may occur. This takes into account both the field of view of the person and also the locus of attention of all other senses for a given body orientation. Since it is typically easier to obtain the head pose rather than the body orientation in crowded environments (due to occlusions), the head pose provides an approximation of the direction of the transactional segment. It is characterized by the direction θ (which is the person's head orientation), aperture γ, and length l. The value γ is directly related to the aperture of the visual field of view. Its range is in $[114°, 190°]$ whenever the binocular or the total field of view is considered [46]. In [98] the parameter γ has been set to $160°$ to consider the binocular aperture plus $20°$ on both sides to include the other senses. Ba and

Odobez [4] used the same measure for approximating the range of possible eye gaze directions. The third parameter is the length l, expressed in cm or in m based on the data, which represents the distance to which interactions may occur. The parameter l can be set following [100], for example, $l \leq 2$ m can be used to detect groups ranging from an intimate zone to a socio-consultive one. These three elements determine the TSbF of a person. The TSbF is based on sampling from two probability distributions, a Gaussian distribution G and a Beta distribution B. The G *distribution* is used to generate samples related to the aperture of the frustum, so it is centered in the head orientation θ of a person with a variance set such that the full width of the Gaussian corresponds to the desired aperture of the frustum. In a Gaussian distribution, 99% of the samples are located in the range of $[-3\sigma, 3\sigma]$; this range will correspond to the full aperture of the frustum.

The B *distribution* is used to generate samples that are dense in close proximity of the person while being few going far away; to achieve this shape, we set the distribution parameters $\alpha = 0.8$ and $\beta = 1.1$. The values returned by the B distribution are in $[0, 1]$ and need to be multiplied by the desired length of the frustum l. The samples obtained using these two distributions are thus in polar coordinates (an angle and a distance), to obtain samples in the 2D space (or in the ground plane assuming that the given positions are in world coordinates), it is sufficient to apply a simple trigonometric rule to each of them. Given a pair of samples from G and B, (G_i, B_i), and the position of a person, (p_x, p_y), the 2D position of each sample is

$$s_x = p_x + \cos(G_i) * B_i * l, \tag{3.1}$$
$$s_y = p_y + \sin(G_i) * B_i * l.$$

Drawing independently n samples from both distributions and applying the above equations for each sample, a set of points that fall in the transactional segment are obtained. Sampling is an important factor because, in general, it allows smoothing noisy data by looking for a statistical consensus.

Moreover, the TSbF model, thanks to its final shape (see Fig. 3.5B), is able to model the *peripheral field of view*, the natural lateral decay of the human perception. The approach in [99] had generated samples from a 2D Gaussian distribution which were subsequently pruned based on the aperture of the human field of view. This process produced sharp edges on the boundaries of the frustum (see Fig. 3.5A) which were not realistic and led to poorer results in terms of quality and performance.

Each person in a scene is thus modeled using his/her frustum represented as a 2D histogram h_i of size $N_c \times N_r$, normalized by the number of samples (n), where N_c and N_r span the area of the scene captured by the camera. Choosing this histogram binning is usually a crucial step in many applications that involve sampling, and indeed to this one. A too coarse binning may lead to an information loss while a thinner binning easily makes the histogram unrepresentative of the underlying sample distribution. In practical

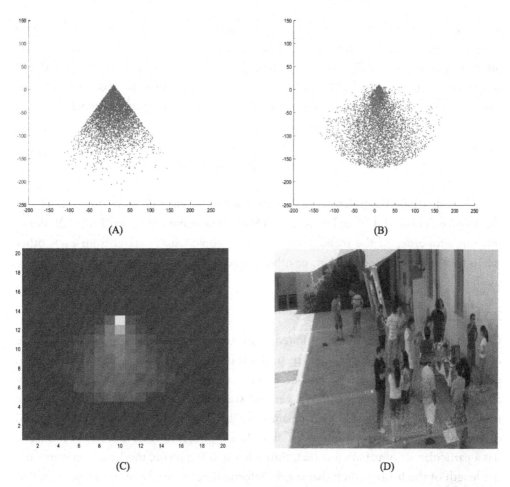

Figure 3.5 (A) The old TSbF model proposed in [99]. (B) TSbF based on sampling from two distributions proposed in [98]. (C) TSbF descriptor, the intensity reflects the number of particles that fall in each bin; the denser, the higher the probability of interaction. (D) An example of frustum projected on the groundplane.

applications with situations similar to a coffee break or a poster session in which the space of the scene is not wide, a binning of 20×20 has been found [99,98] to be the best trade-off in terms of performance and space required. In case of larger spaces this choice could become too coarse, and a thinner binning could be required. Once the 2D histogram of the entire scene has been created, the final feature is obtained by concatenating each row, obtaining a vector of size $1 \times (N_c * N_r)$. The fact that the histogram is created for the entire scene ensures that the spatial relations between each TSbF are not discarded, and it also allows us to easily use any distance function that we want to test.

Like in most clustering techniques, it is necessary here to provide a function that quantifies the distance between pairs of features. For example, in [99,98] the Jensen–Shannon (JS) and the Kullback–Leibler (KL) divergence has been successfully used, but other choices, such as the Earth Mover Distance or the χ^2 could be reasonable alternatives. To obtain a measure of affinity rather than a distance, a Gaussian kernel,

$$\gamma(h_i, h_j) = \exp\left\{-\frac{d(h_i, h_j)}{\sigma}\right\},\qquad(3.2)$$

can be used, where the function "d" refers to some distance measure (like the KL- or the JS-divergence) and h_i and h_j to two TSbFs. The parameter σ in Eq. (3.2) allows intrinsic properties of the scene (e.g., how far people usually stand from each other when they are in an F-formation, or other properties directly connected to the metric used) to be taken into account.

3.3.2.3 External Factors

Other important, yet not fully exploited, features for group detection include the role of the geometry of the scene, e.g., the position of the furniture may affect the creation of groups and thus the performance of the detection and tracking algorithms. The first attempt in this direction appeared in a recent work of Zhang et al. [113] in which the notion of similarity between two persons is related not only to their proximity and orientation, but also to their spatial location. In fact, in [113] each location of the scene has a particular set of parameters that, through a training phase, model the aperture and the length of the frustum such that it gets deformed based on the available space in that location. For example, if there is a wall in an empty room, the persons that are oriented toward it should have smaller and more closed frustums than those in the middle of the room.

3.4 DETECTION MODELS

In this section we present two models for group detection. In Section 3.4.1 we describe the details of how to exploit game theory in the context of group detection using the sociological features presented in the previous section. Section 3.4.2 presents an alternative probabilistic approach which relies on mixture models to represent groups. Note that each model is instantiated with a specific choice of sociological features; however, they are general enough to be implemented with most of the other features presented in the previous section.

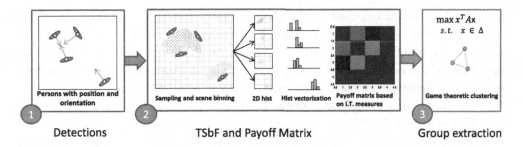

Figure 3.6 The pipeline of the game-theoretic model: (1) detect persons position and orientation, (2) construct the sociological feature and compute the payoff matrix of the game, (3) extract the groups as point of equilibrium (ESS clustering).

3.4.1 Game-Theoretic Conversational Grouping Model

In this section we present a group detection method based on game theory, i.e., the Robust Game-Theoretic Conversational Group detector (R-GTCG) [98].[1] In this method the grouping task is defined as a noncooperative game between two players. The strategies available in the game correspond to the persons in the scene. The payoff obtained by a player choosing strategy i, while the opponent picks strategy j, is directly proportional to the likelihood that the pair of persons (i, j) have to interact. The process iterates until a group emerges. See Fig. 3.6 for the full pipeline.

Being more specific, to build the game-theoretic model we need two ingredients, high-level features and a function to quantify the likelihood of two persons having a conversation. The first ingredient can be obtained through one of the methods in Section 3.3.2, and the latter is strictly related to the desired feature model. For example, in the case of the TSbF, one can use the Kullback–Leibler or Jensen–Shannon divergence to quantify the level of interaction. The game-theoretic model is based on the framework proposed in [92] which is applied to the problem of detecting F-formations as a noncooperative *clustering game*. This game algorithm is a clustering technique which is appealing for a series of desirable properties:

- The similarity function is not required to be a metric, so it is usable with the Kullback–Leibler or other asymmetric measures.
- Setting the number of clusters, like in the k-means procedure, is not needed. This is crucial, since the number of groups in a scene is unknown.
- Is quite fast if the game is composed of fewer strategies ($\simeq 1000$).

Given a set of objects $O = \{1, \ldots, n\}$ and an $n \times n$ affinity matrix $A = (a_{ij})$ which quantifies the pairwise similarities between the objects in O, we consider a situation whereby two players play a game which consists of simultaneously selecting an object

---

[1] Code available at http://xwasco.github.io/GTCG/, checked on October 1, 2016.

from O. The reward obtained by the players is proportional to the similarity of the chosen elements. The *pure strategies* available to both players and the affinity matrix A represent the payoff function. More precisely, each a_{ij} represents the payoff received by a player that chooses strategy i against an opponent playing strategy j. In the task of detecting groups, the objects correspond to the persons detected in a scene, while the payoff function is a similarity measure between the high-level feature of subjects as described in Section 3.3.2.2.

A cardinal point in game theory is the notion of a *mixed strategy*, which is simply a probability distribution $x = (x_1, \ldots, x_n)^T$ over the set of pure strategies O (note that in this case the vector x is not related to the feature of persons in a scene). Mixed strategies clearly belong to the n-dimensional simplex $\Delta = \{x \in \mathbb{R}^n : \sum_{i=1}^{n} x_i = 1 \text{ and } x_i \geq 0, \ i = 1, \ldots, n\}$. Given a mixed strategy $x \in \Delta$, we define its *support* as $\sigma(x) = \{i \in O : x_i > 0\}$.

The expected payoff received by an individual playing a mixed strategy y against an opponent with mixed strategy x is given by $y^T A x$. The set of *best replies* against a mixed strategy x is defined as $\beta(x) = \{y \in \Delta : y^T A x = \max_z z^T A x\}$. A mixed strategy $x \in \Delta$ is said to be a *Nash equilibrium* if it is a best reply to itself, namely, if $x \in \beta(x)$ or, in other words, if

$$x^T A x \geq y^T A x \tag{3.3}$$

for all $y \in \Delta$. Intuitively, at a Nash equilibrium no player has an incentive to unilaterally deviate from it.

The clustering game is played within an evolutionary setting in which two players, each assumed to play a preassigned strategy, are repeatedly drawn at random from a large population. Here, a given mixed strategy $x \in \Delta$, x_j ($j \in O$) is assumed to represent the proportion of players that are programmed to select pure strategy j. A dynamic evolutionary selection process will then make the state x to evolve according to the principle of natural selection, eventually, the better-than-average (pure) strategies will survive while the others will get extinct. Within this context, a mixed strategy $x \in \Delta$ is said to be an *evolutionary stable strategy* (ESS) if it is a Nash equilibrium and if, for each best reply y to x, we have $x^T A y > y^T A y$. Intuitively, ESSs are mixed strategies such that any small deviation from them will lead to an inferior payoff. Interested readers can consult [103] for an introduction to evolutionary game theory. In [92] a combinatorial characterization of ESSs is given, making them plausible candidates for the notion of a cluster (called *ESS-cluster*). The motivation behind this claim resides in the fact that ESS-clusters incorporate the two basic features which characterize all clustering algorithms, i.e.,

- *internal coherency*: elements belonging to the cluster should have high mutual similarities;
- *external incoherency*: the overall cluster internal coherency decreases by introducing external elements which are not similar enough to the elements within the cluster.

To find an ESS-cluster, one can use the classical *replicator dynamics* [103], a class of dynamical systems that mimic a Darwinian selection process over the set of objects (pure strategies):

$$x_i(t+1) = x_i(t) \frac{(A\mathbf{x}(t))_i}{\mathbf{x}(t)^T A \mathbf{x}(t)} \tag{3.4}$$

for all $i \in O$. The process starts from a point $\mathbf{x}(0)$ on the barycenter of the simplex Δ ($\forall i,\ x_i = \frac{1}{n}$), and it is iterated until convergence (typically when the distance between two successive states is smaller than a given threshold, or a desired number of steps is reached). The entire dynamical process is driven by the payoff function which is defined precisely to favor the evolution of highly coherent objects that are mutually similar. Mutual similarity in the specific task of group detection implies a desire of two persons to interact.

The support $\sigma(\mathbf{x})$ of the converged population state $\mathbf{x}$ represents a cluster, that is, a group in our context. In order to extract all the ESS-clusters, and thus detect all the groups, the elements of the current cluster are removed from the original set and Eq. (3.4) is iterated again on the remaining elements until all the elements are grouped. Analyzing the time and space complexity of the R–GTCG method, we can state the following: the time complexity is dominated by the Replicator Dynamics (Eq. (3.4)) which is $\mathcal{O}(n^2)$ per step, where n is the number of elements in the dataset (persons in the scene). Given k, the number of steps required by Eq. (3.4) to converge, the overall time complexity to extract one group is $\mathcal{O}(k * n^2)$. Usually, few steps ($k \simeq 100$) are required for convergence since the number of elements (persons) in a scene is not huge. For the space complexity, the matrix A is in the class of $\mathcal{O}(n^2)$.

3.4.2 The Dirichlet Process Mixture Model

The game-theoretic approach has the advantage of being deterministic, and therefore it is easy to use in practice. However, the drawback is that it is less flexible than the probabilistic approach. In this section, we present an alternative model which is based on probability theory in order to cope with the flexibility issue of the game-theoretic method.

The main idea is to represent the set of individuals in a given frame with a Gaussian mixture model with a potentially infinite number of components. Let $\mathbf{X}_t = \{\mathbf{x}_t^k\}_{k=1}^K$ be the joint state of the K individuals, and $\boldsymbol{\Theta}_t = \{\mathbf{z}_t^k\}_{k=1}^K$ with $\mathbf{z}_t^k \in \{0, 1, \ldots, G\}$ be the joint state of the G groups. $\mathbf{x}_t^k$ can be any combination of the set of features that we discussed in Section 3.3. $\mathbf{z}_t^k$ is the group label for the kth individual. As an example, suppose we have 5 individuals and 2 groups at time t: $\boldsymbol{\Theta}_t = [1, 1, 2, 2, 0]^T$ indicates that the first two individuals belong to the first group, the third and fourth are in the second group, and the fifth individual is a singleton.

Group modeling is addressed as a problem of mixture model fitting, where each group is defined as a mixture component in the chosen feature space. Individuals are

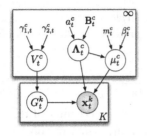
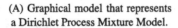
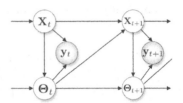

(A) Graphical model that represents a Dirichlet Process Mixture Model.

(B) Model for joint individual-group tracking.

Figure 3.7 Models for detection and tracking.

seen as observations drawn from the mixture. The number of mixture components (i.e., groups) is in general unknown a priori and may change over time. Consequently, standard mixture models like Gaussian mixtures are not suitable since they do not deal with the dynamic inclusion and exclusion of components. Dirichlet Process Mixture Models (DPMMs) [64] have been used to overcome this limitation. DPMMs can represent mixture distributions with an unbounded number of components, where the complexity of the mixture adapts to the observed data.

Each individual $\mathbf{x}_t^k$ is interpreted as an observation coming from one of the infinitely many components of the Dirichlet Process mixture. This component represents the group it belongs to. In order to keep inference tractable, a stick-breaking representation [83] of the Dirichlet Process prior [32] is used. Such a representation is a constructive process generating an infinite set of nonnegative numbers summing to 1, by sequentially sampling from a series of *Beta* distributions. The obtained sequence can be interpreted as the mixing coefficients of a set of components defining a mixture model. The graphical model associated to the generative process linking mixture components c to observations is shown in Fig. 3.7A where, at time step t, we have K points and

$$V_t^c | \gamma_{1;t}^c, \gamma_{2;t}^c \sim Beta\left(\gamma_{1;t}^c, \gamma_{2;t}^c\right), \qquad G_t^k | \{v_{1,t}, v_{2,t}, \ldots\} \sim Discrete\left(\rho_t^c\left(V_t^c\right)\right),$$

$$\mathbf{\Lambda}_t^c | \mathbf{B}_t^c, d_t^c \sim Wi\left(\mathbf{B}_t^c, d_t^c\right), \qquad \mu_t^c | m_t^c, \beta_t^c, \mathbf{\Lambda}_t^c \sim \mathcal{N}\left(m_t^c, \left(\beta_t^c \mathbf{\Lambda}_t^c\right)^{-1}\right),$$

$$\mathbf{x}_t^k | G_t^k \sim \mathcal{N}\left(\mu_{G_t^k}, \mathbf{\Lambda}_{G_t^k}^{-1}\right),$$

where *Discrete* and *Wi* are the Discrete and Wishart distributions, respectively, $\mathbf{x}_t^k$ represents the k-th data point, G^k is an assignment variable relating each kth data point to the mixing components, V^c and the pair $(\mu^c, \mathbf{\Lambda}^c)$ represent the parameters of the cth mixture component in the stick-breaking construction [83], with $(\mu^c, \mathbf{\Lambda}^c)$ representing the location of the component in the parameter space and V^c defining the mixing proportions through ρ_t^c. For convenience, all the parameters of each mixture component c are grouped together as $\boldsymbol{\theta}_t^c = \{\mathbf{v}_t^c, \mu_t^c, \boldsymbol{\Lambda}_t^c, G_t^c\}$, and $\boldsymbol{\Theta}_t = \{\boldsymbol{\theta}_t^1, \boldsymbol{\theta}_t^2, \ldots\}$.

The likelihood of a data point given the model is defined as follows:

$$p\left(\mathbf{x}_t^k | \boldsymbol{\Theta}_t\right) = \sum_{c=1}^{\infty} \rho_t^c \cdot \mathcal{N}\left(\mathbf{x}_t^k | m_t^c, a_t^c \left(\mathbf{B}_t^c\right)^{-1}\right). \tag{3.5}$$

This enables us to determine the probability each person belongs to any of the groups. Such probability depends on how likely the observation associated to the person is under the Gaussian component associated to the group, and on the probability of the mixture component itself. Hard association of each individual to a group can be made by taking the argmax of Eq. (3.5).

Group dynamics is difficult to model especially because groups tend to split and merge, generating new ones. Nevertheless, the grouping configurations of two consecutive frames are highly correlated due to the temporal smoothness of people's trajectories. This observation is exploited in the Bayesian framework as a sequential inference scheme, where the grouping configuration at one time step can be used as a prior belief for the next. We use the sequential variational inference algorithm presented in [111] which allows us to implement online unsupervised learning of group structures. We refer to [111] for further details.

In this model, it is also possible to add constraints on the possible mixture components (i.e., group configurations) inferred by the learning algorithm, considering elements of proxemics [45], which assume that people tend to unconsciously organize the space around them in concentric zones with different degrees of intimacy. The shorter the distance between two persons, the higher the degree of intimacy. Thus, we define a limit distance ($r = 2$ m [100]), beyond which two individuals can be considered not to be interacting with high probability. When this limit is not respected, the corresponding mixture component is discarded and a new one is initialized in a region of the space which is badly modeled by the mixture distribution.

3.5 GROUP TRACKING

In this section we formally define the problem of the joint individual-group tracking. We introduce two methods: one directly exploits simple features such as individual position and velocity, and the second one is built on top of more sophisticated social description, such as the group detection models of Section 3.4.

We model the state space of individual-group tracking as the decomposition of two subspaces that are *conditionally dependent*. The subspaces are represented by the variables $\mathbf{X}_t$ and $\boldsymbol{\Theta}_t$, such that $\xi_t = [\mathbf{X}_t, \boldsymbol{\Theta}_t]^T$, where the subspace of the individuals is $\mathbf{X}_t$ and the subspace of the groups is $\boldsymbol{\Theta}_t$. The tracking problem is defined as a classical nonlinear discrete-time system employed for the generic object tracking as follows:

$$\mathbf{X}_{t+1} = f_t^X\left(\mathbf{X}_t, \boldsymbol{\Theta}_t, \eta_t^X\right), \qquad \boldsymbol{\Theta}_{t+1} = f_t^\Theta\left(\mathbf{X}_{t+1}, \boldsymbol{\Theta}_t, \eta_t^\Theta\right),$$
$$\mathbf{y}_t = h_t\left(\mathbf{X}_t, \boldsymbol{\Theta}_t, \eta_t^y\right).$$

The conditional dependencies between the variables are shown in Fig. 3.7B: the state of the individuals $\mathbf{X}_{t+1}$ depends on its previous state $\mathbf{X}_t$ and the previous state of the groups $\boldsymbol{\Theta}_t$. The state of the groups $\boldsymbol{\Theta}_{t+1}$ depends on the state of the individuals $\mathbf{X}_{t+1}$ and the previous state of the groups $\boldsymbol{\Theta}_t$. Finally, the observation $\mathbf{y}_{t+1}$ depends on both the state of the individuals and groups because both of them generate the current measurements.

Inference is performed using the Decentralized Particle Filter (DPF) originally proposed in [17], which recursively estimates the posterior distribution $p\left(\boldsymbol{\Theta}_t, \mathbf{X}_{0:t}|\mathbf{y}_{0:t}\right)$ through a *decomposition* of the joint state space in two subspaces $\boldsymbol{\Theta}$ and $\mathbf{X}$. This posterior distribution can be factorized by using the definition of conditional probability as follows:

$$p\left(\boldsymbol{\Theta}_t, \mathbf{X}_{0:t}|\mathbf{y}_{0:t}\right) = p\left(\boldsymbol{\Theta}_t|\mathbf{X}_{0:t}, \mathbf{y}_{0:t}\right) p\left(\mathbf{X}_{0:t}|\mathbf{y}_{0:t}\right) \tag{3.6}$$

where $\mathbf{y}_{0:t} = \left(\mathbf{y}_0, \ldots, \mathbf{y}_t\right)$ and $\mathbf{X}_{0:t} = \left(\mathbf{X}_0, \ldots, \mathbf{X}_t\right)$ represent the sequence of observations and states up to the time t, respectively. The factorization adopted by the DPF circumvents both the inefficiency and ineffectiveness of the classical particle filtering [27] when dealing with large state spaces. The main idea is to split the inference in Eq. (3.6) as follows:

$$p\left(\boldsymbol{\Theta}_t|\mathbf{X}_{0:t}, \mathbf{y}_{0:t}\right) \propto p\left(\boldsymbol{\Theta}_t|\mathbf{X}_{0:t}, \mathbf{y}_{0:t-1}\right) p\left(\mathbf{y}_t|\mathbf{X}_t, \boldsymbol{\Theta}_t\right), \tag{3.7}$$
$$p\left(\mathbf{X}_{0:t}|\mathbf{y}_{0:t}\right) \propto p\left(\mathbf{y}_t|\mathbf{X}_{0:t}, \mathbf{y}_{0:t-1}\right) p\left(\mathbf{X}_t|\mathbf{X}_{0:t-1}, \mathbf{y}_{0:t-1}\right) p\left(\mathbf{X}_{0:t-1}|\mathbf{y}_{0:t-1}\right). \tag{3.8}$$

These equations highlight that the inference is online, that is, it is possible estimate $p\left(\boldsymbol{\Theta}_t|\mathbf{X}_{0:t}, \mathbf{y}_{0:t}\right)$ and $p\left(\mathbf{X}_{0:t}|\mathbf{y}_{0:t}\right)$ given the results of inference at the previous step, $p\left(\boldsymbol{\Theta}_{t-1}|\mathbf{X}_{0:t-1}, \mathbf{y}_{0:t-1}\right)$ and $p\left(\mathbf{X}_{0:t-1}|\mathbf{y}_{0:t-1}\right)$, respectively.

Since performing inference of Eqs. (3.7) and (3.8) is usually not feasible because it involves the computation of integrals when expanding the equations (see the supplementary material of [9]), Monte Carlo techniques are used to approximate the solution of the problem. The DPF uses sequential importance sampling [27], a popular online Monte Carlo method, to perform inference for the distributions in Eqs. (3.7) and (3.8). In sequential importance sampling, a set of weighted samples (or hypotheses) are updated as soon as a new observation comes. The weights (corresponding to the importance of each sample) are first updated, given the current observation. New hypotheses are then sampled for the next time step considering their weights. The less important a hypothesis is, the less probable its survival. In DPF the sampling/weighting procedure

---

Algorithm 1: The DPF algorithm [17]. INPUT: samples $\{\mathbf{X}_t^{(i)}\}_{i=1,\dots,N_x}$, samples $\{\mathbf{\Theta}_t^{(i,j)}\}_{i=1,\dots,N_x,j=1,\dots,N_z}$, where i and j are the indices for the space $\mathbf{X}$ and $\mathbf{\Theta}$, respectively. OUTPUT: a new set of particles $\{\mathbf{X}_{t+1}^{(i)}\}_{i=1,\dots,N_x}$ and $\{\mathbf{\Theta}_{t+1}^{(i,j)}\}_{i=1,\dots,N_x,j=1,\dots,N_z}$.

---

1. Approximate $p\left(\mathbf{X}_t|\mathbf{y}_{0:t}\right)$ through the importance weights

$$w_t^{(i)} \propto \frac{p_{N_z}\left(\mathbf{y}_t|\mathbf{X}_{0:t}^{(i)},\mathbf{y}_{0:t-1}\right)p_{N_z}\left(\mathbf{X}_t^{(i)}|\mathbf{X}_{0:t-1}^{(i)},\mathbf{y}_{0:t-1}\right)}{\pi\left(\mathbf{X}_t^{(i)}|\mathbf{X}_{0:t-1}^{(i)},\mathbf{y}_{0:t}\right)}.$$

2. Resample $\left\{\mathbf{X}_t^{(i)},\mathbf{\Theta}_t^{(i,j)}\right\}$ according to $w_t^{(i)}$.

3. Approximate $p\left(\mathbf{\Theta}_t|\mathbf{X}_{0:t},\mathbf{y}_{0:t}\right)$ through the importance weights $\bar{q}_t^{(i,j)} \propto p\left(\mathbf{y}_t|\mathbf{X}_t^{(i)},\mathbf{\Theta}_t^{(i,j)}\right)$.

4. Generate $\mathbf{X}_{t+1}^{(i)}$ according to $\pi\left(\mathbf{X}_{t+1}|\mathbf{X}_{0:t}^{(i)},\mathbf{y}_{0:t+1}\right)$.

5. Approximate $p\left(\mathbf{\Theta}_t|\mathbf{X}_{0:t+1},\mathbf{y}_{0:t}\right)$ through the importance weights

$$q_t^{(i,j)} = \bar{q}_t^{(i,j)} p\left(\mathbf{X}_{t+1}^{(i)}|\mathbf{X}_{0:t}^{(i)},\mathbf{\Theta}_t^{(i,j)}\right).$$

6. Resample $\mathbf{\Theta}_t^{(i,j)}$ according to $q_t^{(i,j)}$.

7. Generate $\mathbf{\Theta}_{t+1}^{(i,j)}$ according to $\pi\left(\mathbf{\Theta}_{t+1}|\mathbf{X}_{0:t+1}^{(i)},\mathbf{\Theta}_t^{(i,j)}\right)$.

---

is applied on each subspace in a nested way (see Algorithm 1): first, the algorithm estimates the importance weights for $\mathbf{X}_t$ and $\mathbf{\Theta}_t$ (steps 1 and 3); second, new sample sets are generated for both distributions using the current sample sets (steps 4 and 7). Note that there is an additional step (step 5) that enables us to estimate $\mathbf{\Theta}_t$ by looking ahead to the hypotheses of $\mathbf{X}_{t+1}$, in the same spirit of [25]. In practice, the time complexity of the algorithm is $\mathcal{O}\left(N_x \cdot N_z \cdot K^2\right)$.

More specifically, in step 1 of Algorithm 1, the standard importance sampling formulation (*Observation · Dynamics*)/(*Proposal*) is applied to approximate $p\left(\mathbf{X}_{0:t}|\mathbf{y}_{0:t}\right)$. The difference with the standard framework lies in the term $\mathbf{y}_{0:t-1}$, the presence of which is formally motivated by a mathematical derivation discussed in [17]. Intuitively, the conditioning of $\mathbf{y}_{0:t-1}$ injects the knowledge acquired by explaining the observations $\mathbf{y}$ in the subspace $\mathbf{\Theta}$ at time $t-1$.

3.5.1 DPF for Group Tracking

In this section, we explicitly define the probability distributions highlighted in gray in Algorithm 1 for the joint individual-group tracking problem. For simplicity, we decided to define the state space of individuals as $\mathbf{x}_t^k = [x_t, y_t, \dot{x}_t, \dot{y}_t]$, which corresponds to using the individual positions and velocities as social features. We want to remark that the presented method is general, and other social features can be included in the state representation.

Individual Proposal $\pi\left(\mathbf{X}_{t+1}|\mathbf{X}_{0:t}, \mathbf{y}_{0:t+1}\right)$

This distribution models the dynamics of the individuals. Inspired by [73], we adopt the notion of composite proposal, incorporating two sources of information:

$$\pi\left(\mathbf{X}_{t+1}|\mathbf{X}_{0:t}, \mathbf{y}_{0:t+1}\right) = \alpha\, \pi_{\mathrm{dyn}}\left(\mathbf{X}_{t+1}|\mathbf{X}_t\right) + (1-\alpha)\, \pi_{\mathrm{det}}\left(\mathbf{X}_{t+1}|\mathbf{X}_t, \mathbf{y}_{t+1}\right),$$

where we assume Markovianity between the $\mathbf{X}$s and conditional independence with respect to the observations $\mathbf{y}_{0:t}$. We adopt a locally linear dynamics with Gaussian noise $\mathbf{x}_{t+1}^k = A\mathbf{x}_t^k + \eta^x$, where T is the sampling interval and $\eta^x \sim \mathcal{N}\left(\mathbf{0}, \Sigma^k\right)$. Therefore, $\mathbf{x}_{t+1}^k \sim \mathcal{N}\left(A\mathbf{x}_t^k, \Sigma^k\right)$, which is easy to evaluate and sample from. We have

$$\pi_{\mathrm{dyn}}\left(\mathbf{X}_{t+1}|\mathbf{X}_t\right) = \prod_{k=1}^K \mathcal{N}\left(\mathbf{x}_{t+1}^k|A\mathbf{x}_t^k, \Sigma^k\right), \tag{3.9}$$

that is, a multivariate Gaussian distribution with block-diagonal covariance matrix, $\mathrm{diag}(\Sigma^1, \Sigma^2, \ldots, \Sigma^K)$. We assumed that $\Sigma^k = \Sigma$ for each $k = 1, \ldots, K$.

The second term $\pi_{\mathrm{det}}\left(\mathbf{X}_{t+1}|\mathbf{X}_t, \mathbf{y}_{t+1}\right)$ generates hypotheses in those regions of the state space where it is more probable to find a person. In practice, we are using an informative proposal that searches for detected people. When a person is detected, the tracker will be more reliable. The distribution is defined as a multivariate Gaussian distribution centered in the detections associated to each target with covariance matrix Σ (same as Eq. (3.9)). The parameter $\alpha = 0.5$ is fixed for all the experiments.

Joint Observation Distribution $p\left(\mathbf{y}_t|\mathbf{X}_t, \Theta_t\right)$

The joint observation distribution is defined to account for both the appearance and the group membership contributions as follows:

$$p\left(\mathbf{y}_t|\mathbf{X}_t, \Theta_t\right) \propto g_{\mathrm{app}}\left(\mathbf{y}_t, \mathbf{X}_t, \Theta_t\right) \cdot g_{\mathrm{mem}}\left(\mathbf{y}_t, \mathbf{X}_t, \Theta_t\right).$$

The individuals contribute to the appearance component as follows:

$$g_{\mathrm{app}}\left(\mathbf{y}_t, \mathbf{X}_t, \Theta_t\right) = \prod_{k=1}^K e^{-\lambda_y\, d_y\left(q(\mathbf{y}_t, \mathbf{x}_t^k), \tau^k\right)},$$

where $q\left(\mathbf{y}_t, \mathbf{x}_t^k\right)$ extracts a descriptor from the current bounding box in the image given by $\mathbf{x}_t^k$, τ^k is the descriptor of the template of the kth individual and d_y is a distance between descriptors. It is easy to notice that the appearance component is defined as the standard template-based technique [15], widely used in particle filtering approaches [49]. In the experiments, we used the Bhattacharyya distance between RGB color histograms, and the template was never updated.

We defined two different membership components depending on the availability or not of the group detector presented in the previous section. In case the detector is not available, we assume independence with respect to $\mathbf{y}_t$, therefore the definition is the following:

$$g_{\mathrm{mem}}\left(\mathbf{y}_t, \mathbf{X}_t, \mathbf{\Theta}_t\right) = e^{-\lambda_d\, d_{cl}(\mathbf{\Theta}_t, \mathbf{X}_t)} \prod_{g=1}^{G} \mathcal{N}\left(S_t^g | \mu, \sigma\right),$$

where $d_{cl}(\mathbf{\Theta}_t, \mathbf{X}_t)$ is a cluster validity measurement, such as the Davies–Bouldin index [24] and S_t^g is the size of the gth group in $\mathbf{\Theta}_t$. The second term of the membership component penalizes the hypotheses of having too-large groups (in the experiments, $\mu = 1$ and $\sigma = 1.5$). However, such measurements are usually heuristics, and they do not ensure that a model is better than another in absolute terms.

When the DPMM group detector is available, the membership function can be defined in a more elegant way by using Eq. (3.5) directly as follows:

$$g_{\mathrm{mem}}\left(\mathbf{y}_t, \mathbf{X}_t, \mathbf{\Theta}_t\right) = \prod_{k=1}^{K} \sum_{c=1}^{C} \rho_t^c \cdot \mathcal{N}\left(\mathbf{y}_t^{x^k} | m_t^c, a_t^c \left(\mathbf{B}_t^c\right)^{-1}\right),$$

where C is the truncation level of the mixture [13],

$$\rho^c = \begin{cases} \nu^c, & \text{if } c = 1, \\ \nu^c \cdot \prod_{j=1}^{c-1}(1 - \nu^j), & \text{if } c > 1, \end{cases}$$

with

$$\nu^c = \begin{cases} \dfrac{\gamma_1^c}{\gamma_1^c + \gamma_2^c}, & \text{if } c < C, \\ 1, & \text{if } c = C, \end{cases}$$

and $\mathbf{y}_t^{x^k}$ being the detection associated to the kth individual. The membership function encourages the hypotheses that fit well with the mixture components, and therefore it is more likely to obtain compact groups.

Joint Individual Distribution $p\left(\mathbf{X}_{t+1}|\mathbf{X}_{0:t}, \boldsymbol{\Theta}_t\right)$

This distribution models the dynamics of the individual, taking into account the presence of the groups as $\mathbf{x}_{t+1}^k = \mathbf{x}_t^k + B\mathbf{g}_t^k + \eta^x$ where

$$B = \begin{bmatrix} 0 & 0 & T & 0 \\ 0 & 0 & 0 & T \\ 0 & 0 & 0 & 0 \\ 0 & 0 & 0 & 0 \end{bmatrix}, \quad \mathbf{g}_t^k = \frac{\sum_{l=1}^K \mathbf{x}_t^l \mathbb{I}\left(\mathbf{z}_t^k == \mathbf{z}_t^l\right)}{\sum_{l=1}^K \mathbb{I}\left(\mathbf{z}_t^k == \mathbf{z}_t^l\right)},$$

$\mathbb{I}(\cdot)$ is the indicator function, and $\mathbf{g}_t^k$ is the position and velocity of the group the kth individual belongs to. Note that matrix B selects only the velocity vector of $\mathbf{g}_t^k$, discarding the positional information. This encourages individuals in the same group to have similar dynamics. The resulting probability distribution is

$$p\left(\mathbf{X}_{t+1}|\mathbf{X}_{0:t}, \boldsymbol{\Theta}_t\right) = \prod_{k=1}^K \mathcal{N}\left(\mathbf{x}_{t+1}^k|\mathbf{x}_t^k + B\mathbf{g}_t^k, \Sigma\right).$$

Joint Group Proposal $\pi\left(\boldsymbol{\Theta}_{t+1}|\mathbf{X}_{0:t+1}, \boldsymbol{\Theta}_t\right)$

The joint group proposal represents how the groups change over time based on the observation of the individuals. For this reason, we need again to distinguish between the case where the group detector is available from the case where it is not.

In case that the group detector is not available, we manually specify the possible events that may happen to a group (namely merge, split, and none). This enables us to define a surrogate distribution which is easier to sample from than the original proposal. The joint group proposal is therefore defined as

$$\pi\left(\boldsymbol{\Theta}_{t+1}|\mathbf{X}_{t+1}, \boldsymbol{\Theta}_t\right) = f\left(\prod_{g=1}^G \pi\left(e_{t+1}^g|\mathbf{X}_{t+1}, g_t, g_t'\right), \boldsymbol{\Theta}_t\right) \tag{3.10}$$

where $\pi\left(e_{t+1}^g|\mathbf{X}_{t+1}, g_t, g_t'\right)$ is the surrogate distribution defined on the set of events $e^g \in$ {Merge, Split, None} related to the gth group, and g' is the group associated to the gth group using the distance between their centroids.

The surrogate distribution is learned offline using the multinomial logistic classifier. To this end, a set of possible scenarios containing events have been simulated and labeled [7]. We use as features (i) the inter-group distance between g and the nearest group g', considering their positions and sizes (d_{KL}, symmetrized Kullback–Leibler distance between Gaussians) and velocities (d_v, Euclidean distance), and (ii) the intra-group variance between the positions of the individuals in the gth group (d_{intra}). Thus, the input of the multinomial logistic regression is a 6-dimensional vector, i.e., ($d_{KL}, d_v, d_{\text{intra}}$)

for time instances t and $t + 1$. The output is the probability values for each class in e^g; we use these values to sample the new action for each group.

When the DPMM detector is available, the group proposal distribution is learned in an online and unsupervised fashion. Assume that the DPMM $\tilde{\Theta}_t$ is trained using the individual state estimate (not the particles) up to the current frame, so that only one model is kept at each time step. To sample the new hypotheses set at time $t + 1$, a copy of the model $\Theta_t^{(i)}$ is instantiated for each new particle as the DPMM learned to date. The model $\tilde{\Theta}_t^{(i)}$ is updated using the ith hypothesis $\mathbf{X}_{t+1}^{(i)}$ in accordance with the method discussed in Section 3.4.2, thus generating N different models $\Theta_{t+1}^{(i)}$, $i = \{i, \ldots, N\}$.

Then, M samples are generated from each model, that is, $\Theta_{t+1}^{(i,j)}$, $i = \{i, \ldots, N\}, j = \{1, \ldots, M\}$. In order to perform this second sampling step, new grouping hypotheses are formulated on the basis of what has been learned by the model. To this end, a new mixture hypothesis is assigned to each particle by performing ancestral sampling from the graphical model in Fig. 3.7A, for each of the mixture components. In practice, the parameters of component c are obtained by fixing the hyperparameters γ_1^c, γ_2^c, $\mathbf{B}^c$, α^c, m^c, β^c to their current estimate and sampling in a top-down manner the random variables in the graphical model, sampling children as soon as all their parents have been sampled.

3.6 EXPERIMENTS

In this section the results obtained in the task of group detection and tracking are reported.

3.6.1 Results of Group Detection

3.6.1.1 Datasets

The methods have been evaluated on the following publicly available datasets: Friends-Meet (FM) [7], Discovering Groups of People in Images (DGPI) [18], PosterData [47], CocktailParty [112], CoffeeBreak [22], Synth [22], GDet [22], and BIWI [79].

3.6.1.2 Evaluation Metrics

The evaluation of the detection models follows the standard method of [84]: a group is considered as correctly detected if at least $\lceil (T \cdot |G|) \rceil$ of its members are correctly grouped by the algorithm, and if no more than $\lceil (1 - T) \cdot |G| \rceil$ false subjects are identified, where $|G|$ is the cardinality of the labeled group G. Variable T was set equal to 2/3 for all the datasets except for the DGPI [18] and FM datasets; for them it was set to $T = 1/2$ and $T = 0.6$, respectively, for a fair comparison with the state-of-the-art. Based on this criterion, the *precision*, *recall*, and *F1-score* per frame can be computed; the final score is obtained averaging all the results on a single frame.

Table 3.1 Results of the GT detector. Only the best results are shown while the parameters are discussed in the results section (σ in Eq. (3.2) and l in Eq. (3.1)). The comparative methods are: IRPM [8], HFF [22], DS [84], MULTISCALE [85], GTCG [99]; "R-GTCG" [98] is the method presented in this chapter

Method*	CoffeeBreak (S1 + S2)*			PosterData			GDet		
	Prec	Rec	F1	Prec	Rec	F1	Prec	Rec	F1
IRPM [8,84]	0.60	0.41	0.49	—	—	—	—	—	—
HFF [22,84]	0.82	0.83	0.82	**0.93**	**0.96**	**0.94**	0.67	0.57	0.62
DS [47,84]*	0.68	0.65	0.66	**0.93**	0.92	0.92	—	—	—
MULTISCALE [85]	0.82	0.77	0.80	—	—	—	—	—	—
GTCG [99] JS	0.83	**0.89**	0.86	0.92	**0.96**	**0.94**	0.76	0.76	0.76
R-GTCG [98]	**0.86**	0.88	**0.87**	0.92	**0.96**	**0.94**	0.76	0.76	0.76
	$\sigma = 0.2, l = 145$			$\sigma = 0.25, l = 115$			$\sigma = 0.7, l = 180$		

Method*	Cocktail Party			Synth		
	Prec	Rec	F1	Prec	Rec	F1
IRPM [8,84]	—	—	—	0.71	0.54	0.61
HFF [22,85]	0.59	0.74	0.66	0.73	0.83	0.78
MULTISCALE [85]	0.69	0.74	0.71	0.86	0.94	0.90
GTCG [99] JS	0.86	**0.82**	**0.84**	**1.00**	**1.00**	**1.00**
R-GTCG [98]	**0.87**	**0.82**	**0.84**	**1.00**	**1.00**	**1.00**
	$\sigma = 0.6, l = 170$			$\sigma = 0.1, l = 75$		

* Note that in [84] the parameters for the DS method were not fully optimized.

* In case of double citations, the first one refers to the original method, while the latter refers to the more recent paper from which the results have been taken.

3.6.1.3 Comparing Methods

The game-theoretic method was compared with the Hough-based approach of [22] in its renewed version of [84] (HFF), the hierarchical extension of the Hough-based approach of [85] (MULTI), the InterRelation Pattern Matrix (IRPM) [8,28] (see Section 3.3.2.1), the Dominant Set-based technique of [47] (DS), the approach of Choi et al. [18] (DGPI), and finally with the DPMM [111] (see Section 3.4.2).

3.6.1.4 Performance Evaluation

The results of the game-theoretic model (R-GTCG) and the corresponding parameters are reported in Table 3.1. We report the performance averaged over 10 runs to evaluate the stability of the method. The R-GTCG detector uses the TSbF presented in Section 3.3.2.2 whose parameters have been found empirically on each single dataset due to the great difference across the datasets in terms of geometry of the scene and level of crowdedness. The distance function that we used is the Jensen–Shannon divergence as suggested in [99].

The only cases where R-GTCG does not outperform the state-of-the-art are on the PosterData, with a difference of 1% in the precision with respect to DS and HHF,

Table 3.2 (Left) The results obtained on the DGPI dataset. $\Diamond$ = using only the facing classes, $\blacklozenge$ = using all group types. (Right) The average results on FriendsMeet. $\star$ = the queueing sequences are removed, # = the queuing and the side-by-side sequences are removed

DGPI dataset				FriendsMeet dataset	
Method	Prec	Rec	F1	Method	GDSR
DGPI [18]	0.59	0.65	0.62	DPMM [111]	0.946
R-GTCG $\Diamond$	0.54	0.57	0.56	R-GTCG $\star$	0.899
R-GTCG $\blacklozenge$	0.99	0.73	0.84	R-GTCG #	0.934

and in the recall of the CoffeeBreak with a difference of 1%. In all the other datasets, the results are definitely superior, saturating, for example, the synthetic benchmark, and outperforming by over 10% the *F1-score* on the GDet and the CocktailParty. The results of the IRPM approach are considerably lower because its F-formation model depends on many heuristics which are not always valid in practice: the length of the period of time that the interactions are accumulated on and the interaction threshold. For example, IRPM is not able to capture brief interactions that last less that the interaction threshold. It is worth noting that the performances across the different runs of the algorithm have been quite stable, with a maximum variance of about 0.6% for both the precision and recall values.

The DGPI and FM dataset were designed for different purposes than the other datasets analyzed so far. The DGPI dataset includes a larger set of groups made of 7 different types, and thus having 7 diverse ground-truths. The FM contains queues of individuals and persons walking side by side. For these reasons, we decide to do a specific evaluation of those datasets, separated from the others.

The method presented in [18] is the state-of-the-art on the DGPI dataset. The comparison with [18] was made in two ways: first taking as ground-truth the union of the ground-truths in which the persons are facing, thus can have a conversation (the second row of Table 3.2, left) and in the latter case by taking the union of all the ground-truths. The obtained results are compared with the average obtained from Table 2 of [18] on the related ground-truth. The parameter has been set to the following values $\sigma = 0.5$ and $l = 1.3$ m (a value in the social space range reported in [45,100]). If all 7 classes are considered as generic groups and used for the evaluation, the method has very high performance with a precision reaching 0.99, a recall of 0.73, and an F1-score of 0.84 (the last row of Table 3.2, left). Unfortunately, in [18] only the average over the detections per type of group is reported (precision 0.50, recall 0.44, and F1-score 0.47), so this comparison is not completely fair, yet it gives an indication of the quality of the game-theoretic approach also in these conditions.

The results on the FM dataset are reported in Table 3.2, right. The parameters used in this dataset have been kept fixed for all the sequences since there is no change of the viewpoint, namely, $l = 1.5$ (a value in the social space range reported in [45,100]) and

$\sigma = 0.5$. The results are compared with those for DPMM in [9], and for the evaluation their approach has been followed when computing the GDSR (group detection success rate): a group is correctly detected if 60% of the persons are matched. Moreover, we analyze two subsets of the available sequences: (i) the sequences with queuing persons being removed from the evaluation (Seq 11-12-13) with the results reported in the second row of Table 3.2, right; and (ii) the sequences containing queuing persons or those that are walking side-by-side (Seq 4-8-11-12-13) being removed (see the last row of Table 3.2, right). These cases represent conditions that cannot be detected by the R-GTCG method because these are not conversational or facing groups. The first case corresponds to the same subset evaluated in [9] and in the next section. The average performance is reported in the second row of Table 3.2, right. As one can note, the performance is worse than that obtained by the DPMM. This is motivated by the fact that the two sequences (Seq 4-8) containing persons that are walking side-by-side are still present in the evaluation. If these sequences are also removed from the set (third row) the performance becomes very good and comparable with that of DPMM. It is worth noting that DPMM uses the temporal information to infer the groups at time $t + 1$ from time t; this information is not available to R-GTCG. On the other hand, R-GTCG uses the head/body orientation which is not needed by DPMM. Again, as for DGPI, this comparison is not completely fair, yet it gives an indication of the goodness of the method. An interesting failure case for many group detection methods, like R-GTCG and DPMM, is the problem of the *associates* investigated in [113]. An associate is a person that is detected as a part of a group even if he/she is simply in close proximity with it and oriented toward it, but is not taking part in the conversation. This condition typically occurs, for example, during poster sessions and can potentially lower the performance of a group detector if not properly handled.

After this empirical evidence we can provide an overall final analysis. The proposed approach is to be preferred over the others under a wide variety of different scenarios. In general, the performance is incredibly stable under both noisy (real) and ideal (synthetic) dataset. For example, the highest performance in the CoffeeBreak has been reached even if it is a very noisy dataset in terms of head orientation since only 4 orientations are possible. From previous results [99] authors suggested to use symmetric measures (Jensen–Shannon) to quantify the pairwise interaction while modeling a social interaction. This is reasonable because it assumes that both individuals want to maintain a connection with the same strength, implying a symmetric affinity. This is reflected in terms of stable and good performance for this detection task. From the computational point of view, if near real-time detection is required, the approach is able to perform group detection at a minimum of 15 fps given the person detections when using nonoptimized Matlab code.

3.6.2 Results of Group Tracking

3.6.2.1 Datasets

The methods presented in Section 3.5 were evaluated on two challenging datasets: FriendsMeet (FM) [7] and BIWI [79].

In the experiments, we decided to decouple the evaluation of the tracking models from the specific choice of a person detector in order to evaluate the theoretical value of the conditional dependency relations introduced between groups and individuals. To this end, we simulated the person detector by generating detections from the ground-truth with a false positive and negative rates of 20% and adding a spatial Gaussian noise.

We present here two variants of the method which were presented in Section 3.5: DEEPER-JIGT directly uses the raw position of the individuals, while DP2-JIGT uses the group detector presented in Section 3.4.2. In this way, it is possible to investigate the impact of the group detector for tracking applications.

In this section, we will prove that: (i) both DEEPER-JIGT and DP2-JIGT perform joint tracking with some differences between them, (ii) methods that resemble multiperson trackers perform poorly on groups, and (iii) the proposed methods allow obtaining an optimal compromise between individual and group tracking performance.

3.6.2.2 Evaluation Metrics

The evaluation is done in terms of individual and group accuracy using standard tracking metrics, such as False Positive (FP) and False Negative (FN) rates [88] for detection, Mean Square Error (MSE) of the estimated positions and its standard deviation, MultiObject Tracking Precision (MOTP) and Accuracy (MOTA) [11] for tracking and id-switch (ID) [71]. In all these metrics, intersection operations among bounding boxes of individuals translate naturally to intersections among convex hulls of groups. We also introduced the Group Detection Success Rate (GDSR) as the rate of the correctly detected groups. In this case we consider that a group is *correctly detected* if at least 60% of its members are detected.

Given the difficulties of real scenarios, the tracker of the individuals can lose some targets because of occlusions, low resolution, and due to the fact that their appearance model is not updated over time. Particular attention has to be paid to the initialization issue. When an individual target is lost (distance above 0.6 m) we reinitialize the individual track, using the detections simulated by the ground truth (as described in Section 3.6.2). This allows us not to use any specific algorithm to perform reinitialization and reacquisition of the targets. The average reinitialization rate per track is reported in Table 3.4. It is worth noticing that there exist many reinitialization strategies that can be adopted in this step, see [14] as an example.

Table 3.3 Results on the real FM dataset excluding the queue sequences (see the text for the details). Group detection (columns 2–4) and group tracking (columns 5–6). For MOTP (in meters), the lower the better

	1-FP	1-FN	GDSR	MOTP [m]	MOTA
DP2-JIGT	**97.81%**	**97.54%**	**94.65%**	0.92	**73.85%**
DEEPER-JIGT	95.72%	89.99%	85.78%	**0.87**	65.18%
VAR-JIGT	75.02%	32.51%	21.53%	3.12	3.14%
Data ass. (baseline)	97.92%	98.21%	94.89%	1.02	70.35%

3.6.2.3 Comparing Methods

The proposed methods are compared to two baselines. The first is called VAR-JIGT which is a variant of the proposed DEEPER-JIGT, where the dynamics contributions of the groups are independent from the one of the individuals. The second method named "data ass." is a group tracking method based on data association that models groups as atomic entities, acting as a generic multiple-target tracker [14,19,109,70,51]. This data association-based tracker uses the detections of groups given by the DPMM detector [111].

3.6.2.4 Performance Analysis

Group tracking is performed by associating the groups at time t with the groups at time $t − 1$ through nearest neighbor on the position–velocity vector.

Table 3.3 reports the statistics on the real FM dataset. DEEPER-JIGT performs consistently better than VAR-JIGT, and DP2-JIGT outperforms DEEPER-JIGT, especially in terms of false negative rate and GDSR. This suggests that there are fewer false negatives, and that we have a more accurate localization of the groups when detected. In terms of tracking accuracy (last column), DP2-JIGT is still better than DEEPER-JIGT, but is less precise (slightly higher MOTP).

The statistics reported in Table 3.3 (last row) show that while the detection accuracy measures obtained with the described data association technique (FP, FN, and GDSR) are comparable with those obtained by DP2-JIGT, DP2-JIGT outperforms it in terms of tracking (MOTP and MOTA). This is due to the fact that the data association-based tracker is not able to deal correctly with splitting and merging events.

We also analyzed the individual tracking performance reported in Table 3.4. One can notice that DP2-JIGT, DEEPER-JIGT, and VAR-JIGT trackers are comparable in terms of MSE and MOTP, while VAR-JIGT outperforms DP2-JIGT and DEEPER-JIGT in terms of the other measures (FP, FN, reinit, and ID). This is due to the fact that VAR-JIGT models the individual tracking and group tracking as two separated processes. On the one hand, the individual component is not influenced by the group tracking, thus favoring a more accurate estimate of individuals. On the other hand, this

Table 3.4 Results of individual tracking for the real FM dataset excluding the queue sequences (see the text for the details)

	1-FP	1-FN	MSE [px]	MOTP [px]	Re-init	ID
DP2-JIGT	81.25%	78.11%	0.25	0.71	3.2%	156
DEEPER-JIGT	82.87%	79.82%	0.24	**0.71**	3.3%	148
VAR-JIGT	**88.12%**	**84.05%**	**0.22**	0.72	**1.6%**	**132**

Table 3.5 Group results on the BIWI dataset

	1-FP	1-FN	GDSR	MOTP [m]	MOTA
DP2-JIGT	37.66%	**89.43%**	51.86%	0.47	22.94%
DEEPER-JIGT	53.77%	78.00%	**53.59%**	**0.44**	**29.43%**
VAR-JIGT	**60.55%**	51.57%	29.60%	1.03	9.58%

independence causes poor group tracking performance because group tracking does not consider the individual component. In practice, we observed that VAR-JIGT produces either big group estimates (low FP rate) or no group estimate at all (low FN rate). Since the group estimates of VAR-JIGT vary consistently over time, new group identifiers are generated, resulting in high group MOTP and low MOTA (see Table 3.3). Therefore, when considering the results of Tables 3.3 and 3.4 jointly, DP2-JIGT and DEEPER-JIGT provide the best compromise between individual and group tracking performance if compared with VAR-JIGT. The compromise is a small loss in the individual tracking performance with respect to VAR-JIGT which will result in a big boost of the group tracking metrics.

The BIWI dataset [79] presents different challenges for group tracking: first, group dynamics is poor, i.e., group events are very rare, and second, group events are not annotated in the ground truth. Unfortunately, the literature lacks other datasets where group events are present and annotated for the tracking purpose (i.e., keeping consistence of group labels across different frames). The two sequences of the BIWI have individuals who walk alone in one direction and stay in the field of view of the camera for a few frames.

The results reported in Table 3.5 were computed for the annotated frames of the sequences where the ground-truth was available from [79]. DP2-JIGT outperformed DEEPER-JIGT in terms of the false negative rate, which means that it is less conservative than DEEPER-JIGT which generates fewer groups. Despite this benefit, DP2-JIGT pays more in terms of the false positive rate and tracking accuracy. The main reason for this shortcoming is that the lifespan of a group in this dataset is really limited to a few tens of frames. The online DPMM needs a bootstrap period to learn plausible configurations of new groups, which, in this case, is greater than their lifespan. This generates more false detections and tracks.

Table 3.6 Comparison between different types of features and group detection models

| Method | Low level | | High level |
	Position	Orientation	
IRPM [8]			✓
HFF [22]	✓	✓	
DS [47]	✓	✓	
Multiscale [85]	✓	✓	
R-GTCG [98]			✓

Some qualitative results that compare DEEPER-JIGT and DP2-JIGT and some failing conditions are reported in a video at http://youtu.be/TOYm060sZDc. In particular, an advantage of DP2-JIGT is that it is able to initialize groups faster than DEEPER-JIGT. This is due to the fact that DEEPER-JIGT tries to merge pairs of individuals and/or groups, while DP2-JIGT uses all the data simultaneously. We also noticed that DEEPER-JIGT tends to merge groups with singletons and sometimes with other groups even if they are far away (e.g., S07 $t = 202$ and $t = 366$), while DP2-JIGT uses the social constraint to avoid them.

3.7 DISCUSSION

Before concluding the chapter, we want to carry out an analysis on the features that the different state-of-the-art methods use. Table 3.6 summarizes the group detection methods with respect to the type of features that are involved (low-level vs. high-level). The best performing method in terms of the results of Table 3.1 is R-GTCG. The feature used by R-GTCG is TSbF (Section 3.3.2.2), which is considered a high-level feature that combines the position and the orientation of a person using notion from sociology (transactional segment) and from biology (the human field of view). TSbF is similar in nature to SVF of IRPM (Section 3.3.2.1) but with an important difference in quantifying the likelihood that two persons have to have an interaction. TSbF thus implicitly takes into account the uncertainty of an interaction and the noise carried by the low-level features. This explains the good performance of TSbF and its plasticity to different scenarios. On the other hand, one can ask why SVF in IRPM performs poorly with respect to the low-level feature. The answer is that SVF relies on the strong assumption that the position and orientation provided are correct, which is not always the case. Moreover, the IRPM group detection method is also not capturing transitions between different groups and short group interactions. One can notice that by introducing the high-level feature of R-GTCG, the detection performance increases compared with the other low-level features used in HFF, DS, and Multiscale. In the tracking methods, the high-level features are not used, still only relying on the low-level information (position and orientation). Adding the high-level features also in the tracking of groups could lead to a more stable model. We will leave this last point as future work.

3.8 CONCLUSIONS

In this chapter, we described a subset of the existing features that can be used for group analysis. We divided them into low-level features and high-level features, with the distinction being based on the level of sociological information embedded into them. Given the features which describe the individuals of interest, two ways to perform group analysis have been analyzed, namely, group detection and group tracking. Among the literature of group detection methods, we described two successful approaches. The game-theoretic approach has the advantage of being deterministic and therefore it easy to use in practice. However, the drawback is that it is less flexible than a probabilistic approach. The main idea is to represent the set of individuals in a given frame with a Gaussian mixture model with a potentially infinite number of components. Finally, we described a group tracking framework that can be used directly on top of the sociological features or can be also integrated with the probabilistic detector. The experiments on diverse and challenging datasets show the effectiveness of the group detection and tracking methods.

Group detection and tracking are two relatively new research fields of computer vision with many interesting applications. Moreover, there is still a wide gap between computer vision and sociology, and therefore there is a lot of research that has to be done to close the gap between these two fields.

REFERENCES

[1] Andriluka Mykhaylo, Roth Stefan, Schiele Bernt. People-tracking-by-detection and people-detection-by-tracking. In: IEEE conference on computer vision and pattern recognition. IEEE; 2008. p. 1–8.

[2] Arrow H, McGrath JE, Berdahl JL. Small groups as complex systems: formation, coordination, development, and adaptation. SAGE Publications; 2000.

[3] Arulampalam M Sanjeev, Maskell Simon, Gordon Neil, Clapp Tim. A tutorial on particle filters for online nonlinear/non-Gaussian Bayesian tracking. IEEE Trans Signal Process 2002;50(2):174–88.

[4] Ba Sileye O, Odobez J. Multiperson visual focus of attention from head pose and meeting contextual cues. IEEE Trans Pattern Anal Mach Intell 2011;33(1):101–16.

[5] Ba SO, Odobez JM. A study on visual focus of attention recognition from head pose in a meeting room. In: MLMI. 2006. p. 75–87.

[6] Bazzani L, Cristani M, Murino V. Collaborative particle filters for group tracking. In: IEEE international conference on image processing. 2010.

[7] Bazzani L, Murino V, Cristani M. Decentralized particle filter for joint individual-group tracking. In: IEEE conference on computer vision and pattern recognition. June 2012.

[8] Bazzani L, Tosato D, Cristani M, Farenzena M, Pagetti G, Menegaz G, Murino V. Social interactions by visual focus of attention in a three-dimensional environment. Expert Syst 2013.

[9] Bazzani L, Zanotto M, Cristani M, Murino V. Joint individual-group modeling for tracking. IEEE Trans Pattern Anal Mach Intell 2015;37(4):746–59.

[10] Benfold Ben, Reid Ian. Stable multi-target tracking in real-time surveillance video. In: CVPR. June 2011. p. 3457–64.

[11] Bernardin Keni, Stiefelhagen Rainer. Evaluating multiple object tracking performance: the CLEAR MOT metrics. Int J Image Video Process 2008:1–10.

[12] Beyan Cigdem, Carissimi Niccolo, Capozzi Francesca, Vascon Sebastiano, Bustreo Matteo, Pierro Andrea, Becchio Cristina, Murino Vittorio. Detecting emergent leader in a meeting environment using nonverbal visual features only. In: Proc. international conference on multimodal interaction. ACM; 2016.

[13] Blei DM, Jordan MI. Variational inference for Dirichlet process mixtures. Bayesian Anal 2006;1(1):121–44.

[14] Breitenstein Michael D, Reichlin Fabian, Leibe Bastian, Koller-Meier Esther, Van Gool Luc. Online multiperson tracking-by-detection from a single, uncalibrated camera. IEEE Trans Pattern Anal Mach Intell 2011;33:1820–33.

[15] Brown LG. A survey of image registration techniques. ACM Comput Surv 1992;24:325–76.

[16] Chang MC, Krahnstoever N, Ge W. Probabilistic group-level motion analysis and scenario recognition. In: IEEE ICCV. 2011.

[17] Chen Tianshi, Schon TB, Ohlsson H, Ljung L. Decentralized particle filter with arbitrary state decomposition. IEEE Trans Signal Process 2011;59(2):465–78.

[18] Choi W, Chao YW, Pantofaru C, Savarese S. Discovering groups of people in images. In: ECCV. 2014.

[19] Choi W, Pantofaru C, Savarese S. A general framework for tracking multiple people from a moving camera. IEEE Trans Pattern Anal Mach Intell 2012.

[20] Choi Wongun, Savarese Silvio. A unified framework for multi-target tracking and collective activity recognition. In: Computer vision – ECCV 2012. Springer; 2012. p. 215–30.

[21] Ciolek T Matthew, Kendon Adam. Environment and the spatial arrangement of conversational encounters. Sociol Inq 1980;50(3–4):237–71.

[22] Cristani M, Bazzani L, Paggetti G, Fossati A, Tosato D, Del Bue A, Menegaz G, Murino V. Social interaction discovery by statistical analysis of F-formations. In: Proc. of BMVC. BMVA Press; 2011. p. 23.1–12.

[23] Cupillard Freédéric, Brémond Francois, Thonnat Monique. Tracking groups of people for video surveillance. In: Video-based surveillance systems: computer vision and distributed processing. 2002. p. 89–100.

[24] Davies DL, Bouldin DW. A cluster separation measure. IEEE Trans Pattern Anal Mach Intell 1979(2):224–7.

[25] de Freitas Nando, Dearden R, Hutter F, Morales-Menendez R, Mutch J, Poole D. Diagnosis by a waiter and a Mars explorer. Proc IEEE 2004;92(3):455–68.

[26] Dollar Piotr, Wojek Christian, Schiele Bernt, Perona Pietro. Pedestrian detection: an evaluation of the state of the art. IEEE Trans Pattern Anal Mach Intell 2012;34(4):743–61.

[27] Doucet A, De Freitas N, Gordon N. Sequential Monte Carlo methods in practice. Springer-Verlag; 2001.

[28] Farenzena M, Tavano A, Bazzani L, Tosato D, Pagetti G, Menegaz G, Murino V, Cristani M. Social interaction by visual focus of attention in a three-dimensional environment. In: Workshop on pattern recognition and artificial intelligence for human behavior analysis at AI*IA. 2009.

[29] Farenzena Michela, Bazzani Loris, Murino Vittorio, Cristani Marco. Towards a subject-centered analysis for automated video surveillance. In: Proceedings of the 15th international conference on image analysis and processing. Berlin, Heidelberg: Springer-Verlag; 2009. p. 481–9.

[30] Feldmann M, Fränken D, Koch W. Tracking of extended objects and group targets using random matrices. IEEE Trans Signal Process 2011;59:1409–20.

[31] Felzenszwalb Pedro F, Girshick Ross B, McAllester David, Ramanan Deva. Object detection with discriminatively trained part-based models. IEEE Trans Pattern Anal Mach Intell 2010;32(9):1627–45.

[32] Ferguson TS. A Bayesian analysis of some nonparametric problems. Ann Stat 1973;1(2):209–30.

[33] Forsyth DR. Group dynamics. Wadsworth/Cengage Learning; 2010.

[34] Freeman L. Social networks and the structure experiment. In: Research methods in social network analysis. 1989. p. 11–40.

[35] Gan Tian, Wong Yongkang, Zhang Daqing, Kankanhalli Mohan S. Temporal encoded F-formation system for social interaction detection. In: Proceedings of the 21st ACM international conference on multimedia. New York: ACM; 2013. p. 937–46.

[36] Ge Weina, Collins Robert T, Ruback R Barry. Vision-based analysis of small groups in pedestrian crowds. IEEE Trans Pattern Anal Mach Intell 2012;34(5):1003–16.

[37] Gennari G, Hager GD. Probabilistic data association methods in visual tracking of groups. In: IEEE conference on computer vision and pattern recognition. 2004.

[38] Gilholm Kevin, Godsill Simon, Maskell Simon, Salmond David. Poisson models for extended target and group tracking. In: SPIE conference: signal and data processing of small targets. 2005.

[39] Girshick Ross. Fast R-CNN. In: The IEEE international conference on computer vision. December 2015.

[40] Girshick Ross, Donahue Jeff, Darrell Trevor, Malik Jitendra. Rich feature hierarchies for accurate object detection and semantic segmentation. In: Proceedings of the IEEE conference on computer vision and pattern recognition. 2014.

[41] Gning A, Mihaylova L, Maskell S, Pang SK, Godsill S. Group object structure and state estimation with evolving networks and Monte Carlo methods. IEEE Trans Signal Process 2011;59(4):1383–96.

[42] Goffman E. Encounters: two studies in the sociology of interaction. The advanced studies in sociology series. Bobbs–Merrill; 1961.

[43] Goffman Erving. Behavior in public places: notes on the social organization of gatherings. Free Press; 1966.

[44] Groh Georg, Lehmann Alexander, Reimers Jonas, Frieß Marc René, Schwarz Loren. Detecting social situations from interaction geometry. In: 2010 IEEE second international conference on social computing. IEEE; 2010. p. 1–8.

[45] Hall Edward T. The hidden dimension. Garden City (NY): A Doubleday Anchor book; 1966.

[46] Howard IP, Rogers BJ. Binocular vision and stereopsis. Oxford psychology series. Oxford University Press; 1995.

[47] Hung H, Kröse B. Detecting F-formations as dominant sets. In: ICMI. 2011.

[48] Hüttenrauch Helge, Severinson Eklundh Kerstin, Green Anders, Topp Elin Anna. Investigating spatial relationships in human–robot interaction. In: 2006 IEEE/RSJ international conference on intelligent robots and systems. IEEE; 2006. p. 5052–9.

[49] Isard M, Blake A. Condensation: conditional density propagation for visual tracking. Int J Comput Vis 1998;29:5–28.

[50] Jabarin B, Wu J, Vertegaal R, Grigorov L. Establishing remote conversations through eye contact with physical awareness proxies. In: CHI '03 extended abstracts. 2003.

[51] Kalal Zdenek, Mikolajczyk Krystian, Matas Jiri. Tracking–learning–detection. IEEE Trans Pattern Anal Mach Intell 2012;34:1409–22.

[52] Kendon Adam. Conducting interaction: patterns of behavior in focused encounters, vol. 7. CUP Archive; 1990.

[53] Khamis Sameh, Morariu Vlad I, Davis Larry S. A flow model for joint action recognition and identity maintenance. In: 2012 IEEE conference on computer vision and pattern recognition. IEEE; 2012. p. 1218–25.

[54] Lablack A, Djeraba C. Analysis of human behaviour in front of a target scene. In: IEEE international conference on pattern recognition. 2008. p. 1–4.

[55] Lan Tian, Wang Yang, Yang Weilong, Robinovitch Stephen N, Mori Greg. Discriminative latent models for recognizing contextual group activities. IEEE Trans Pattern Anal Mach Intell 2012;34(8):1549–62.

[56] Langton SHR, Watt RJ, Bruce V. Do the eyes have it? Cues to the direction of social attention. Trends Cogn Neurosci 2000;4(2):50–8.

[57] Lau B, Arras KO, Burgard W. Multi-model hypothesis group tracking and group size estimation. Int J Soc Robot 2010;2(1):19–30.

[58] Leal-Taixé Laura, Milan Anton, Reid Ian, Roth Stefan, Schindler Konrad. Motchallenge 2015: towards a benchmark for multi-target tracking. Available from arXiv:1504.01942, 2015.

[59] Leal-Taixé Laura, Pons-Moll Gerard, Rosenhahn Bodo. Everybody needs somebody: modeling social and grouping behavior on a linear programming multiple people tracker. In: IEEE international conference on computer vision workshops, 1st workshop on modeling, simulation and visual analysis of large crowds. 2011.

[60] Li Ruonan, Porfilio Parker, Zickler Todd. Finding group interactions in social clutter. In: IEEE conference on computer vision and pattern recognition. June 2013.

[61] Li Stan Z, Zhu Long, Zhang ZhenQiu, Blake Andrew, Zhang HongJiang, Shum Harry. Statistical learning of multi-view face detection. In: Proceedings of the 7th European conference on computer vision. London: Springer-Verlag; 2002. p. 67–81.

[62] Lin Wen-Chieh, Liu Yanxi. A lattice-based MRF model for dynamic near-regular texture tracking. IEEE Trans Pattern Anal Mach Intell 2007;29(5):777–92.

[63] Liu X, Krahnstoever N, Ting Y, Tu P. What are customers looking at? In: Advanced video and signal based surveillance. 2007. p. 405–10.

[64] Lo AY. On a class of Bayesian nonparametric estimates: I. Density estimates. Ann Stat 1984;1(12):351–7.

[65] Marin-Jimenez M, Zisserman A, Ferrari V. Here's looking at you, kid. Detecting people looking at each other in videos. In: British machine vision conference. 2011.

[66] Marques Jorge S, Jorge Pedro M, Abrantes Arnaldo J, Lemos JM. Tracking groups of pedestrians in video sequences. In: IEEE conference on computer vision and patter recognition workshops, vol. 9. 2003. p. 101.

[67] Matsumoto Y, Ogasawara T, Zelinsky A. Behavior recognition based on head-pose and gaze direction measurement. In: Proc. international conference on intelligent robots and systems, vol. 4. 2002. p. 2127–32.

[68] Mauthner T, Donoser M, Bischof H. Robust tracking of spatial related components. In: International conference on pattern recognition. 2008. p. 1–4.

[69] Mckenna Stephen J, Jabri Sumer, Duric Zoran, Wechsler Harry, Rosenfeld Azriel. Tracking groups of people. In: Computer vision and image understanding. 2000.

[70] Milan Anton, Roth Stefan, Schindler Konrad. Continuous energy minimization for multi-target tracking. IEEE Trans Pattern Anal Mach Intell 2013.

[71] Milan Anton, Schindler Konrad, Roth Stefan. Challenges of ground truth evaluation of multi-target tracking. In: 2013 IEEE conference on computer vision and pattern recognition workshops. IEEE; 2013. p. 735–42.

[72] Murphy-Chutorian Erik, Trivedi Mohan Manubhai. Head pose estimation in computer vision: a survey. IEEE Trans Pattern Anal Mach Intell 2009;31:607–26.

[73] Okuma Kenji, Taleghani Ali, De Freitas Nando, Little James J, Lowe David G. A boosted particle filter: multitarget detection and tracking. In: Computer vision – ECCV 2004. Springer; 2004. p. 28–39.

[74] Paisitkriangkrai S, Shen CH, Zhang J. Performance evaluation of local features in human classification and detection. IET Comput Vis 2008;2(4):236–46.

[75] Panero J, Zelnik M. Human dimension & interior space: a source book of design reference standards. Whitney Library of Design. 1979.

[76] Pang Sze Kim, Li Jack, Godsill Simon. Models and algorithms for detection and tracking of coordinated groups. In: Symposium of image and signal processing and analysis. 2007.

[77] Pavan M, Pelillo M. Dominant sets and pairwise clustering. IEEE Trans Pattern Anal Mach Intell 2007;29(1):167–72.

[78] Pellegrini Stefano, Ess Andreas, Van Gool Luc. Improving data association by joint modeling of pedestrian trajectories and groupings. In: European conference on computer vision. 2010. p. 452–65.

[79] Pellegrini Stefano, Ess Andreas, Schindler Konrad, van Gool Luc. You'll never walk alone: modeling social behavior for multi-target tracking. In: International conference on computer vision. 2009.

[80] Preparata FP, Shamos MI. Computational geometry: an introduction. Springer; 1985.

[81] Qin Zhen, Shelton Christian R. Improving multi-target tracking via social grouping. In: IEEE conference on computer vision and pattern recognition. 2012.

[82] Rodriguez Mikel, Laptev Ivan, Sivic Josef, Audibert Jean-Yves. Density-aware person detection and tracking in crowds. In: 2011 IEEE international conference on computer vision. IEEE; 2011. p. 2423–30.

[83] Sethuraman J. A constructive definition of Dirichlet priors. Stat Sin 1994;4:639–50.

[84] Setti Francesco, Hung Hayley, Cristani Marco. Group detection in still images by F-formation modeling: a comparative study. In: WIAMIS. 2013.

[85] Setti Francesco, Lanz Oswald, Ferrario Roberta, Murino Vittorio, Cristani Marco. Multi-scale F-formation discovery for group detection. In: International conference on image processing. 2013.

[86] Setti Francesco, Russell Chris, Bassetti Chiara, Cristani Marco. F-formation detection: individuating free-standing conversational groups in images. PLoS ONE 2015;10(5).

[87] Smith K, Ba S, Odobez J, Gatica-Perez D. Tracking the visual focus of attention for a varying number of wandering people. IEEE Trans Pattern Anal Mach Intell 2008;30(7):1–18.

[88] Smith Kevin, Gatica-Perez Daniel, Odobez Jean-Marc, Ba Sileye. Evaluating multi-object tracking. In: CVPR. 2005. p. 36.

[89] Smith P, Shah M, da Vitoria Lobo N. Determining driver visual attention with one camera. IEEE Trans Intell Transp Syst 2003;4(4):205–18.

[90] Sochman J, Hogg DC. Who knows who – inverting the social force model for finding groups. In: IEEE international conference on computer vision workshops. Nov. 2011. p. 830–7.

[91] Stiefelhagen Rainer, Finke Michael, Yang Jie, Waibel Alex. From gaze to focus of attention. In: Visual information and information systems. 1999. p. 761–8.

[92] Torsello A, Rota Bulò S, Pelillo M. Grouping with asymmetric affinities: a game-theoretic perspective. In: IEEE Computer Society conference on computer vision and pattern recognition, vol. 1. 2006. p. 292–9.

[93] Tosato D, Farenzena M, Spera M, Cristani M, Murino V. Multi-class classification on Riemannian manifolds for video surveillance. In: IEEE European conference on computer vision. 2010.

[94] Tran KN, Gala A, Kakadiaris IA, Shah SK. Activity analysis in crowded environments using social cues for group discovery and human interaction modeling. Pattern Recognit Lett 2013.

[95] Tsai Roger Y. A versatile camera calibration technique for high-accuracy 3D machine vision metrology using off-the-shelf TV cameras and lenses. IEEE J Robot Autom 1987;3(4):323–44.

[96] Tuzel O, Porikli F, Meer P. Region covariance: a fast descriptor for detection and classification. In: Proceedings of the European conference on computer vision. 2006. p. 589–600.

[97] Tuzel Oncel, Porikli Fatih, Meer Peter. Pedestrian detection via classification on Riemannian manifolds. IEEE Trans Pattern Anal Mach Intell 2008;30:1713–27.

[98] Vascon Sebastiano, Mequanint Eyasu Z, Cristani Marco, Hung Hayley, Pelillo Marcello, Murino Vittorio. Detecting conversational groups in images and sequences: a robust game-theoretic approach. Comput Vis Image Underst 2016;143:11–24 [Inference and learning of graphical models: theory and applications in computer vision and image analysis].

[99] Vascon Sebastiano, Mequanint Zemene Eyasu, Cristani Marco, Hung Hayley, Pelillo Marcello, Murino Vittorio. A game-theoretic probabilistic approach for detecting conversational groups. In: Proceedings, Asian conference on computer vision. Lect Notes Comput Sci. Heidelberg: Springer; 2014.

[100] Vinciarelli A, Pantic M, Bourlard H. Social signal processing: survey of an emerging domain. Image Vis Comput 2009;27(12):1743–59.

[101] Viola M, Jones MJ, Viola P. Fast multi-view face detection. In: Proc. of computer vision and pattern recognition. 2003.

[102] Wang Ya-Dong, Wu Jian-Kang, Kassim Ashraf A, Huang Wei-Min. Tracking a variable number of human groups in video using probability hypothesis density. In: International conference on pattern recognition. 2006.

[103] Weibull JW. Evolutionary game theory. Cambridge (MA): MIT Press; 2005.

[104] Whittaker S, Frohlich D, Daly-Jones O. Informal workplace communication: what is it like and how might we support it? In: CHI '94. 1994. p. 208.

[105] Wu B, Nevatia R. Optimizing discrimination-efficiency tradeoff in integrating heterogeneous local features for object detection. In: Proceedings of the international conference of computer vision and pattern recognition. 2008.

[106] Wu B, Nevatia R. Detection and segmentation of multiple, partially occluded objects by grouping, merging, assigning part detection responses. Int J Comput Vis 2009;82(2).

[107] Wu Bo, Ai Haizhou, Huang Chang, Lao Shihong. Fast rotation invariant multi-view face detection based on real AdaBoost. In: Proceedings of the sixth IEEE international conference on automatic face and gesture recognition. Washington (DC): IEEE Computer Society; 2004. p. 79–84.

[108] Yamaguchi K, Berg AC, Ortiz LE, Berg TL. Who are you with and where are you going? In: IEEE conference on computer vision and patter recognition. 2011.

[109] Yang Bo, Nevatia R. Multi-target tracking by online learning of non-linear motion patterns and robust appearance models. In: IEEE conference on computer vision and pattern recognition. June 2012. p. 1918–25.

[110] Yu T, Lim S, Patwardhan KA, Krahnstoever N. Monitoring, recognizing and discovering social networks. In: CVPR. 2009.

[111] Zanotto M, Bazzani L, Cristani M, Murino V. Online Bayesian non-parametrics for social group detection. In: British machine vision conference. 2012.

[112] Zen Gloria, Lepri Bruno, Ricci Elisa, Lanz Oswald. Space speaks: towards socially and personality aware visual surveillance. In: 1st ACM international workshop on multimodal pervasive video analysis. 2010. p. 37–42.

[113] Zhang Lu, Hung Hayley. Beyond F-formations: determining social involvement in free standing conversing groups from static images. In: IEEE conference on computer vision and pattern recognition. June 2016.

[114] Zhang Zhengyou. A flexible new technique for camera calibration. IEEE Trans Pattern Anal Mach Intell 2000;22(11):1330–4.

CHAPTER 4

Exploring Multitask and Transfer Learning Algorithms for Head Pose Estimation in Dynamic Multiview Scenarios

Elisa Ricci*,†, Yan Yan‡, Anoop K. Rajagopal§, Ramanathan Subramanian¶, Radu L. Vieriu‡, Oswald Lanz*, Nicu Sebe‡

*Center for Information and Communication Technology, Fondazione Bruno Kessler, Trento, Italy
†Department of Engineering, University of Perugia, Perugia, Italy
‡Department of Information Engineering and Computer Science, University of Trento, Trento, Italy
§Department of Electrical Engineering, Indian Institute of Science, Bangalore, India
¶Centre for Visual Information Technology, International Institute of Information Technology, Hyderabad, India

Contents

4.1 INTRODUCTION

Motivated by several applications such as video surveillance, human–computer interaction, behavior analysis and social computing, extensive research has been devoted to head pose estimation recently [26]. Several approaches precisely compute the head pose when the target is close to the camera, as high resolution images enable accurate facial feature extraction and depth information can be additionally integrated [14]. Nevertheless, HPE from surveillance videos is challenging despite recent advancements [6,27,33], as faces are captured at very low resolution and appear blurred.

Group and Crowd Behavior for Computer Vision
DOI: 10.1016/B978-0-12-809276-7.00005-9

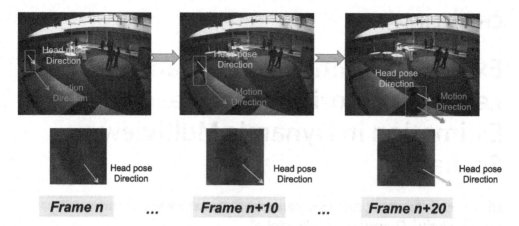

Figure 4.1 HPE under target motion. Automatically extracted face crops corresponding to three different positions of a target exhibiting the same 3D head pose are shown. Note significant changes in facial appearance as the target moves closer to the camera – these appearance differences severely impede the performance of traditional pose estimation (classification) methods.

HPE accuracy on surveillance video data can be improved by fusing information from multiple cameras, as monocular systems are often insufficient for analyzing human behavior in large environments. Surprisingly, only a few HPE methods consider a multiview setting [25,29,35,43] and typically compute head pose as a target rotates in-place [25,35]. However, the ability to estimate head pose of moving targets is key as head orientation is primarily employed as a surrogate for gaze direction to infer social interactions [32]. HPE of moving targets is a challenging problem as illustrated in Fig. 4.1: facial appearance of a person exhibiting identical 3D head pose at three different scene locations varies considerably due to the perspective and scale. These appearance changes severely impede HPE performance using traditional approaches.

This chapter explicitly examines the problem of **multiview head pose estimation under target motion**. Since precise head pose estimates are superfluous for most real-life applications and also onerous to compute from low-resolution faces, most approaches simplify HPE to classification of head pose into a finite number of classes, with each class denoting a quantized range of head poses. A number of approaches [27,33] have attempted HPE when the target facial appearance is more or less consistent with respect to the camera – this is the case when the camera is located at infinity, or when the target is stationary. However, very few algorithms [43] are designed to estimate the head pose of a moving target in smart-room environments.

In this chapter we describe two methodologies for HPE under target motion, namely, transfer learning and multitask learning. The rationale behind exploring these two approaches for our problem is the following:

1. Since existing HPE datasets involve stationary targets (and are typically acquired from meeting scenes), it may be economical to build a model on these datasets and adapt the learned knowledge to moving targets by utilizing a few exemplars thereof. **Transfer Learning** (TL) is a popular machine learning methodology when the training and test data have differing attributes. For the multiview HPE problem, the train and test datasets can differ with respect to low-level scene properties (scene dimension, camera placement, and illumination), range of head poses exhibited by targets (meeting vs museum scene), as well as target motion. We explore how a variety of transfer learning approaches can be employed to address the above discrepancies.

2. Another problem closely related to transfer learning is **Multitask Learning** (MTL), where classification/regression models corresponding to a number of related tasks are learned together, so that the joint learning leads to more efficient models as compared to learning each task independently. MTL is inherently designed to address HPE under target motion. If the scene of interest is partitioned into a uniform grid, one can expect some similarities as well as differences in facial appearance across grid partitions. If modeling the pose–appearance relationship within each segment is considered as a task, MTL can be invoked for jointly learning a set of segment-specific head pose classifiers. Following this intuition, we propose FlExible GrAph-guided Multitask Learning (FEGA-MTL) for multiview HPE.

The remainder of this chapter is organized as follows. First, we review related works on HPE and transfer and multitask learning. Then, we describe the proposed methodologies for addressing the problem of HPE under target motion and we show the results of our experimental evaluation. In Section 4.4 conclusions are drawn.

4.2 RELATED WORK

In this section we discuss related work for (i) head pose estimation from low-resolution imagery, (ii) transfer learning approaches, and (iii) multitask learning techniques.

4.2.1 Head Pose Estimation from Low-Resolution Images

Much work has been dedicated to estimating people's head pose due to its relevance in areas such as human–computer interaction, security, advertising, cognitive psychology, etc. The head pose is used as a proxy for attention in applications related to driving [11], human–robot interaction [20], or in structured/unstructured social interaction settings [32]. While efforts have been generally channeled on HPE from high-resolution images and videos, little attention has been paid to inferring head pose from surveillance data, especially when multiple cameras are available. Pioneering attempts along this line of research include the work of Voit et al. [34,35]. They typically use Neural Networks to infer the pose from single views separately and then fuse the results under a Bayesian framework. Common benchmarks on which their approaches are validated include CLEAR06 and CLEAR07 [31]. In [35] a system that combines a head tracking module, an HOG-based camera reliability measure, and a Neural Net classifier for the two main head pose parameters (i.e., pan and tilt) is presented. Mean pose errors of 7.2° and 9.3°, respectively, are reported on CLEAR07.

Given a large field-of-view camera capturing a number of moving subjects, GMMs and HMMs incorporating location and head pose information are used to determine the number of persons who attend to an outdoor advertisement in [30]. A novel KL distance-based descriptor is introduced in [27] for coarse head pose estimation of moving people in crowded scenes. The performance reported in [27], however, is surpassed in [33] by the use of ARCO descriptors, claimed to be less sensitive to scale/lighting variations as well as occlusions. In [25] view-specific probability distributions for pose classification are computed using SVMs. Consequently, these are fused to obtain a more precise pose estimate. In [43] nine cameras are used to project the views onto a sphere model. A face detection algorithm infers the center of the face and thus the rotational angles that characterize the head pose. Although it has not been tested on known benchmarks, the method is claimed to be robust to occlusions and able to achieve very small angular errors. More recently, [37] has addressed multiview head pose classification using knowledge transfer and multitask learning approaches, respectively.

4.2.2 Transfer Learning

In general, transfer learning describes the ability of leveraging knowledge from a source domain, to help solve a problem in a related target domain. It relaxes the assumption of having labeled data both for training and testing drawn from similar distributions and sharing similar feature spaces, as standard machine learning algorithms require. Transfer learning tackles the meticulous and most of the time expensive work of labeling data when a new task is to be solved, by allowing domains, tasks, and distributions used to be different. One of the most elaborate surveys for TL approaches is [28]. According to [28], methods employing transfer learning can be categorized, based on the means used for transferring knowledge from source to target, into four groups. *Instance-based* transfer learning [8,18] involves reusing samples from the source, along with some target samples, assuming that the former samples are still useful. A TL approach based on the well-known AdaBoost is presented in [8], which leverages on extensive labeled source data in addition to a few target samples to derive an accurate target classifier. Jiang et al. [18] run a relevance test on the source samples to discover and remove the potentially harmful ones from training. *Feature-based* TL tries to find a "good" representation that reduces the differences between the two domains. Labeled source and target data features are copied to synthesize an augmented feature space in [9], on which supervised learning is employed while jointly optimizing source and target feature weights to maximize prediction accuracy. In *parameter-based* transfer learning [3], the goal is to exploit model parameters that can be shared between source and target. Finally, *relational knowledge transfer* [10,24], source and target are treated as relational domains and i.i.d. assumptions are relaxed.

Recently, there has been an increased interest in TL approaches applied to vision tasks, in particular to human activity recognition. Zheng et al. [45] describe a cross-view action recognition method that uses a sparse transferable dictionary pair learned from source and target views. Farhadi et al. [15] transfer discriminative hyperplanes, called splits, between source and target views of the same activity, while Yang et al. [41] train a transferable distance function for action detection with sparse data.

4.2.3 Multitask Learning

Multitask learning [4] aims at simultaneously learning classifiers for multiple tasks, by assuming the existence of a shared representation between them. MTL has recently been successfully employed in image classification [42], visual tracking [44], and multiview action recognition [39]. The intuition behind MTL is that a joint learning procedure accounting for task relationships is more efficient than learning each task separately. Traditionally, MTL methods [1] assume that all the tasks are related and their dependencies can be modeled by a set of latent variables. However, in many real world applications, such as HPE under target motion, not all tasks are related, and enforcing erroneous (or nonexistent) dependencies may lead to *negative knowledge transfer*. To some extent, MTL can be seen as a form of *inductive transfer*, therefore being prone to *negative transfer*. Recently, sophisticated methods have been introduced to counter this problem. These methods assume prior knowledge (e.g., in the form of a graph) defining task dependencies [7], or learn task relationships in conjunction with task-specific parameters [19,46, 17,47,16].

4.3 TL AND MTL FOR MULTIVIEW HEAD POSE ESTIMATION

In this section we describe the proposed approaches for multiview head pose estimation under target motion. All the techniques we present comprise two main steps: a training phase where head pose classifiers are learned and a test phase where head pose classification is performed on novel instances. In this work, we are mainly interested in determining the *head pan* for detecting face-to-face interactions and seek to assign the target's head pan to one of eight classes, each denoting a quantized 45° range. Our approaches also rely on a multiview particle filter tracker [22] for target position estimation and head localization. Features describing the face regions are subsequently extracted from the obtained cropped regions. This preprocessing phase is described in detail in Section 4.3.1. The output of the tracker, the head localization, and the features are used both in the training and in the test phases. In the following we describe the preprocessing step and the proposed learning algorithms.

4.3.1 Preprocessing

In the preprocessing step, multiview face crops are extracted using a visual tracker. A multiview, color-based particle filter [22] is used to compute the 3D body centroid of moving targets. A $30 \times 30 \times 20$ cm-sized dense 3D grid (with 1 cm resolution) of hypothetic head locations is then placed around the estimated 3D head position provided by the particle filter. The grid size accounts for tracker's variance and horizontal/vertical offsets of the head from the body centroid due to pan, tilt, and roll. Assuming a spherical model of the head, a contour likelihood is computed for each grid point by projecting a 3D sphere onto each view using camera calibration information. The grid point with the highest likelihood sum is determined as the head location.

The head is then cropped and resized to 20×20 pixels in each view. Head crops from the different views are concatenated to generate the multiview face crops, and similar to previous works [2,6,33] we employ powerful covariance or HOG descriptors to effectively describe the face appearance for head pose classification.

4.3.2 Transfer Learning for HPE

In this section, we describe two different transfer learning solutions to overcome the adverse impact of changing attributes between the source and target data on HPE performance. Firstly, to address the problem of head-pan classification under varying head-tilt, we propose a domain adaptive version of the ARCO pose classifier [33] based on the instance-based transfer learning technique described in [8]. However, this adaptation is still not effective for determining the head pose of moving targets. For solving this problem, we propose a second transfer learning approach where a set of face patch weights are learned from the source data, with each patch weight indicating the saliency of the face patch for pose classification. These weights are then adapted to the target scenario, incorporating a patch *reliability score* measuring the face patch's appearance distortion under target motion.

4.3.2.1 Head-Pan Classification Under Varying Head-Tilt

In this subsection we focus on problem of predicting head pan in the target upon learning from many source and a few target examples. Apart from varying image acquisition conditions, often facial appearance in the source and target data differs due to the range of head poses exhibited by subjects (e.g., all source examples correspond to a frontal tilt, while target tilts are in the range $[-90°, 90°]$). An effective method for transferring knowledge across datasets through induction of a few target examples in the learning process is proposed in Tradaboost [8]. Tradaboost is modeled on AdaBoost where, given a training set comprising source and target samples, a set of weak learners are learned such that misclassified target samples are given priority at each step. In this way, the resulting model is tuned to effectively predict target samples. Analogously, the ARCO

Algorithm 1 ARCO-Xboost.

Input: Combined source $(x_i, y_i \in \mathcal{T}_s)$, target $(x_i, y_i \in \mathcal{T}_t)$ training set $\mathcal{T} = \{(x_1, y_1), \ldots, (x_N, y_N), (x_{N+1}, y_{N+1}), \ldots, (x_{N+M}, y_{N+M})\}$, $\{y_i\}, \{y_i\} = 1, \ldots, J$, number of learners L.

$\forall i$ initialize weights $w_i = \frac{1}{N+M}$ and posterior probabilities $P_j(x_i) = \frac{1}{J}$.

Set $\alpha_s = \frac{1}{2}\ln(1 + \sqrt{2\ln\frac{N}{L}})$

   for $l = 1, \ldots, L$

     Initialize learner $F_l = 0$.

     Compute response values z_i and weights w_i from $P_j(x_i)$

     if $L > 1$

       Normalize the weight vector $w_1, \ldots, w_{N+M}$

       Compute the error on target, $\epsilon_t = \sum_{j=1}^{J} \frac{w_j[y_j \neq h(x_j)]}{\sum_{i=1}^{N+M} w_i}$,

       where $x_j = \{x_i \in j\}$, $w_j = \{w_i, \forall x_i \in j\}$, $h(\cdot)$ is the classified label.

       Set $\alpha_t = \frac{1}{2}\ln(\frac{1-\epsilon_t}{\epsilon_t})$, $\epsilon_t < \frac{1}{2}$

       Update weights

       $w_i \leftarrow w_i e^{-\alpha_s(y_j \neq h(x_j))}$ (modify misclassified source weights)

       $w_i \leftarrow w_i e^{\alpha_t(y_j \neq h(x_j))}$ (modify misclassified target weights)

     end if

     Compute learner F_l using least-squares regression from computed z_{ij}'s and w_i's.

     Compute new $P_j(x_i)$'s and $h(x_i)$'s.

   end for

Output: Set of learners $\{F_l\}$

---

framework [33] also employs a multiclass Logitboost classifier $\{F_l\}$ for each image patch, comprising $l = 1, \ldots, L$ weak classifiers. Given a training set $\{x_i\}$ with N samples corresponding to class labels $1, \ldots, J$, the Logitboost algorithm iteratively learns training samples most difficult to classify through a set of weights w_i and posterior probabilities, $P_j(x_i)$. Each weak learner solves a weighted-regression problem, whose goodness of fit is measured by the response value vector for the ith training sample, $z_i = \{z_{ij}\}_{j=1}^{J}$.

Following [8,33], we designed ARCO-Xboost, an inductive transfer learning approach for the ARCO Logitboost classifier as follows. Given $N + M$ training data comprising N source and M target samples, with $N \gg M$, the error on target (ϵ_t) is computed at every step upon normalizing the w_i's. Also, α_s and α_t, which are respectively the attenuating and boosting factors for misclassified source and target samples, are determined. Finally, the weights of misclassified target data are boosted by a factor of e^{α_t}, so that the model incorporates more target-specific information, while the weights for misclassified source are attenuated by a factor of $e^{-\alpha_s}$ to discourage learning of these samples. ARCO-Xboost is summarized in Algorithm 1.

Experimental Results

We consider two datasets in our experiments: the CLEAR and the DPOSE datasets. The CLEAR dataset [31] is a popular multiview dataset used for evaluating multiview HPE. Acquired from 15 targets rotating in-place in the middle of a room, the dataset comprises over 30,000 synchronously acquired images from four cameras with head-pose measurements. The DPOSE dataset [29] contains sequences acquired from 16 targets, where the target is either rotating in-place at the room center or moving around freely in a room and moving their head in all possible directions. The dataset consists of over 50,000 images. Head pan, tilt, and roll measurements for various poses are recorded using an accelerometer, gyro, magnetometer platform strapped onto the head using an elastic band running down from the back of the head to the chin. Target head movements are captured using cameras mounted at the corners of a 6×4.8 m^2 room. The CLEAR and DPOSE datasets differ with respect to camera configurations, illumination conditions, and targets.

For our experiments, the source training set comprised 300 CLEAR images for each of the eight frontal tilt classes. The classification accuracies reported correspond to the mean value obtained from four independent trials involving randomly chosen target training sets. For the sake of evaluating how ARCO-Xboost improves classification performance over ARCO, we used covariance features derived from the 12-dimensional feature set $\phi = [x, y, R, G, B, I_x, I_y, OG, Gabor_{\{0, \pi/6, \pi/3, 4\pi/3\}}, KL]$. Here, x, y and R, G, B denote spatial positions and color values, while Ix, Iy, and OG respectively denote intensity gradients and gradient orientation of pixels. *Gabor* is the set of coefficients obtained from Gabor filtering at aforementioned orientations (frequency = 16 Hz), while KL denotes maximal divergence between corresponding patches in the target face image and each of the pose-class templates computed as described in [27]. The presented results correspond to two covariance features, namely, $Cov(d = 12)$, which denotes covariance descriptors computed from all features in ϕ, and $Cov(d = 7)$, where covariances are computed only for color and Gabor features.

We compare ARCO-Xboost with other state-of-the-art transfer learning methods. More specifically, we consider the Feature Replication (FR) method proposed in [9], the Adaptive Support Vector Machine (A-SVM) approach presented in [40], the Domain Adaptation Machine (DAM) algorithm [13], the Domain Adaptive Metric Learning (DAML) [21], and the Domain Transfer Multiple Kernel Learning (DTMKL) described in [12]. Fig. 4.2 (right) shows the results of our evaluation when $Cov(d = 12)$ features are used. For SVM-like methods, we considered a Gaussian kernel. The regularization parameters of all considered methods were tuned upon cross-validation. Three auxiliary classifiers are used in A-SVM and DAM, while 20 prelearned base kernels are adopted in DTMKL. From Fig. 4.2, we observe that all transfer learning approaches achieve very similar performance, with DTMKL achieving a slightly superior accuracy. The

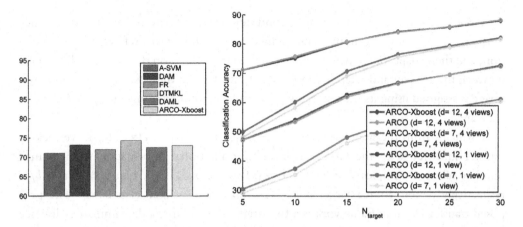

Figure 4.2 (Left) Comparison with state-of-the-art transfer learning approaches. Results are plotted for models trained with source+5 target samples/class, and with 4-view $Cov(d = 12)$ features. (Right) Comparison with ARCO at varying size of the target set.

improved performance of DTMKL can be attributed to the use of multiple kernels in the learning framework.

Classification accuracy trends upon increasing the number of inducted target samples from 5–30 samples/class, with $L = 12$, are shown in Fig. 4.2 (right). Here, we also compare the classification performance achieved with 4-view and single-view features – mean value of the accuracies achieved with each of the four views is considered for the single-view case. Considerably higher accuracies are obtained when features extracted from all four views are employed for pose classification, implying that multiview information improves robustness of pose classification on low-resolution images. Also, larger gains with ARCO-Xboost are obtained with single-view features and when fewer target examples are inducted in the training set.

4.3.2.2 Head-Pan Classification Under Target Motion

In this subsection we describe how knowledge from source images capturing stationary targets can be effectively used to determine head pan in target images involving freely moving persons but exhibiting the same range of head poses as in the source. The challenge in this scenario is that facial appearance for a given pose changes with the target's position due to varying camera perspective and scale.

To this end, we propose a two-step, adaptive weights learning technique. First, upon dividing of the multiview facial appearance image into a number of overlapping patches, the weight of each patch denoting its saliency for pose classification is learned from source images. These patch weights can be directly applied to the target dataset if it also involves stationary targets. However, since the target dataset involves moving persons, visibility of face patches and their reliability for pose classification would vary based on the target's position. Therefore in the second step, we transform the person's appearance in the target dataset to a canonical appearance corresponding to a reference spatial position, and then adapt source patch weights to the target based on the visibility differences between the current and reference target positions. Finally, the pose class of a target test image is assigned using its nearest training example, computed using a weighted distance measure.

Formally, from the large source set $\mathcal{T}_s = \{(\mathbf{x}_1, l_1), (\mathbf{x}_2, l_2), \ldots, (\mathbf{x}_{N_s}, l_{N_s})\}$, we seek to transfer knowledge to the target incorporating additional information from a small number of target samples $\mathcal{T}_t = \{(\mathbf{x}_1, 1_1), (\mathbf{x}_2, 1_2), \ldots, (\mathbf{x}_{N_t}, 1_{N_t})\}$. Here, $\mathbf{x}_i/\mathbf{x}_i$ and $l_i/1_i$ respectively denote source/target image features and associated class labels. The proposed transfer learning framework is a two-step process. First, a discriminative distance function is learned on the source. Given that each image consists of Q patches, we learn a weighted-distance on the source, $D_{W_s}(\mathbf{x}_i, \mathbf{x}_j)$, as a parameterized linear function, i.e., $D_{W_s}(\mathbf{x}_i, \mathbf{x}_j) = W_s^T \mathbf{d}_{ij}$, where $\mathbf{d}_{ij}$ is the distance (we use Euclidean distance) between corresponding patches in images. W_s is the source patch weight vector, which encodes the saliency of each face patch for pose classification. Details can be found in the original paper [29]. In the second step, a distance function $D_{W_t}(\cdot)$ is learned on target data $\mathcal{T}_t$. W_s is used in this phase, in order to transfer the source knowledge onto the target. The reliability score for each target patch as computed from the canonical transformation [29] is also considered. We formulate the adaptation problem as:

$$\min_{W_t, \xi_i, \alpha, \beta} \quad \gamma_1(\alpha, \beta) \|\mathbf{B} W_t\|^2 - \gamma_2(\alpha, \beta) W_s^T W_t - \gamma_3(\alpha, \beta) \|W_s\|^2 + \frac{1}{N_t} \sum_{i=1}^{N_t} \xi_i, \quad (4.1)$$

$$\text{s.t.} \quad \min_{1_i \neq 1_k} W_t^T \mathbf{d}_{ik} - \max_{1_i = 1_j} W_t^T \mathbf{d}_{ij} \geq 1 - \xi_i, \quad \forall i,$$

$$W_t \geq 0, \quad \xi_i \geq 0, \quad \alpha(1 - \alpha) - \beta^2 > 0,$$

where $\gamma_1(\alpha, \beta) = \lambda_1 + \frac{\lambda_2 \alpha}{\Delta(\alpha, \beta)}$, $\gamma_2(\alpha, \beta) = \frac{2\lambda_2 \beta}{\Delta(\alpha, \beta)}$, $\gamma_3(\alpha, \beta) = \frac{\lambda_2(1-\alpha)}{\Delta(\alpha, \beta)}$, and $\Delta(\alpha, \beta) = \alpha(1 - \alpha) - \beta^2$.

The matrix $\mathbf{B}$ takes into account patch reliability. We assume that the room is divided into R distinctive regions, and to effectively learn appearance variation with position, we have K_r target training samples for each region $r \in R$. The patch reliability score vector, $\hat{\rho} = [\rho_q]$, $q = 1, \ldots, Q$, is determined from the mean reliability score of the P patch pixels, i.e., $\rho_q = \frac{1}{P} \sum_{p=1}^{P} r_p$ and the expected patch reliability for region r, $r = 1, \ldots, R$, is computed as $\hat{\rho}_r = \frac{1}{K_r} \sum_{i=1}^{K_r} \hat{\rho}_i$. Given $\hat{\rho}_r$, a diagonal matrix $\mathbf{B} \in \mathbb{R}^{Q \times Q}$ for region r is defined such that $B_{pq} = e^{-(1 - \frac{\eta}{\hat{\rho}_r})}$ if $p = q$ and 0 otherwise. Further details on the proposed approach can be found in the original paper [29].

Table 4.1 Performance comparison for 8-class head-pan classification under target motion. The room is divided into 4 quadrants (R1–R4). NWD$_s$ classification accuracies obtained with the different features are indicated in braces

	R1	R2	R3	R4
ARCO-Xboost $Cov(d=7)$	41.1	43.6	45.9	41.7
ARCO-Xboost $Cov(d=12)$	66.1	67.6	66.2	59.1
WD $Cov(d=7)$	65.8 (33.1)	67.4 (41.5)	59.6 (51.2)	60.6 (37.8)
WD $Cov(d=12)$	69.8 (45.0)	72.4 (51.6)	63.0 (59.6)	62.4 (42.3)
WD LBP	**74.7** (60.9)	**77.6** (61.3)	**66.9** (58.7)	**64.5** (58.3)
Multiview SVM	47.6	51.3	41.0	41.6

Experimental Results

The proposed framework for pose classification under target motion is evaluated against the ARCO-Xboost and the Multiview SVM (MSVM)-based method in [25]. MSVM-based pose estimation feeds gradient features from the target appearance image in each camera view to a multiclass SVM classifier, the output of which is used to compute a probability distribution over all pose classes. Then, a combined distribution fusing the multiview information is computed for determining the pose class. We divide the scene of interest into $R = 4$ nonoverlapping regions and assume that a few target training examples are available per quadrant. Apart from $Cov(d = 7)$ and $Cov(d = 12)$ features, we also employ 64 bin-indexed local binary pattern (LBP) descriptors [36] to learn face patch weights using the proposed framework.

Table 4.1 presents the region-wise classification results. We also analyze how learning of patch weights is beneficial by comparing classification accuracies achieved with a nearest-neighbor (NN) classifier employing the weighted (WD) and unweighted (NWD) distance measures.[1] Considering the mean classification accuracy over all quadrants, learning of face patch weights through the proposed adaptive framework is immensely beneficial under target motion.

We also compare WD performance with state-of-the-art transfer learning methods in Fig. 4.3. The adaptive weights learning approach, which explicitly incorporates camera geometry information into learning, outperforms most competing approaches. However, DTMKL, which is a powerful framework employing 20 prelearned kernel classifiers for domain adaptation, produces the highest classification accuracies. It is worth noting that while some methods use multiple auxiliary (source) classifiers for knowledge transfer (e.g., DAM uses three classifiers), our approach employs only a single source classifier. With the adoption of multiple source models we also expect that the performance of our approach improves.

---

[1] The NN classifier assigns the class label of the nearest target training example to the test image.

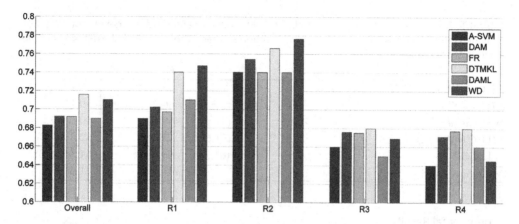

Figure 4.3 Comparison with state-of-the-art transfer learning approaches with moving subjects in target set. Experiments are performed using source+5 target samples/class/quadrant, and with *LBP* features.

4.3.3 Multitask Learning for HPE

In this section we also address the problem of multiview head pose classification under target motion and describe FEGA-MTL. The intuition behind FEGA-MTL is simple. Upon dividing the scene ground plane into a uniform grid, one can expect some similarities as well as differences in facial appearance for a given head pose across grid partitions. To model this phenomenon we propose to use multitask learning for constructing a set of region-specific head pose classifiers. To eliminate the need for training data, FEGA-MTL is also designed to operate in an unsupervised manner, i.e., by employing the motion direction of targets as weak labels to signify their head orientation. Post training, FEGA-MTL outputs (i) pose classifiers for each region, and (ii) the optimal scene partitioning, where grid regions with similar facial appearance for a given head pose constituting a cluster. During classification, the tracker provides target position based on which the appropriate region-based pose classifier is invoked to output the head pan class. We now describe FEGA-MTL in detail.

4.3.3.1 FEGA-MTL

To apply FEGA-MTL, we initially divide the scene ground plane into a uniform 5×5 grid. We seek to learn the pose–appearance relationship in each partition. The algorithm learns from a training set $\mathcal{R}_t = \{(\mathbf{x}_i^t, y_i^t) : i = 1, 2, \ldots, N_t\}$ for each region $t = 1, 2, \ldots, R$, where $\mathbf{x}_i^t \in \mathbb{R}^D$ denote D-dimensional feature vectors and $y_i^t \in \{1, 2, \ldots, C\}$ are the head pose labels ($C = 8$ classes in our setting).

We define two graphs guiding the learning process. One graph specifies the similarity in appearance for a given head pose across the grid regions based on camera geometry. If grid partitions form the graph nodes, we determine the edge set $\mathcal{E}_1$ and the associated edge weights γ_{mn} quantifying the appearance distortion between $\mathcal{R}_m$ and $\mathcal{R}_n$ due to positional change from region m to region n (these edge weights indicate whether knowledge sharing between regions m and n is beneficial or not). The details of the construction of the region graph can be found in the original paper [38]. A second graph models the fact that facial appearances should be more similar for neighboring head pose classes. Exploiting this information, a pose graph $\mathcal{E}_2$ is defined with associated edge weights $\beta_{ij} = 1$ if i and j correspond to neighboring pose classes c_i, c_j, and $\beta_{ij} = 0$ otherwise.

Formally, modeling spatial regions as separate tasks, for each task t we define a training set $\mathcal{R}_t$ and a matrix $\mathbf{X}_t \in \mathbb{R}^{N_t \times D}$, $\mathbf{X}_t = [\mathbf{x}_1^t, \ldots, \mathbf{x}_{N_t}^t]'$, where $(\cdot)'$ indicates the transposition operator. We also define the matrix $\mathbf{X} \in \mathbb{R}^{N \times D}$, $\mathbf{X} = [\mathbf{X}_1', \ldots, \mathbf{X}_R']'$, where $N = \sum_{t=1}^R N_t$ denotes the total number of training samples. For each training sample, we construct a binary label vector $\mathbf{y}_i^t \in \mathbb{R}^{RC}$ as

$$\mathbf{y}_i^t = [\underbrace{0, 0, \ldots, 0}_{\text{Task 1}}, \underbrace{0, 1, \ldots, 0}_{\text{Task 2}}, \ldots, \underbrace{0, 0, \ldots, 0}_{\text{Task R}}],$$

i.e., the position of the nonzero element indicates the task and class membership of the corresponding training sample. A label matrix $\mathbf{Y} \in \mathbb{R}^{N \times RC}$ is then obtained concatenating the $\mathbf{y}_i^t$'s for all training samples. For each region t and pose class c, we consider the weight vectors $\mathbf{s}_{t,c}, \boldsymbol{\theta}_{t,c}, \mathbf{w}_{t,c} \in \mathbb{R}^D$ and define the associated matrices $\mathbf{S}, \boldsymbol{\Theta}, \mathbf{W} \in \mathbb{R}^{D \times RC}$,

$$\mathbf{S} = [\underbrace{\mathbf{s}_{1,1}, \ldots, \mathbf{s}_{1,C}}_{\text{Task 1}}, \ldots, \underbrace{\mathbf{s}_{R,1}, \ldots, \mathbf{s}_{R,C}}_{\text{Task R}}], \quad \boldsymbol{\Theta} = [\underbrace{\boldsymbol{\theta}_{1,1}, \ldots, \boldsymbol{\theta}_{1,C}}_{\text{Task 1}}, \ldots, \underbrace{\boldsymbol{\theta}_{R,1}, \ldots, \boldsymbol{\theta}_{R,C}}_{\text{Task R}}] \text{ and } \mathbf{W} = \mathbf{S} + \boldsymbol{\Theta}.$$

FEGA-MTL learns a set of region-specific weight vectors for pose classification $\mathbf{w}_{t,c} \in \mathbb{R}^D$, $\mathbf{w}_{t,c} = \mathbf{s}_{t,c} + \boldsymbol{\theta}_{t,c}$. Each weight vector is obtained by summing up two components, $\mathbf{s}_{t,c}$ which models the appearance relationships among regions and $\boldsymbol{\theta}_{t,c}$ accounting for region-specific appearance variations. Using a matrix notation for the sake of clarity, we propose to solve the following optimization problem:

$$\min_{\mathbf{S}, \boldsymbol{\Theta}} \left\| \mathbf{U}^{\frac{1}{2}} (\mathbf{Y} - \mathbf{X}(\mathbf{S} + \boldsymbol{\Theta})) \right\|_F^2 + r(\mathbf{S}, \boldsymbol{\Theta}) \tag{4.2}$$

where

$$r(\mathbf{S}, \boldsymbol{\Theta}) = \lambda_\theta \|\boldsymbol{\Theta}\|_F^2 + \lambda_s \|\mathbf{S}\|_F^2 + \lambda_s \lambda_1 \sum_{(i,j) \in \mathcal{E}_1} \gamma_{ij} \|\mathbf{s}_{t_i,c} - \mathbf{s}_{t_j,c}\|_1 + \lambda_s \lambda_2 \sum_{(i,j) \in \mathcal{E}_2} \beta_{ij} \|\mathbf{s}_{t,c_i} - \mathbf{s}_{t,c_j}\|_1.$$

In the loss function the matrix $\mathbf{U} \in \mathbb{R}^{N \times N}$, $\mathbf{U} = \mathbf{N}(\mathbf{Y}\mathbf{Y}')^{-1}$ is obtained multiplying two terms. The normalization factor $(\mathbf{Y}\mathbf{Y}')^{-1}$ compensates for different number of samples per task, while the matrix $\mathbf{N} = \text{diag}(v_i^t)$ aims to weight differently samples labeled by a human annotator and those automatically obtained by exploiting the information about the walking direction. Specifically, we assign a weight $v_i^l = 1$ for samples with a true label (i.e., human annotation), while v_i^l is set to a value $\rho \leq 1$ for weakly labeled data.

The regularization function $r(\cdot)$ is made of several components. The first term penalizes large region-specific appearance variations, the second regulates model complexity, and the ℓ_1 norm terms impose the weights $\mathbf{s}_{t,c}$ of appearance-wise related regions and neighboring classes to be close together. Specifically, γ_{ij}'s and β_{ij}'s are the appearance similarity-based weights of region graph edges $\mathcal{E}_1$ and pose graph edges $\mathcal{E}_2$, respectively. Similar parameters $\mathbf{s}_{t,c}$ for neighboring head orientations are obtained as λ_2 increases. Region clusters are formed as $\lambda_1 \to \infty$. Importantly, this effect is feature-specific: cluster structure varies from feature to feature. Less important features are used similarly by all tasks, while discriminative features are used differently by different tasks. This is one of the main reasons why our method is termed flexible.

The optimization problem in Eq. (4.2) is convex. To solve it, we propose an algorithm based on smoothing proximal gradient method [7]. After the training phase, the computed weights $\mathbf{w}_{t,c}$ are used for classification. While testing, upon determining the region $\bar{t}$ associated to a test sample $\mathbf{x}_{test}$ using the person tracker, the corresponding weights vectors are used to compute the head pose label, i.e., $y_{test} = \arg\max_{c=1,...,C} \mathbf{w}_{\bar{t},c}^l \mathbf{x}_{test}$.

As stated above, FEGA-MTL can operate both with labeled and unlabeled data. This is important as obtaining a large repository of annotated data for HPE under target motion is costly. In this work, we exploit the fact that people usually tend to look in the direction of their motion to collect a large set of weakly-labeled exemplars without any human intervention. We use walking direction, which can be conveniently extracted from the ground locations output by the tracker, as a proxy for head pose. Specifically, we exploit the tracker output both in terms of estimated target position and particle-spread. Given the tracker estimates for each target, we first employ a smoothing spline approximation to interpolate the trajectory. To filter out noisy samples, we compute the Euclidean distance between tracker estimates and their smoothed counterparts $f_l(\cdot)$, and retain those samples with distance below threshold θ_D. Furthermore, as tracking failures can also contribute to noisy labeling, we monitor the entropy of the target position distribution propagated by the particle filter which, up to a certain extent, indicates the accuracy of the target position estimate. We reject position estimates that result from large localization uncertainty, i.e., where the volume of the typical set approximated from the particle set via kernel density estimation [23] is above a threshold θ_P. Finally, a filtering technique is applied to detect short segments where head appearance is consistent with the observed motion employing this procedure. The filtering process aims to reject samples corresponding to static positions, tracking failures, and sudden changes in direction, where the face may appear blurred and the walking direction may not correspond to the head orientation. The result of the filtering process is a set of short image sequences that can be used to learn head pose classifiers customized to a specific multiview environment and lighting condition.

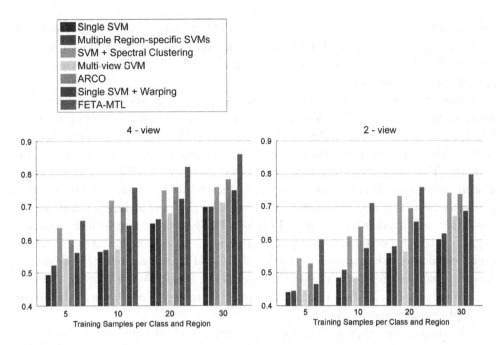

Figure 4.4 Comparison with state-of-the-art head pose classification methods.

Experimental Results

To quantitatively assess the performance of our method, we conduct our experiments on the publicly available DPOSE dataset [29]. We consider an initial, uniformly spaced grid with $R = 25$ regions and define mutually exclusive training/validation/test sets. For all considered classification methods, the regularization parameters are tuned using the validation set. In particular, we set $\lambda_s = 2$, $\lambda_\theta = 2^2$, $\lambda_1 = 2^2$, $\lambda_2 = 1$ for FEGA-MTL in our experiments. To extract short sequences with consistent head and body motion when annotated training data are unavailable, thresholds are set to $\theta_P = 0.5$ m^2 using a Gaussian kernel with variance 0.15 m for entropy estimation.

Fig. 4.4 compares FEGA-MTL with competing HPE methods. The mean classification accuracies obtained from five independent trials are reported, where a randomly chosen training set is employed in each trial. In this series of experiments, we consider annotated training data. To underline the usability of FEGA-MTL with arbitrary camera configuration, we show the results obtained with both four (Fig. 4.4, left) and two camera views (Fig. 4.4, right). As expected, all the considered methods perform better when information from four cameras is used. As baselines, we consider recent multiview approaches, namely, the warping algorithm in [29] combined with RBF-SVMs for classification and presented in the previous section (no transfer learning is required in this case), the approach in [25] which probabilistically fuses the output of multiple

SVMs, and the monocular ARCO [33] and SVM+Spectral Clustering [5] methods. As shown in Fig. 4.4, both ARCO and the method in [25] perform poorly with respect to FEGA-MTL as they are not designed to account for facial distortions due to scale/perspective changes.

Considering baselines that have explicitly accounted for motion-based facial appearance variations while predicting head pose, the warping method in [29] achieves lower accuracy with respect to FEGA-MTL, despite considerably outperforming Single SVM. Here, it is also important to point out two differences between our approach and [29]. The scene is a priori divided into four quadrants in [29], which is not necessarily optimal for describing the pose–appearance relationship under arbitrary camera geometry. Secondly, task dependencies are ignored in [29], and an independent classifier is used for each quadrant. In contrast, FEGA-MTL discovers the optimal configuration of grid clusters that best describes the pose–appearance relationship given camera geometry. Considering task relationships enables FEGA-MTL to achieve higher classification accuracy than a single global classifier (Single SVM), Single SVM+Warping and separate region-specific classifiers that do not consider interregion appearance relationships (Multiple Region-Specific SVMs).

We compare our approach against SVM+Spectral Clustering adopted as a proxy for [5] (a rigorous comparison is not possible as the approach in [5] is monocular). In our implementation of SVM+Spectral Clustering, we use the training images and the spectral clustering algorithm described in [5] to learn a set of spatial regions according to facial appearance similarity. The number of clusters is set to five. Then, five independent SVM classifiers are trained (one for each learned region). As shown in Fig. 4.4, by learning the optimal region partitioning and the classifiers simultaneously, we achieve higher accuracy than SVM+Spectral Clustering.

Table 4.2 compares HPE performance of various MTL methods. The advantage of employing MTL is obvious since all MTL approaches greatly outperform a single SVM. However, assuming that all tasks share a common component, i.e., using the ℓ_{21} MTL approach [1] is suboptimal, and having a flexible learning algorithm which is able to infer appearance relationships among regions improves classification accuracy. This is confirmed by the fact that in all situations (varying training set sizes and number of camera views), FTC MTL [46], Clustered MTL [47], and FEGA-MTL achieve superior performance. FEGA-MTL, which independently considers features and employs graphs to explicitly model region and head pose-based appearance relationships, achieves the best performance. The usefulness of modeling both region and pose-based task dependencies through FEGA-MTL is evident when observing the results in Table 4.2. Using the region graph alone is beneficial as such, while employing the region and pose graphs in conjunction produces the best classification performance.

Fig. 4.5 shows some qualitative results obtained with FEGA-MTL and the optimal spatial partitioning learned for a 3-camera system with 5 training images/class/region.

Table 4.2 Comparing head pose classification accuracy with competing MTL methods

	5 training samples/class/region			10 training samples/class/region		
	2-view	3-view	4-view	2-view	3-view	4-view
Single SVM	0.441 ± 0.011	0.494 ± 0.024	0.523 ± 0.016	0.486 ± 0.012	0.549 ± 0.008	0.564 ± 0.013
Dirty model MTL [17]	0.546 ± 0.006	0.585 ± 0.008	0.603 ± 0.011	0.655 ± 0.011	0.686 ± 0.009	0.696 ± 0.008
ℓ_{21} MTL [1]	0.525 ± 0.010	0.567 ± 0.009	0.589 ± 0.012	0.642 ± 0.012	0.675 ± 0.015	0.696 ± 0.014
Robust MTL [16]	0.550 ± 0.012	0.580 ± 0.011	0.581 ± 0.009	0.655 ± 0.005	0.689 ± 0.004	0.705 ± 0.008
Clustered MTL [47]	0.540 ± 0.007	0.590 ± 0.007	0.619 ± 0.009	0.639 ± 0.014	0.682 ± 0.011	0.711 ± 0.010
Flexible Task Clusters MTL [46]	0.555 ± 0.008	0.598 ± 0.009	0.621 ± 0.007	0.65 ± 0.005	0.681 ± 0.008	0.715 ± 0.006
FEGA-MTL (pose graph only, $\lambda_1 = 0$)	0.564 ± 0.006	0.605 ± 0.006	0.637 ± 0.007	0.661 ± 0.009	0.699 ± 0.011	0.728 ± 0.005
FEGA-MTL (region graph only, $\lambda_2 = 0$)	0.581 ± 0.002	0.623 ± 0.004	0.643 ± 0.006	0.677 ± 0.006	0.718 ± 0.003	0.733 ± 0.007
FEGA-MTL (region graph + pose graph)	**0.602 ± 0.002**	**0.643 ± 0.003**	**0.660 ± 0.004**	**0.711 ± 0.003**	**0.748 ± 0.004**	**0.759 ± 0.005**

Figure 4.5 (Left) Head pose classification results for a target moving freely in a room. (Right) The learned clusters of spatial regions. Cluster corresponding to the target position (denoted using a stick model) is highlighted.

Clustered regions correspond to identical columns of the task similarity matrix $\mathbf{S}$, i.e., two regions t_i and t_j merge if $\mathbf{s}_{t_i,c} = \mathbf{s}_{t_j,c}$ $\forall c$. Constrained by the appearance similarity graph weights, spatially adjacent regions tend to cluster together. While regions closer to the camera-less room corner tend to form large clusters, smaller clusters are observed as one moves closer to the cameras owing to larger facial appearance distortions caused by perspective and scale changes.

We now evaluate FEGA-MTL performance when head pose labels extracted using motion trajectories are used for learning. We consider three different settings: supervised (as in [37]), semisupervised and unsupervised. For unsupervised learning, we train FEGA-MTL exclusively using 1000 images (5 images/class/region) with head pose labels computed using motion direction. For supervised learning, we train the classifier only using annotated examples. We also evaluate FEGA-MTL performance in the semisupervised case where the training set comprises the above annotated-plus-weakly-labeled examples. Table 4.3 shows the results of our evaluation (note that the case corresponding to zero annotated samples in the semisupervised setting exemplifies the unsupervised setting). While accuracy achieved with unsupervised learning is expectedly lower than with supervised learning, weakly labeled examples, nevertheless, boost performance when used in conjunction with annotated ones. Finally, the filtering approach employed for weak labeling is also found to enhance classification performance. Overall, the obtained empirical results confirm the efficacy of FEGA-MTL when unlabeled examples are used for training, and the usefulness of the proposed filtering procedure. As a reference, we also compute the accuracy obtained when motion direction is used as a label and no learning and no filtering are performed. This corresponds to estimating the level of noise of the weakly annotated samples. As expected performance significantly degrade.

Table 4.3 FEGA-MTL classification accuracy obtained with different training sets

No. annotated samples/class/region	Supervised		Semisupervised						No learning
			with filtering			no filtering			motion direction
	5	10	0	5	10	0	5	10	—
1-view	0.49	0.58	0.45	0.59	0.67	0.44	0.59	0.63	
2-view	0.60	0.71	0.58	0.66	0.76	0.55	0.64	0.70	0.45
3-view	0.64	0.74	0.61	0.69	0.78	0.61	0.69	0.74	
4-view	0.66	0.75	0.64	0.74	0.82	0.61	0.73	0.78	

4.4 CONCLUSIONS

This chapter addressed the problem of head pose estimation under target motion, when the scene is captured by multiple, wide-angle cameras. Challenges in this set-up include low-resolution of faces, and change in facial appearance with motion due to varying camera perspective and scale. We described two different frameworks to address HPE under target motion, namely, transfer learning and multitask learning. We also presented exhaustive experimental results obtained using both approaches and demonstrating the effectiveness of the described solutions. We believe that our methods represent effective solutions for addressing different high level tasks in the areas of human behavior analysis, social signal processing, and social robotics. Due to the overwhelming success of deep learning approaches, a natural follow up of this work is to extend the proposed frameworks adopting convolutional neural networks.

REFERENCES

[1] Argyriou Andreas, Evgeniou Theodoros, Pontil Massimiliano. Multi-task feature learning. In: NIPS. 2007.
[2] Benfold Ben, Reid Ian. Unsupervised learning of a scene-specific coarse gaze estimator. In: ICCV. 2011.
[3] Bonilla Edwin V, Chai Kian M, Williams Christopher. Multi-task Gaussian process prediction. In: NIPS. 2007.
[4] Caruana Rich. Multitask learning. Mach Learn 1997;28(1):41–75.
[5] Chamveha Isarun, Sugano Yusuke, Sugimura Daisuke, Siriteerakul Teera, Okabe Takahiro, Sato Yoichi, Sugimoto Akihiro. Head direction estimation from low resolution images with scene adaptation. Comput Vis Image Underst 2013;117(10):1502–11.
[6] Chen Cheng, Odobez Jean-Marc. We are not contortionists: coupled adaptive learning for head and body orientation estimation in surveillance video. In: CVPR. 2012.
[7] Chen Xi, Lin Qihang, Kim Seyoung, Carbonell Jaime G, Xing Eric P, et al. Smoothing proximal gradient method for general structured sparse regression. Ann Appl Stat 2012;6(2):719–52.
[8] Dai Wenyuan, Yang Qiang, Xue Gui R, Yu Yong. Boosting for transfer learning. In: ICML. 2007.
[9] Daumé III Hal. Frustratingly easy domain adaptation. In: ACL. 2009. p. 256–63.

[10] Davis Jesse, Domingos Pedro. Deep transfer via second-order Markov logic. In: ICML. ACM; 2009. p. 217–24.

[11] Doshi Anup, Trivedi Mohan M. Head and eye gaze dynamics during visual attention shifts in complex environments. J Vis 2012;12(2):9.

[12] Duan Lixin, Tsang Ivor W, Xu Dong. Domain transfer multiple kernel learning. IEEE Trans Pattern Anal Mach Intell 2012;34(3):465–79.

[13] Duan Lixin, Tsang Ivor W, Xu Dong, Chua Tat-Seng. Domain adaptation from multiple sources via auxiliary classifiers. In: ICML. 2009.

[14] Fanelli Gabriele, Gall Juergen, Van Gool Luc. Real time head pose estimation with random regression forests. In: CVPR. 2011.

[15] Farhadi Ali, Tabrizi Mostafa Kamali. Learning to recognize activities from the wrong view point. In: ECCV. Springer; 2008. p. 154–66.

[16] Gong Pinghua, Ye Jieping, Zhang Changshui. Robust multi-task feature learning. In: KDD. 2012.

[17] Jalali Ali, Sanghavi Sujay, Ruan Chao, Ravikumar Pradeep K. A dirty model for multi-task learning. In: NIPS. 2010.

[18] Jiang Jing, Zhai ChengXiang. Instance weighting for domain adaptation in NLP. In: ACL. 2007.

[19] Kang Zhuoliang, Grauman Kristen, Sha Fei. Learning with whom to share in multi-task feature learning. In: ICML. 2011.

[20] Katzenmaier Michael, Stiefelhagen Rainer, Schultz Tanja. Identifying the addressee in human–human–robot interactions based on head pose and speech. In: ICMI. 2004.

[21] Kulis Brian, Saenko Kate, Darrell Trevor. What you saw is not what you get: domain adaptation using asymmetric kernel transforms. In: CVPR. 2011.

[22] Lanz Oswald. Approximate Bayesian multibody tracking. IEEE Trans Pattern Anal Mach Intell 2006;28:1436–49.

[23] Lanz Oswald. An information theoretic rule for sample size adaptation in particle filtering. In: ICIAP. 2007.

[24] Mihalkova Lilyana, Mooney Raymond J. Transfer learning by mapping with minimal target data. In: AAAIW. 2008.

[25] Muñoz-Salinas Rafael, Yeguas-Bolivar Enrique, Saffiotti Alessandro, Medina-Carnicer R. Multi-camera head pose estimation. Mach Vis Appl 2012;23(3):479–90.

[26] Murphy-Chutorian Erik, Trivedi Mohan Manubhai. Head pose estimation in computer vision: a survey. IEEE Trans Pattern Anal Mach Intell 2009;31:607–26.

[27] Orozco Javier, Gong Shaogang, Xiang Tao. Head pose classification in crowded scenes. In: BMVC. 2009.

[28] Pan Sinno Jialin, Yang Qiang. A survey on transfer learning. IEEE Trans Knowl Data Eng 2010;22(10):1345–59.

[29] Rajagopal Anoop Kolar, Subramanian Ramanathan, Ricci Elisa, Vieriu Radu L, Lanz Oswald, Sebe Nicu. Exploring transfer learning approaches for head pose classification from multi-view surveillance images. Int J Comput Vis 2014;109(1–2):146–67.

[30] Smith Kevin, Ba Sileye O, Odobez Jean-Marc, Gatica-Perez Daniel. Tracking the visual focus of attention for a varying number of wandering people. IEEE Trans Pattern Anal Mach Intell 2008;30(7):1212–29.

[31] Stiefelhagen Rainer, Bowers Rachel, Fiscus Jonathan. In: Multimodal technologies for perception of humans: international evaluation workshops. 2008.

[32] Subramanian Ramanathan, Yan Yan, Staiano Jacopo, Lanz Oswald, Sebe Nicu. On the relationship between head pose, social attention and personality prediction for unstructured and dynamic group interactions. In: ICMI. 2013.

[33] Tosato Diego, Farenzena Michela, Cristani Marco, Spera Mauro, Murino Vittorio. Multi-class classification on Riemannian manifolds for video surveillance. In: ECCV. 2010.

[34] Voit Michael, Nickel Kai, Stiefelhagen Rainer. Neural network-based head pose estimation and multi-view fusion. In: Multimodal technologies for perception of humans. 2006. p. 291–8.

[35] Voit Michael, Stiefelhagen Rainer. A system for probabilistic joint 3D head tracking and pose estimation in low-resolution, multi-view environments. In: Computer vision systems. 2009.

[36] Wang Xiaoyu, Han Tony, Yan Shuicheng. An HOG–LBP human detector with partial occlusion handling. In: ICCV. 2009.

[37] Yan Yan, Ricci Elisa, Subramanian Ramanathan, Lanz Oswald, Sebe Nicu. No matter where you are: flexible graph-guided multi-task learning for multiview head pose classification under target motion. In: ICCV. 2013.

[38] Yan Yan, Ricci Elisa, Subramanian Ramanathan, Liu Gaowen, Lanz Oswald, Sebe Nicu. A multi-task learning framework for head pose estimation under target motion. IEEE Trans Pattern Anal Mach Intell 2016;38(6):1070–83.

[39] Yan Yan, Ricci Elisa, Subramanian Ramanathan, Liu Gaowen, Sebe Nicu. Multitask linear discriminant analysis for view invariant action recognition. IEEE Trans Image Process 2014;23(12):5599–611.

[40] Yang Jun, Yan Rong, Hauptmann Alexander G. Cross-domain video concept detection using adaptive SVMs. In: ACM multimedia. 2007.

[41] Yang Weilong, Wang Yang, Mori Greg. Efficient human action detection using a transferable distance function. In: ACCV. 2009.

[42] Yuan Xiao-Tong, Liu Xiaobai, Yan Shuicheng. Visual classification with multitask joint sparse representation. IEEE Trans Image Process 2012;21(10):4349–60.

[43] Zabulis Xenophon, Sarmis Thomas, Argyros Antonis A. 3D head pose estimation from multiple distant views. In: BMVC. 2009.

[44] Zhang Tianzhu, Ghanem Bernard, Liu Si, Ahuja Narendra. Robust visual tracking via structured multi-task sparse learning. Int J Comput Vis 2013;101(2):367–83.

[45] Zheng Jingjing, Jiang Zhuolin, Phillips P Jonathon, Chellappa Rama. Crossview action recognition via a transferable dictionary pair. In: BMVC. 2012.

[46] Zhong Leon Wenliang, Kwok James Tin Yau. Convex multitask learning with flexible task clusters. In: ICML. 2012.

[47] Zhou Jiayu, Chen Jianhui, Ye Jieping. Clustered multi-task learning via alternating structure optimization. In: NIPS. 2011.

CHAPTER 5

The Analysis of High Density Crowds in Videos

Mikel Rodriguez*, Josef Sivic[†], Ivan Laptev[†]
*MITRE, Computer Vision Research, McLean, VA, USA
[†]Laboratoire d'Informatique de l'École Normale Supérieure, ENS/INRIA/CNRS, Paris, France

Contents

Group and Crowd Behavior for Computer Vision
DOI: 10.1016/B978-0-12-809276-7.00006-0

Figure 5.1 Examples of high density crowded scenes.

5.1 INTRODUCTION

In recent years, video surveillance of public areas has proliferated at an ever increasing rate, from CCTV systems that monitor individuals in subway systems, sporting events and airport facilities to networks of cameras that cover key locations within large cities. Along with the growing ubiquity of video surveillance, computer vision algorithms have recently begun to play a growing role in these monitoring systems. Up until recently, this type of video analysis has, for the most part, been limited to the domain of sparse and medium person density scenes primarily due to the limitations of person detection and tracking. As the density of people in the scene increases, a significant degradation in the performance is usually observed in terms of object detection, tracking, and events modeling, given that many existing methods depend on being able to separate people from the background. This inability to deal with crowded scenes such as those depicted in Fig. 5.1 represents a significant problem; given that these crowded scenes need the most of video analysis as that they host a large population of pedestrians.

This chapter first reviews (Section 5.2) recent studies that have begun to address the various challenges associated with the analysis of crowded scenes focusing on: (i) learning typical motion patterns of crowded scenes and segmenting the motion of the agents in a crowd; (ii) determining the density of people in a crowded scene; (iii) tracking the motion of individuals in crowded scenes; and (iv) crowd event modeling and anomaly detection. After reviewing related work, we describe our three contributions to crowd analysis in a video.

In particular, in Section 5.3, we present a crowd analysis algorithm powered by behavior priors that are learned on a large database of crowd videos gathered from the Internet [47]. The algorithm works by first learning a set of crowd behavior priors offline. During testing, crowd patches are matched to the database, and behavior priors are transferred. The proposed algorithm performs like state-of-the-art methods for tracking people having common crowd behaviors and outperforms the methods when the tracked individual behaves in an unusual way.

In Section 5.4, we address the problem of person detection and tracking in crowded video scenes. We propose to leverage information on the global structure of the scene and to resolve all detections simultaneously. In particular, we explore constraints imposed by the crowd density and formulate person detection as the optimization of a joint energy function combining crowd density estimation and the localization of individual people [46]. We demonstrate how the optimization of such an energy function significantly improves person detection and tracking in crowds. We validate our approach on a challenging video dataset of crowded scenes.

Section 5.5 explores the notion of learning a representation that is tailored for high density crowded scenes. In particular, the section explores the use of convolutional neural net features learned using crowd tracking as a "free" source of supervision.

Finally, the chapter concludes by describing ongoing and future research directions in crowd analysis.

5.2 LITERATURE REVIEW

The problem of crowd analysis in videos comprises a wide range of subproblems. In the subsections that follow, we describe a representative subset of studies that address the major tasks associated with analyzing high density crowded scenes. These studies are grouped into four commonly studied problems within crowd analysis: modeling and segmenting the motion of a crowd, the estimation of crowd density, detecting and tracking individuals in a crowded scene, and modeling collective crowd events and behaviors.

5.2.1 Crowd Motion Modeling and Segmentation

Learning typical motion patterns of moving objects in the scene from videos is an important visual surveillance task given that it provides algorithms with motion priors that can be used to improve tracking accuracy and allow for anomalous behavior detection. Typically, given an input video, the goal is to partition the video into segments with coherent motion of the crowd, or alternatively find (multiple) dominant motion directions at each location in the video.

A significant amount of effort has been placed on studying this problem in the context of typical surveillance scenarios which contain low-to-medium person densities. More recently, a number of studies have begun to focus on segmenting motion patterns of high density scenes. Several crowd flow segmentation works represent crowd motion patterns using low-level features computed over short temporal extents [2,25,45,20], such as an optical flow. These features are then combined with Lagrangian particle dynamics [2] or simple agglomerative clustering algorithm [25,53] to partition a crowd video sequence into segments with single coherent motion. Multiple dominant motions at each location of the crowd video can be found using latent variable topic models [6] applied to optic flow vectors clustered into a motion vocabulary [45].

An alternative representation of crowd motion patterns forgoes directly incorporating low-level motion features in favor of mid-level features such as object tracks. The main thrust behind these approaches resides in the fact that they allow for long-term analysis of a scene and can capture behaviors that occur over long spatiotemporal extents. For example, point trajectories of pedestrians or traffic within a scene (such as a cross-road) can be clustered into coherent motion clusters [57,33,52]. Trajectories which do not match any of the clusters can be then flagged as abnormal events.

5.2.2 Estimating Density of People in a Crowded Scene

Determining the density of objects in a scene has been studied by a number of works. The objective of most of the studies that focus on this problem is to provide accurate estimates of person densities in the form of people per squared meter or person counts within a given spatiotemporal region of a video.

A significant number of density estimation methods are based on aggregating counts obtained from local object detectors. In these approaches an object detector is employed to localize individual person instances in an image. Having obtained the localizations of all person instances, density estimation can proceed in a straightforward manner. A number of these methods are not particularly well suited for crowded scenes, given they assume that pedestrians are disconnected from each other by the distinct background color such that it may be possible to detect individual instances via a Monte Carlo process [13], morphological analysis [4], or variational optimization [42]. This class of methods tends to generate accurate density estimation within the bounds of the previously mentioned assumptions.

Another density estimation paradigm is based on regression. This class of methods forgoes the challenges of detecting individual agents and instead focuses on learning a mapping between density and a set of global features. Lempitsky and Zisserman [35] cast the problem of density estimation as that of estimating an image density whose integral over any image region gives the count of objects within that region. Learning to infer such a density is formulated as a minimization of a regularized risk quadratic cost function. A linear transformation of feature responses that approximates the density function at each pixel is learned. Once trained, an estimate for object counts can be obtained at every pixel or in a given region by integrating across the area of interest.

A number of regression-based methods begin by segmenting the scene into clusters of objects and then proceed to regress on each of the clusters separately. For example, Chan et al. [9] segment crowd video using a mixture of dynamic textures. For each crowd segment, various features are extracted, while applying a perspective map

Figure 5.2 Not all actors in a crowd are equal. Depending on their roles, be it a protester, police or marathon runner, individual behavior is shaped by these factors.

to weight each image location according to its approximate size in the real scene. Finally, the number of people per segment is estimated with Gaussian process regression. Ryan et al. [50] use foreground/background segmenter to localize crowd segments in the video and estimate the count of people within each segment using local rather than global features.

However, most of the above discussed methods have been evaluated in low/medium density crowds, and it is not clear how they would perform in heavily crowded scenes.

5.2.3 Crowd Event Modeling and Recognition

Over the years event modeling has traditionally been limited to scenes containing low density of people. However, recently, the computer vision community has begun to focus on crowd behavior analysis. There are several complementary approaches to solving the problem of understanding crowd behaviors.

The most conventional approach to modeling crowd events is the "object based" paradigm, in which a crowd is considered as a collection of individuals (bounding boxes, segmented regions, etc.). Ke et al. [27] propose a part-based shape template representation that involves sliding the template across all possible locations and measuring the shape matching distance between a subregion of the input sequence and the manually generated template (Fig. 5.2).

The work of Kratz et al. [31] focuses on recognizing anomalous behaviors in high density crowded scenes by learning motion-pattern distributions that capture the variations of local spatiotemporal motion patterns to compactly represent the video volume. To this effect, this work employs a coupled HMM that models the spatial relationship of motion patterns surrounding each video region. Each spatial location in the video is modeled separately, creating a single HMM for each tube of observations.

Another study that focuses on detecting abnormal crowd behavior is the work of Mehran et al. [41]. Instead of explicitly modeling a set of distinct locations within the video as in Kratz et al., this work takes a holistic approach that uses optical flow to compute the social force between moving particles to extract interaction forces. The interaction forces are then used to model the normal behaviors using a bag of words.

5.2.4 Detecting and Tracking in a Crowded Scene

Person detection and tracking is one of the most researched areas in computer vision, and a substantial body of work has been devoted to the problem. In general, the goal of these works is to determine the location of individuals as they move within crowded scenes.

Tracking in crowded scenes has been addressed in a variety of contexts, including the study of dense clouds of bats [5] and biological cells in microscopy images [36], as well as medium to high density gatherings of people in monocular video sequences [21,37,8,34,7,3,60] and multiple camera configurations [19,28].

In medium-density crowded scenes, research has been done on tracking-by-detection methods [34,7] in multiobject tracking. Such approaches involve the continuous application of a detection algorithm in individual frames and the association of detections across frames.

Another approach followed by several studies centers around learning scene-specific motion patterns, which are then used to constrain the tracking problem. In [3] global motion patterns are learned, and participants of the crowd are assumed to behave in a manner similar to the global crowd behavior. Overlapping motion patterns have been studied [45] as a means of coping with multimodal crowd behaviors. These types of approaches operate in the off-line (or *batch*) mode (i.e., when the entire test sequence is available during training and testing) and are usually tied to a *specific scene*. Furthermore, they are not well suited for tracking rare events that do not conform to the global behavior patterns of the same video.

In the following section, we describe a crowd tracking algorithm that builds on the progress in large database driven methods, which have demonstrated great promise in providing priors on object locations in complex scenes for object recognition [38,49,48] or scene completion [23], and recognizing actions of people at a distance [15], as well as predicting and transferring motion from a video to a single image [39,59].

5.3 DATA-DRIVEN CROWD ANALYSIS IN VIDEOS

Here we wish to use a large collection of crowd videos to learn crowd motion patterns by performing long-term analysis in an off-line manner. The learned motion patterns can be used in a range of application domains such as crowd event detection or anomalous behavior recognition. In this particular work, we choose to use the motion patterns learned on the database to drive a tracking algorithm. The idea is that any given crowd video can be thought of as being a mixture of previously observed videos. For example,

Figure 5.3 A crowd as a combination of previously observed crowd patches. Each crowd patch contains a particular combination of crowd behavior patterns.

a crowded marathon video, such as the one depicted in the middle of Fig. 5.3, contains regions that are similar to other crowd videos. In it we observe a region of people running in a downward direction, similar to the video depicted in the top left, as well as a region containing people running toward the right, as in the video depicted in the bottom left. These different videos can provide us with strong cues as to how people behave in a particular region of a crowd. By learning motion patterns from a large collection of crowded scenes, we should be able to better predict the motion of individuals in a crowd.

Our data-driven tracking algorithm is composed of three components: We start by learning a set of motion patterns off-line from a large database of crowd videos. Subsequently, given an input video, we proceed to obtain a set of coarsely matching crowd videos retrieved from the large crowd database. Having obtained a subset of videos which roughly match the scale and orientation of our testing sequence, in the second phase of our algorithm, we use this subset of videos to match patches of the input crowded scene. Our goal is to explain input video by the collection of space-time patches of many other videos and to transfer learned patterns of crowd behavior from videos in the database. The final component of our algorithm pertains to how we incorporate local motion priors into a tracking framework. The three components of the approach are described next.

5.3.1 Off-Line Analysis of Crowd Video Database

A crowd motion pattern refers to a set of dominant displacements observed in a crowded scene over a given time scale. These observed motion patterns can be represented either directly, using low-level motion features such as optical flow, or they can be modeled at a higher level, by a statistical model of flow direction obtained from a long term analysis of a video. In this section we describe each of these representations.

5.3.1.1 Low-Level Representation

Examples of low-level motion features include sparse or dense optical flows, spatiotemporal gradients, and feature trajectories obtained using Kanade–Lucas–Tomasi feature tracking. In this work, a low-level crowd pattern representation is a motion flow field which consists of a set of independent flow vectors representing the instantaneous motion present in a frame of a video. The motion flow field is obtained by first using an existing optical flow method [40] to compute the optical flow vectors in each frame, and then combining the optical flow vectors from a temporal window of frames of the video into a single global motion field.

5.3.1.2 Mid-Level Representation

An alternative representation of crowd motion patterns forgoes directly incorporating low-level motion features in favor of a hierarchical Bayesian model of the features. The main thrust behind the use of an unsupervised hierarchical model within this domain is that it allows for long-term analysis of a scene and can capture both overlapping behaviors at any given location in a scene and spatial dependencies between behaviors. For this purpose, we adopt the representation used in [45] which employs a correlated topic model (CTM) [6] based on a logistic normal distribution, a distribution that is capable of modeling dependence between its components. CTM allows for an unsupervised framework for modeling the dynamics of crowded and complex scenes as a mixture of behaviors by capturing spatial dependencies between different behaviors in the same scene.

5.3.2 Matching

Given a query test video, our goal here is to find similar crowded videos in the database with the purpose of using them as behavior priors. The approach consists of a two-stage matching procedure depicted in Fig. 5.4, which we describe in the remainder of this section.

5.3.2.1 Global Crowded Scene Matching

Our aim in this phase is to select a subset of videos from our dataset that share similar global attributes (Fig. 5.4B). Given an input video in which we wish to track an individual, we first compute the gist descriptor of the first frame. We then select the top 40 nearest neighbors from our database. By searching for similar crowded scenes first, instead of directly looking for local matching regions in a crowd video, we avoid searching among the several million crowd patches in our database and thus dramatically reduce the memory and computational requirements of our approach.

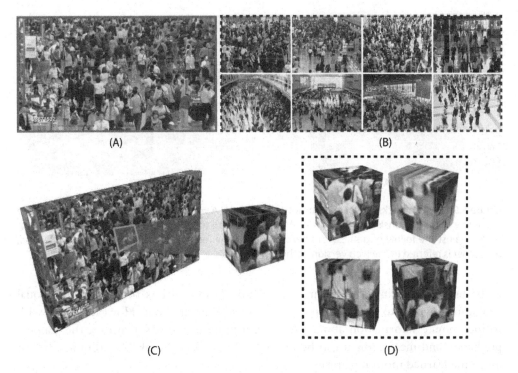

Figure 5.4 Global and local crowd matching. (A) Testing video. (B) Nearest neighbors retrieved from the database of crowd videos using global matching. (C) A query crowd patch from the testing video. (D) Matching crowd patches from the pool of global nearest neighbor matches.

5.3.2.2 Local Crowd Patch Matching

Given a set of crowded scenes which roughly match a testing video, we proceed to retrieve local regions that exhibit similar spatiotemporal motion patterns from this subset of videos. A number of different space-time feature descriptors have been proposed. Most feature descriptors capture local shape and motion in a neighborhood of interest using spatiotemporal image gradients and/or optical flow. In our experiments we employ the HOG3D descriptor [29], which has demonstrated excellent performance in action recognition [55]. Given a region of interest in our testing video (i.e., current tracker position), we compute HOG3D of the corresponding spatiotemporal region of the video. We then proceed to obtain a set of similar crowd patches from the preselected pool of global matching crowd scenes by retrieving the k nearest neighbors (Fig. 5.4D) from the crowd patch that belong to the global matching set.

5.3.3 Transferring Learned Crowd Behaviors

We incorporate the precomputed motion patterns associated with matching crowd patches as priors over a standard Kalman filter tracker. When there is no behavior prior

Figure 5.5 Example tracks of rare and abrupt behaviors. (A) A woman (red) walks perpendicular to the main flow of traffic, consisting of a crowd of people walking toward the left (blue). (B) A cameraman walks against the follow of traffic. (For interpretation of the references to color in this figure legend, the reader is referred to the web version of this chapter.)

to be used in tracking, the linear motion model alone drives the tracker and equal weighting is given to the Kalman prediction and measurement. However, if we wish to incorporate information about the learned motion patterns as priors, the Kalman prediction and measurement can be reweighted to reflect the likelihood of a behavior given the learned motion patterns.

5.3.4 Experiments and Results

This section evaluates our approach on a challenging video dataset collected from the web and spanning a wide range of crowded scenes. In order to track individuals in a wide range of crowd scenes, we aim to sample the set of crowd videos as broadly as possible. To this end, we construct our crowd video collection by crawling and downloading videos from search engines and stock footage websites (such as Getty Images, Google Videos, and BBC Motion Gallery) using text queries such as "crosswalk," "political rally," "festival," and "marathon." We discard duplicate videos, as well as videos taken using alternative imaging methods such as time-lapse videos and videos taken with tilt-shift lenses. Our database contains 520 unique videos varying from two to five minutes (624 min in total) and resized to 720×480 resolution.

The main testing scenario of this work focuses on tracking rare and abrupt behaviors of individuals in a crowd. This class of behaviors refers to motions of an individual within a crowd that do not conform to the global behavior patterns of the same video, such as an individual walking against the flow of traffic. Fig. 5.5 depicts examples of such relatively rare crowd events. In order to assess the performance of the proposed data-driven model in tracking this class of events, we selected a set of 21 videos containing instances of relatively rare events. A first baseline tracking algorithm consisted of the linear Kalman tracker with no additional behavior prior. The second baseline

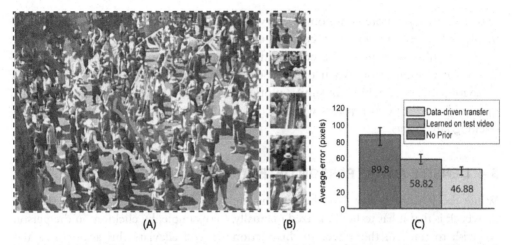

Figure 5.6 (A) Data-driven track of a person walking across a crowded demonstration (green), ground-truth (red), batch mode tracking (yellow). (B) Matching crowd patches from the database. (C) Comparison of average tracking errors when tracking people in rare crowd events. (For interpretation of the references to color in this figure legend, the reader is referred to the web version of this chapter.)

learned motion priors on the testing video itself (batch mode) using the CTM motion representation [45]. Finally, the proposed data-driven approach transferred motion priors from the top k matching database videos, for which motion patterns had been learned off-line using the CTM motion representation.

The rare events are not common in most videos. Therefore, there may only exist a few examples throughout the course of a video sequence. In these scenarios, the data-driven tracking approach is expected work better than batch mode methods, which learn motion priors from the testing video itself. This is due to the fact that the test videos alone are not likely to contain sufficient repetitions of rare events in order to effectively learn motion priors for this class of events.

The tracking errors are depicted in Fig. 5.6C. It can be seen that batch mode tracking is unable to effectively capture strong motion priors for temporally-short events that only occur once throughout a video (with a mean tracking error of 58.82 pixels), whereas data-driven tracking (with a mean tracking error of 46.88 pixels) is able to draw motion priors from crowd patches that both roughly match the appearance of the tracked agent, and exhibit a strongly defined motion pattern. This is evident in Fig. 5.6, which depicts a successfully tracked individual moving perpendicular to the dominant flow of traffic in a political rally scene. The corresponding nearest neighbors (Fig. 5.6B) are crowd patches that, for the most part, contain upward-moving behaviors from the crowd database. Besides, it can be noted that the retrieved crowd patches belong to behaviors which are commonly repeated throughout the course of a clip, such as crossing a busy intersection in the upward direction. By matching a rare event in a testing video with a similar (yet more commonly observed) behavior in our database, we are able to incorporate these strong motion cues as a means of improving tracking performance.

The results above provide a compelling reason for searching a large collection of videos for motion priors when tracking events that do not follow the global crowd behavior pattern. Searching for similar motion patterns in our large database has proven to provide better motion priors which act as strong cues that improve accuracy when tracking rare events.

5.4 DENSITY-AWARE PERSON DETECTION AND TRACKING IN CROWDS

While the person tracker described in the previous section works relatively well, its drawback is that it has to be initialized manually, for example, by clicking on the person we wish to track in the video. In this section we will alleviate this assumption and consider joint detection and tracking of people in crowded videos.

In recent years significant progress has been made in the field of object detection and recognition [18,11,17]. While standard "scanning-window" methods attempt to localize objects independently, several recent approaches extend this work and exploit scene context as well as relations among objects for improved object recognition [12,58,44, 54]. Related ideas have been investigated for human motion analysis where incorporating scene-level and behavioral factors effecting the spatial arrangement and movement of people have been shown beneficial for achieving improved detection and tracking accuracy. Examples of explored cues include: the destination of a pedestrian within the scene [43], repulsion from nearby agents due to the preservation of personal space and social grouping behavior [7], as well as the speed of an agent in the group [26].

We follow this line of work and extend it to the detection and tracking of people in high-density crowds. Rather than modeling individual interactions of people, this work exploits information at the global scene level provided by the crowd density and scene geometry. Crowd density estimation has been addressed in a number of recent works which often pose it as a regression problem [35,10,30] (see Section 5.2). Such methods avoid the hard detection task and attempt to infer person counts directly from low-level image measurements, e.g., histograms of feature responses. Such methods hence provide person counts in image regions but are uncertain about the location of people in these regions. This information is complementary to the output of standard person detectors which optimize precise localization of individual people but lack the global knowledge on the crowd structure. Our precise goal and contribution is to combine these two sources of complementary information for improved person detection and tracking. The intuition behind our method is illustrated in Fig. 5.7 where the constraints of person counts in local image regions help improving the standard head detector.

We formulate our method in the energy minimization framework which combines crowd density estimates with the strength of individual person detections. We minimize

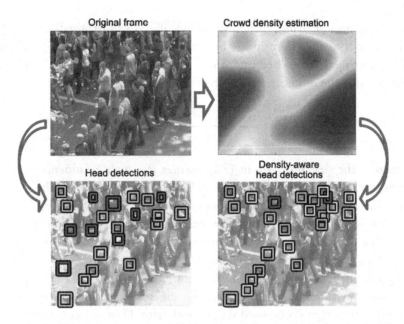

Figure 5.7 Individual head detections provided by state-of-the-art object detector [18] (bottom-left; green, true positives; red, false positives) are improved significantly by our method (bottom-right; yellow, new true positives) using the crowd density estimate (top-right) obtained from the original frame (top-left). (For interpretation of the references to color in this figure legend, the reader is referred to the web version of this chapter.)

this energy by jointly optimizing the density and the location of individual people in the crowd. We demonstrate how such optimization leads to significant improvements of state-of-the-art person detection in crowded scenes with varying densities. In addition to crowd density cues, we explore constraints provided by scene geometry and temporal continuity of person tracks in the video and demonstrate further improvements for person tracking in crowds. We validate our approach on challenging crowded scenes from multiple video datasets.

5.4.1 Crowd Model

We formulate the density-informed person detection as follows. We assume to have a confidence score $s(p)$ of a person detector for each location p_i, $i = 1, \ldots, N$ in an image. In addition, we assume we are given a person density, i.e., the number of people per pixel, $D(p_i)$ estimated in a window of size σ at each location p_i. In this work density estimation is carried out using the regression-based method outlined in [35].

The goal is to identify locations of people in the image such that the sum of detector confidence scores at those locations is maximized while respecting the density of people given by D and preventing significantly overlapping detections, i.e., detections with the area overlap greater than a certain threshold. Using similar notation as in [12], we encode detections in the entire image by a single N-vector $\mathbf{x} \in \{0, 1\}^N$, where $x_i = 1$ if the detection at p_i is "switched on" and 0 otherwise. The detection problem can be then formulated as the minimization of the following cost function

$$\min_{\mathbf{x} \in \{0,1\}^N} \; -\underbrace{s^\top \mathbf{x}}_{E_S} + \underbrace{\mathbf{x}^\top W \mathbf{x}}_{E_P} + \underbrace{\alpha \|D - A\mathbf{x}\|_2^2}_{E_D}. \tag{5.1}$$

Minimizing the first term E_S in (5.1) ensures the high confidence values of the person detector at locations of detected people (indicated by $x_i = 1$). The second, pairwise, term E_P ensures that only valid configurations of nonoverlapping detections are selected. This is achieved by setting $W_{ij} = \infty$, if detections at locations p_i and p_j have significant area overlap ratio, and 0 otherwise. The first two terms of the cost function are similar in spirit to the formulation used in [12] and implement a variation of the standard nonmaximum suppression. In addition, we introduce a new term E_D that concerns the crowd density and penalizes the difference in the density values either (i) measured with a regression-based density estimator D or (ii) obtained by counting "switched on" (or *active*) detections in $\mathbf{x}$. The evaluation of the density of active detections in $\mathbf{x}$ is performed by matrix multiplication $A\mathbf{x}$, where A is an $N \times N$ matrix with rows A_i

$$A_i(q_j) = \frac{1}{2\pi\sigma^2} \exp\left(-\frac{\|p_i - q_j\|^2}{2\sigma^2}\right) \tag{5.2}$$

corresponding to Gaussian windows of size σ centered at positions p_i.

The idea of minimizing the term E_D is illustrated in Fig. 5.8. Intuitively, optimizing the cost (5.1) including the third, density, term E_D enables improving person detection by penalizing confident detections in low person density image regions while promoting low-confidence detections in high person density regions.

5.4.1.1 Tracking Detections

The objective here is to associate head detections in individual frames into a set of head tracks corresponding to the same person within the crowd across time, thereby generating tracks of individuals that conform to the crowd.

We follow the tracking by detection approach of [16], which demonstrated excellent performance in tracking faces in TV footage, but here apply it to track heads in crowded video scenes. The method uses local point tracks throughout the video to associate detections of the same person obtained in individual frames. For each crowd video sequence, we obtain point tracks using the Kanade–Lucas–Tomasi tracker [51]. The point tracks are used to establish correspondence between pairs of heads that have been detected within the crowd.

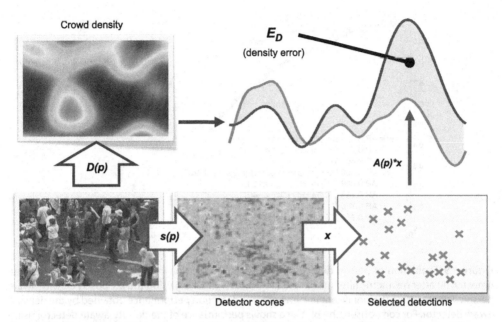

Figure 5.8 Illustration of the energy term E_D of (5.1). Minimizing E_D implies reducing the difference in person density estimates obtained by the estimator $D(p)$ (blue color coding) and by local counting person detections (red color coding). (For interpretation of the references to color in this figure legend, the reader is referred to the web version of this chapter.)

This simple tracking procedure is extremely robust and can establish matches between head detections where the head has not been continuously detected due to pose variation or partial occlusions due to other members of the crowd. In the next section we demonstrate the improvement in detection performance using this type of tracking by association: missing detection below detection threshold can be filled-in, and short tracks corresponding to false positive detections can be discarded. While not done here, the data driven priors described in Section 5.3 could be also incorporated into this tracking by detection framework, for example, to help resolve ambiguities due to occlusions.

5.4.2 Evaluation

In order to test and compare the detection performance, we follow the PASCAL VOC evaluation protocol [17]. To demonstrate the advantage of our method on the detection task, we have compared it to three alternative detectors. Our first *baseline detector* is [18] which was trained on our training data. The second detector integrates the baseline detector with geometric filtering imposing the prior on the size of detections [24,46]. The third detector integrates temporal consistency constraints using tracking. Finally, our density-aware detector optimizes the introduced cost function (5.1) and integrates

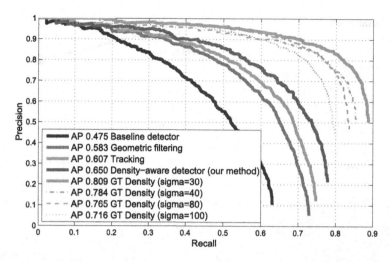

Figure 5.9 Evaluation of person detection performance. Precision–recall curves for the (i) baseline detector, (ii) after geometric filtering, (iii) tracking, and (iv) tracking using the proposed density-aware person detector. Note the significant improvement in detection performance obtained by the density-aware detector. For comparison, the plot also shows performance of the density-aware detector using the ground-truth density, obtained by smoothing ground-truth detections by a Gaussian with different sigmas. Note the improvement in performance for smaller sigmas. In this case, for sigma approaching zero, the density would approach the ground-truth and hence the perfect performance.

geometric filtering and temporal consistency constraints as in the case of other detectors. The comparative evaluation is presented in Fig. 5.9. As can be observed, the density-aware detector (red curve) outperforms all three other detectors by a large margin.

To gain understanding of the density prior introduced in this work, Fig. 5.9 also shows detection results for the density-aware detector using ground truth density estimation (green curves). Interestingly, the detection performance increases significantly in this case, suggesting that our detector can benefit much from future better performing methods of density estimation. As expected, the performance of the detector increases for the more localized ground truth density estimator with small values of σ.

5.4.2.1 Tracking

The objective of this set of experiments is to assess the improvement that can be attained in tracking accuracy using the proposed density-aware crowd model in the presence of a range of crowd densities. In our evaluation we employed a collection of 13 video clips captured at a large political rally. On average, each video clip is roughly two minutes long with a frame size of 720×480. As can be seen, this collection of crowd video clips spans a wide range of densities, viewpoints and zoom levels.

Quantitative analysis of the proposed tracking algorithm was performed by generating ground-truth trajectories for 122 people, which were selected randomly from the set of all people in the crowd. The ground-truth was generated by manually tracking the centroid of each selected person across the video. In our experiments we evaluate tracks independently, by measuring tracking error (measured in pixels) which is achieved by comparing the tracker position at each frame with respect to the position indicated by the ground truth. When our system does not detect a person that has been labeled in the ground truth, this corresponding track is considered lost. In total, our system was able to detect and track 89 out of the 122 labeled individuals.

The average tracking error obtained using the proposed model was 52.61 pixels. In order to assess the contribution of density estimation in tracking accuracy, a baseline tracking procedure consisting of detection, geometric filtering and agglomerative clustering was evaluated. The mean tracking error of this baseline algorithm was 64.64 pixels.

We further evaluated the ability to track people over a span of frames by measuring the difference in the length of the generated tracks in relation to the manually annotated tracks. The mean absolute difference between the length of the ground-truth tracks and the tracks generated by our system was 18.31 frames, whereas the baseline (which does not incorporate density information) resulted in a mean difference of 30.48 frames. It can be observed from these results that our tracking was very accurate, in most cases, and able to maintain correct track labels over time.

5.5 CROWDNET: LEARNING A REPRESENTATION FOR HIGH DENSITY CROWDS IN VIDEOS

5.5.1 Introduction

The previous sections of this chapter employed manually designed representations of crowd behavior in video. Recently, across the field in computer vision, researches have leveraged massive datasets of labeled examples to learn rich, high-performance visual representations [32]. Yet efforts to incorporate some of these approaches to high density crowded scenes are usually limited by the sheer expense of the human annotation required. A natural way to address this limitation would be to employ unsupervised learning, which aims to use data without any annotation. Unfortunately, despite several decades of sustained effort, unsupervised methods have not yet been shown to extract useful information from large collections of full-sized, real images.

Interestingly, a number of recent image classification approaches [14,1] have begun to ameliorate the need for massive labeled datasets by adopting "self-supervised" formulations for learning effective representations of various data types and problems. In this work we closely follow the image representation learning approach introduced in [56] and adapt it to video and high density crowded scenes.

In this section of the chapter we present a simple, yet surprisingly powerful, unsupervised approach to learning a representation for high density crowded scenes using unlabeled crowd videos that span a wide range of scenarios. Much of the recent progress in high density crowd analysis has been focused on visual tracking of people in high density scenes. Leveraging the progress that has been made on this front, we propose to exploit visual tracking in high density scenes as way of automatically learning a representation in the form of a convolutional neural net (CrowdNet) encoding in an un-supervised manner. Specifically, we track thousands of people in high density crowded videos and use these tracks as a source of supervision in order to learn a representation that is tailored to high density crowded scenes. Our key idea is that two crowd patches connected by a track should have similar visual representation in deep feature space since they probably belong to the same person.

In order to leverage the advances in crowd tracking to we employ a Siamese-triplet network with a ranking loss function in order to train CrowdNet representation. This ranking loss function enforces that in the final deep feature space the first crowd patch should be much closer to the tracked patch than any other randomly sampled patch in the crowded scene. In order to assess the effectiveness of this approach we evaluate the proposed representation on a range of high density crowded scene analysis tasks.

5.5.2 Overview of the Approach

Our goal is to learn a video representation that is tailored toward high density crowded scenes. Recent advances in computer vision have shown that visual features learned by neural networks trained for the task of object recognition using more than a million labeled images are useful for many computer vision tasks like semantic segmentation, object detection and action classification. However, these representations which are typically learned on images from datasets such as ImageNet are not representative of the conditions typically seen in high density crowded scenarios. Furthermore, there is no equivalent to ImageNet for high density crowd videos in regards to the sheer number of annotated videos.

We aim to learn a high density crowd video representation. Specifically, our goal is to employ a crowd tracking along with Convolutional Neural Networks (ConvNets) to learn a feature embedding for individual crowd patches such that crowd patches which are visually similar (across different frames in a video) would be close in the embedding space.

We follow the AlexNet [32] architecture to design our base network. However, since our crowd videos do not have labels, it is not clear what should be the loss function and how we should optimize it. However, we have another source of supervision: person tracks across the high density crowded video. The underlying assumption in this work is that a crowd patch does not change drastically within a short time in a video and that a person within a crowd appears in multiple frames of the video.

We sample thousands of crowd patches in these high density crowded scenes and track them over time. Since we are tracking these patches, we know that the first and last tracked frames correspond to the same instance of the moving object or object part. Therefore, any visual representation that we learn should keep these two data points close in the feature space. But just using this constraint is not sufficient: all points can be mapped to a single point in feature space. Therefore, for training our CrowdNet, we sample a third crowd patch from another random spatiotemporal location within the scene which creates a triplet. During training, we use a loss function that explicitly enforces that the first two crowd patches that have been automatically associated by the particle filter tracker are closer in feature space than the first one the random crowd patch.

5.5.3 Crowd Patch Mining in Videos

Given a crowd video, we want to extract crowd patches that contain motion and track these patches to create training instances. Given that all of the videos in our dataset have a static camera one obvious way to find patches of interest is to compute optical flow and use the high magnitude flow regions. Given the flow trajectories, we classify a region as "moving" if the flow magnitude is more than 0.5 pixels. In the experiments, we set height and width of the crowd patch to be 227 pixels.

5.5.4 Tracking

Given the initial crowd patch bounding box, we proceed to track that crowd patch using the density-aware tracking method described in the previous section. After tracking the crowd patch along 30 frames in the video, we obtain the second patch. This patch acts as the similar patch to the query patch in the triplet.

5.5.5 Learning a Representation for High Density Crowds

In the previous section, we discussed how we can use crowd tracking to generate pairs of crowd patches. We use this procedure to generate thousands of such pairs. We now describe how we use these as training instances for our visual representation.

Our goal is to employ crowd patch tracking to learn a feature space such that a given query crowd patch is closer to the tracked patch as compared to any other randomly sampled crowd patch in a scene. To learn this feature space we design a Siamese-triplet network. A Siamese-triplet network consists of three base networks which share the same parameters (see Fig. 5.10). For our experiments, we take the image with size 227×227 as input. The base network is based on the AlexNet architecture [32] for the convolutional layers. Then we stack two fully connected layers on the pool5 outputs, whose neuron numbers are 4096 and 1024, respectively. Thus the final output of each single network is 1024 dimensional feature space $f(\cdot)$.

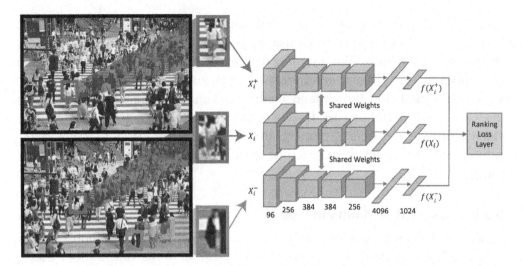

Figure 5.10 CrowdNet is a Siamese-triplet network. Each base network shares the same architecture and parameter weights.

Given a crowd patch X from a video we can obtain its representation from the final later of the CrowdNet network $f(X)$. We can define the distance between two crowd patches X_1, X_2 based on the cosine distances in feature space.

In order to tailor this representation to high density crowded scenes, we want to train CrowdNet to obtain a representation $f(\cdot)$ such that the distance between a query crowd patch and another random crowd patch from the video is large. Therefore, given the set of crowd patches $\mathbb{S}$, where X_i is the crowd patch in the first frame of a video, X_i^+ is the tracked crowd patch and X_i^- is a random crowd patch from the scene, we want to enforce $D(X_i, X_i^-) > D(X_i, X_i^+)$. In order to accomplish this we employ a hinge loss function,

$$L(X_i, X_i^+, X_i^-) = \max\{0, D(X_i, X_i^+) - D(X_i, X_i^-) + G\}, \tag{5.3}$$

where G represents the gap parameters between two distances (set to 0.5 in the experimental section). The objective function employed to train the crowd patch representation is given by:

$$\min_w \frac{\lambda}{2}\|W\|_2^2 + \sum_{i=1}^{N}\max\{0, D(X_i, X_i^+) - D(X_i, X_i^-) + G\}, \tag{5.4}$$

where W represents the weights of CrowdNet, N is the number of triplet crowd patch samples and λ is the weight decay parameter (set to 0.001 in the experiments).

In order to learn the network parameters, we employ mini-batch stochastic gradient descent during training. Given that the three networks share weights, we perform the forward propagation for the whole batch by a single network and calculate the loss based on the output feature. Given a pair of patches X_i, X_i^+, we randomly select another patch X_i^- from another crowd scene.

For unsupervised learning, we obtained over one million crowd frames using YouTube in conjunction with hundreds of keywords to retrieve relevant high density crowd videos. By performing our patch mining method on the videos, we obtain 6 million crowd patches. In order to train our Siamese-triplet networks, we set the batch size as 100, the learning rate starts at 0.001 and we reduce the learning rate by a factor of 10 at every 100K iterations and train for 300K iterations.

5.5.6 Evaluation

First, we demonstrate that CrowdNet learned using tracking across videos represents a reasonably powerful representation for crowd scenes. We perform Nearest Neighbors (NN) using ground-truth (GT) windows of the dataset introduced in the previous section of the chapter. The retrieval-database consists of all selective search windows (more than 0.5 overlap with GT windows) in the data-driven crowd dataset. Our learned crowd representation is far superior to a random AlexNet architecture, and the results are also better than AlexNet trained on ImageNet and applied to this high density crowd scenario.

Quantitatively, we measure the retrieval rate by counting number of correct retrievals in top-K ($K = 10$) retrievals. A retrieval is correct if the semantic class for retrieved patch and query patch are the same. Using the cosine distance we obtain 40% retrieval rate. Our performance is significantly better as compared to 18% by HOG, 21% for HOG/HOF, and 26% by AlexNet with random parameters. This demonstrates that our unsupervised network learns a good visual representation for high density crowded scenes compared to a random parameter CNN.

We also evaluate our learned crowd representation for the crowd event/scene classification task based on the Violence in Crowds dataset introduced in [22]. We train a linear classifier using softmax loss. Using pool5 features from CrowdNet results in superior classification accuracy over traditional crowd representation baselines as can be seen in Fig. 5.11.

5.6 CONCLUSIONS AND DIRECTIONS FOR FUTURE RESEARCH

We have approached the challenge of high density crowd analysis from a number of different new directions. The first sections of this chapter focussed on exploring the various ways of learning a set of collective motion patterns which are geared toward constraining the likely motions of individuals from a specific testing scene. We demonstrated that there are several advantages to searching for similar behaviors among crowd motion patterns in other videos. We have also shown that automatically obtained person

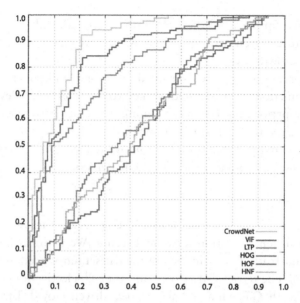

Figure 5.11 Evaluation of the learned crowd representation for the crowd event/scene classification task based on the Violence in Crowds dataset.

density estimates can be used to improve person localization and tracking performance. We also explored the concept of learning a representation that is specifically tailored toward high density crowded scenes. Our experiments indicate that crowd tracking can represent a useful source of intrinsic supervision for high density crowd feature learning.

There are several challenges and open problems in analysis of crowded scenes. First, modeling and recognition of events involving interactions between people and objects still remains a challenging problem. Examples include a person pushing a baby carriage or a fight between multiple people. Second, suitable priors for person detection, tracking, as well as behavior and activity recognition, are also an open problem. Such priors would enable predicting likely events in the scene. At the same time, detected but unlikely events under the prior may be classified as unusual. Finally, the recent progress in visual object and scene recognition has been enabled by the availability of large-scale annotated image databases. Examples include Pascal VOC, LabelMe, or ImageNet datasets. We believe similar data collection and annotation efforts are important to enable progress in visual analysis of crowd videos and more broadly surveillance.

REFERENCES

[1] Agrawal P, Carreira J, Malik J. Learning to see by moving. In: ICCV. 2015.
[2] Ali S, Shah M. A Lagrangian particle dynamics approach for crowd flow segmentation and stability analysis. In: IEEE conference on computer vision and pattern recognition. IEEE; 2007. p. 1–6.
[3] Ali S, Shah M. Floor fields for tracking in high density crowd scenes. In: ECCV. 2008.

[4] Anoraganingrum D. Cell segmentation with median filter and mathematical morphology operation. In: International conference on image analysis and processing, proceedings. IEEE; 1999. p. 1043–6.

[5] Betke M, Hirsh D, Bagchi A, Hristov N, Makris N, Kunz T. Tracking large variable numbers of objects in clutter. In: CVPR. 2007.

[6] Blei D, Lafferty J. A correlated topic model of science. Ann Appl Stat 2007;1(1):17–35.

[7] Breitenstein M, Reichlin F, Leibe B, Koller-Meier E, Van Gool L. Robust tracking-by-detection using a detector confidence particle filter. In: ICCV. 2009.

[8] Brostow G, Cipolla R. Unsupervised Bayesian detection of independent motion in crowds. In: CVPR. 2006.

[9] Chan A, Liang Z, Vasconcelos N. Privacy preserving crowd monitoring: counting people without people models or tracking. In: IEEE conference on computer vision and pattern recognition. IEEE; 2008. p. 1–7.

[10] Chan A, Liang Z, Vasconcelos N. Privacy preserving crowd monitoring: counting people without people models or tracking. In: CVPR. 2008.

[11] Dalal N, Triggs B. Histograms of oriented gradients for human detection. In: CVPR. 2005. p. I:886–93.

[12] Desai C, Ramanan D, Fowlkes C. Discriminative models for multi-class object layout. In: ICCV. 2009.

[13] Descombes X, Minlos R, Zhizhina E. Object extraction using a stochastic birth-and-death dynamics in continuum. J Math Imaging Vis 2009;33(3):347–59.

[14] Doersch C, Gupta A, Efros AA. Unsupervised visual representation learning by context prediction. In: ICCV. 2015.

[15] Efros AA, Berg AC, Mori G, Malik J. Recognizing action at a distance. In: ICCV. 2003.

[16] Everingham M, Sivic J, Zisserman A. "Hello! My name is. . . Buffy" – automatic naming of characters in TV video. In: Proc. BMVC. 2006.

[17] Everingham M, Van Gool L, Williams CKI, Winn J, Zisserman A. The PASCAL visual object classes challenge 2008 (VOC2008) results. Available from http://www.pascal-network.org/challenges/VOC/voc2008/workshop/index.html, 2008.

[18] Felzenszwalb P, Girshick R, McAllester D, Ramanan D. Object detection with discriminatively trained part based models. IEEE Trans Pattern Anal Mach Intell 2010;32(9).

[19] Fleuret F, Berclaz J, Lengagne R, Fua P. Multicamera people tracking with a probabilistic occupancy map. IEEE Trans Pattern Anal Mach Intell 2007;30(2):267–82.

[20] Ge W, Collins RT, Ruback RB. Vision-based analysis of small groups in pedestrian crowds. IEEE Trans Pattern Anal Mach Intell 2012;34(5):1003–16.

[21] Gennari G, Hager G. Probabilistic data association methods in visual tracking of groups. In: CVPR. 2007.

[22] Hassner T, Itcher Y, Kliper-Gross O. Violent flows: real-time detection of violent crowd behavior. In: CVPRW. IEEE; 2012.

[23] Hays J, Efros A. Scene completion using millions of photographs. SIGGRAPH 2007.

[24] Hoiem D, Efros A, Hebert M. Putting objects in perspective. Int J Comput Vis 2008.

[25] Hu M, Ali S, Shah M. Learning motion patterns in crowded scenes using motion flow field. In: IEEE international conference on pattern recognition (ICPR). 2008. p. 1–5.

[26] Johansson A, Helbing D, Shukla P. Specification of the social force pedestrian model by evolutionary adjustment to video tracking data. Adv Complex Syst 2007;10(2):271–88.

[27] Ke Y, Sukthankar R, Hebert M. Event detection in crowded videos. In: IEEE 11th international conference on computer vision. IEEE; 2007. p. 1–8.

[28] Khan S, Shah M. A multiview approach to tracking people in crowded scenes using a planar homography constraint. In: ECCV. 2006.

[29] Klaser A, Marszałek M, Schmid C. A spatiotemporal descriptor based on 3D-gradients. In: BMVC. 2008.

[30] Kong D, Gray D, Tao H. A viewpoint invariant approach for crowd counting. In: Proc. ICPR. 2006.

[31] Kratz L, Nishino K. Anomaly detection in extremely crowded scenes using spatiotemporal motion pattern models. In: IEEE conference on computer vision and pattern recognition. IEEE; 2009. p. 1446–53.

[32] Krizhevsky A, Sutskever I, Hinton GE. ImageNet classification with deep convolutional neural networks. In: Advances in neural information processing systems. 2012. p. 1097–105.

[33] Kuettel D, Breitenstein M, van Gool L, Ferrari V. What's going on? Discovering spatiotemporal dependencies in dynamic scenes. In: CVPR. 2010.

[34] Leibe B, Schindler K, Van Gool L. Coupled detection and trajectory estimation for multi-object tracking. In: ICCV. 2007.

[35] Lempitsky V, Zisserman A. Learning to count objects in images. In: NIPS. 2010.

[36] Li K, Kanade T. Cell population tracking and lineage construction using multiple-model dynamics filters and spatiotemporal optimization. In: IWMIAAB. 2007.

[37] Lin W, Liu Y. Tracking dynamic near-regular texture under occlusion and rapid movements. In: ECCV. 2006.

[38] Liu C, Yuen J, Torralba A. Nonparametric scene parsing: label transfer via dense scene alignment. In: CVPR. 2009.

[39] Liu C, Yuen J, Torralba A, Sivic J, Freeman WT. Sift flow: dense correspondence across different scenes. In: ECCV. 2008.

[40] Lucas B, Kanade T. An iterative image registration technique with an application to stereo vision. In: IJCAI, vol. 3. 1981. p. 674–9.

[41] Mehran R, Oyama A, Shah M. Abnormal crowd behavior detection using social force model. In: IEEE conference on computer vision and pattern recognition. IEEE; 2009. p. 935–42.

[42] Nath S, Palaniappan K, Bunyak F. Cell segmentation using coupled level sets and graph-vertex coloring. In: Medical image computing and computer-assisted intervention. 2006. p. 101–8.

[43] Pellegrini S, Ess A, Schindler K, Van Gool L. You'll never walk alone: modeling social behavior for multi-target tracking. In: ICCV. 2010.

[44] Rabinovich A, Vedaldi A, Galleguillos C, Wiewiora E, Belongie S. Objects in context. In: CVPR. 2007.

[45] Rodriguez M, Ali S, Kanade T. Tracking in unstructured crowded scenes. In: ICCV. 2009.

[46] Rodriguez M, Laptev I, Sivic J, Audibert J-Y. Density-aware person detection and tracking in crowds. In: ICCV. IEEE; 2011.

[47] Rodriguez M, Sivic J, Laptev I, Audibert J-Y. Data-driven crowd analysis in videos. In: ICCV. 2011.

[48] Russell B, Efros AA, Sivic J, Freeman WT, Zisserman A. Segmenting scenes by matching image composites. In: NIPS. 2009.

[49] Russell B, Torralba A, Liu C, Fergus R, Freeman WT. Object recognition by scene alignment. In: NIPS. 2007.

[50] Ryan D, Denman S, Fookes C, Sridharan S. Crowd counting using multiple local features. In: Digital image computing: techniques and applications. IEEE; 2009. p. 81–8.

[51] Shi J, Tomasi C. Good features to track. In: CVPR. 1994.

[52] Solera F, Calderara S, Cucchiara R. Socially constrained structural learning for groups detection in crowd. IEEE Trans Pattern Anal Mach Intell 2016;38(5):995–1008.

[53] Solmaz B, Moore BE, Shah M. Identifying behaviors in crowd scenes using stability analysis for dynamical systems. IEEE Trans Pattern Anal Mach Intell 2012;34(10):2064–70.

[54] Torralba A. Contextual priming for object detection. Int J Comput Vis 2003;53(2).

[55] Wang H, Ullah M, Klaser A, Laptev I, Schmid C. Evaluation of local spatiotemporal features for action recognition. In: Proc. BMVC. 2009.

[56] Wang X, Gupta A. Unsupervised learning of visual representations using videos. In: ICCV. 2015.

[57] Wang X, Ma K, Ng G, Grimson W. Trajectory analysis and semantic region modeling using a non-parametric Bayesian model. In: CVPR. 2008.

[58] Yao B, Fei-Fei L. Modeling mutual context of object and human pose in human–object interaction activities. In: CVPR. 2010.

[59] Yuen J, Torralba A. A data-driven approach for event prediction. In: ECCV. 2010.

[60] Zhao T, Nevatia R, Wu B. Segmentation and tracking of multiple humans in crowded environments. IEEE Trans Pattern Anal Mach Intell 2008;30(7):1198–211.

CHAPTER 6

Tracking Millions of Humans in Crowded Spaces

Alexandre Alahi, Vignesh Ramanathan, Li Fei-Fei
Stanford University, Stanford, CA, USA

Contents

6.1 INTRODUCTION

As Aristotle noted, "man is by nature a social animal." We do not live in isolation. On a daily basis, thousands of individuals walk in terminals, malls, or city centers. They consciously or unconsciously interact with each other. They make decisions on where to go, and how to get to their destination. Their mobility is often influenced by the surrounding. Understanding human social dynamics plays a central role in the design of safer and smarter spaces. It enables the development of ambient intelligence, i.e., spaces that are sensitive and responsive to human behavior. For instance, many sites such as train terminals were constructed several years ago to serve an estimated traffic demand. However, this estimated demand is greatly exceeded by forecasted traffic within a span of one decade. Sensing how individuals move through these large spaces provides insights

Group and Crowd Behavior for Computer Vision
DOI: 10.1016/B978-0-12-809276-7.00007-2

Figure 6.1 Real-world setup. Illustration of one of the monitored corridors in a train terminal. More than 30 cameras are deployed in the presented corridor, whereas 132 cameras are deployed in the terminal. At any given time, the occupancy of the corridor can reach more than 1000 of pedestrians. The label "OD" represents entry/exit zones.

needed to modify the space or design new ones to accommodate increased traffic. This enables reduced congestion and smooth flow of people.

In this chapter, we present computer vision techniques behind understanding the behavior of more than a hundred million individuals in crowded urban spaces. We cover the full spectrum of an intelligent system that detects and tracks humans in high density crowds using a camera network. To the best of our knowledge, we have deployed one of the largest networks of cameras (more than 100 cameras per site) to capture the trajectories of pedestrians in crowded train terminals over the course of two years. At any given time, up to a thousand pedestrians need to be tracked simultaneously (see Fig. 6.1). The captured dataset is publicly available to enable various research communities, from psychology to computer vision, to dive into a large-scale analysis of human mobility in crowded environments.[1] In the remaining of the chapter, we will share all the technical details that lead to a successful analysis of millions of individuals.

While computer vision has made great progress in detecting humans in isolation [1–4], tracking people in high density crowds is very challenging. Individuals highly occlude each other, and their motion behavior is not independent. We present detailed insights on how to address these challenges with sparsity promoting priors, and discrete combinatorial optimization that models social interactions.

Understanding the behavior of pedestrians using a network of cameras comprises the following three steps: (i) human detection in 3D space, (ii) tracklet generation, and (iii) tracklet association. We define *tracklet* as the short trajectory of a human limited to the field-of-view of a single camera. Each camera extracts tracklets corresponding to multiple people. The resulting tracklets are linked across cameras to obtain the long-term trajectories of humans in the full space. In the reminder of this chapter, we expand on each of the three steps and provide more details required to reach real-time performance with high accuracy. First, we cast the human detection problem as an inverse problem with a sparse prior, which can be solved in an efficient optimization framework. Then, we formulate the tracking problem as a linear integer program and use the social affinity of individuals to effectively associate tracklets for long-term tracking.

---

[1] www.ivpe.com/crowddata.htm.

6.2 RELATED WORK

We present an overview of relevant works to solve each of the three steps presented in the introduction: (i) detection, (ii) tracklet generation, and (iii) tracklet association in a camera network.

Human detection. Pedestrians in isolation are accurately detected using a single image and robust classification techniques such as R-CNNs or deformable parts models [1–4]. Individuals are detected in the image plane as opposed to 3D coordinates of people in the real world. With a calibrated camera, it is possible to map detected bounding boxes to the real world coordinates [5,6]. Algorithms with high levels of confidence have been proposed to locate crowded people with a single top view or several head-level overlapping field-of-views [7–10]. For instance, Khan and Shah locate people on the ground by taking the intersection of projected foreground silhouettes on the ground plane. Fleuret et al. [10] use a generative model with a probabilistic framework to outperform previous work. Alahi et al. [6] propose a sparsity driven framework to handle noisy observations and reduce the number of false positives. Golbabaee et al. [11] propose a real-time solver to the sparsity driven framework inspired by the set cover problem. In the next section, we will present more details on the sparsity driven formulations.

Tracklet generation. Once individuals are located on the ground, various graph-based algorithms can be used to track them. Each node represents a detection and the edges measure the similarity cost to link the detections. It is possible to find the global optimum using linear programming to solve the data association problem [12,13]. It outperforms previous works based on Markov Chain Monte Carlo [14] or inference in Bayesian networks [15]. The data association problem is expressed as a graph-theoretic problem for finding the best paths/flows over the graph. The main challenge is to find a robust similarity measure. Recently, Xiang et al. [16] have shown that tracking multiple humans can be formulated as a Markov Decision Process instead of a graph-based formulation. They have learned an appearance-based similarity function to outperform previous ones based on color histograms. But their approach will fail if limited information on the appearance of the pedestrian is available or if all pedestrians look the same (e.g., when only the back of their heads is visible). In this chapter, we present a generic graph-based framework to track multiple humans since both simple and complex similarity measure can be modeled.

Tracklet association. A large body of work models visual appearance to link tracklets across cameras [17–20]. Andriluka et al. [21] use person detection as a cue to perform tracking and vice versa. Javed et al. [22] use travel time and the similarity of appearance features. Song et al. [23] use a stochastic graph evolution strategy. Tracklets extracted by each camera are linked with the Hungarian algorithm [24], MCMC [25], or globally optimal greedy approaches [20]. These approaches have not addressed the linking of tracklets that are dozens of meters away in a highly crowded scene. Alahi et al. [26] propose to model social interactions and, more precisely, social affinities to solve the tracklet association step. In Section 6.6.1 we will present more details on their method.

Tracking with a social prior. Social behavior has recently been incorporated into existing tracking frameworks by modeling the well-known social forces [27] with Kalman filters [28], extended Kalman filters [29], or Linear Programming [13,30]. Antonini et al. [31] use discrete choice models to simulate the walking behavior of people. These approaches improve the operational-level tracking when a few frames are missing (e.g., when given a low-frame rate, or short occlusion cases). They also often model a grouping cue to solve the data association problem [32,13,30]. They model it as a set of pedestrians with similar velocities and spatial proximity. Similarly, the authors of [33] use grouping cues in a hierarchical framework to identify sports player roles. The grouping cue is typically handled as a binary variable indicating group similarity. However, the key challenge is to use a finer representation to capture group association and integrate it into the problem of tracklet association. Yang et al. [34,32] use a conditional random field framework to jointly estimate group membership and tracks. Leal et al. [13] iteratively compute the minimum cost flow for various velocity and grouping assignments until convergence or when a maximum number of iterations is reached. Qin et al. [30] use the Hungarian algorithm to jointly group and link tracklets. However, the Hungarian algorithm does not solve the global minimization over the full long-term track, whereas the minimum network flow formulation does. In this chapter, we present more details on a descriptor representing the grouping cue as a feature to efficiently match behavior across pedestrians.

6.3 SYSTEM OVERVIEW

We work with 132 RGB/Thermal/Depth cameras which monitor 20,000 m^2 with human density reaching 1 individual/m^2. This introduces new challenges in designing detection and tracking algorithms which can work at such a scale. The camera network constantly collects a large volume of visual data. Here are some facts regarding the dataset: individual's travel time in the monitoring area is 50 s and spans 70 m on average. At a given time, up to 1000 people can be in the same area. Typically, after 1 min, more than a million people are detected, and the detections need to be linked to each other. To handle such a deluge of data, we distribute the processing as follows: first, every camera independently locates people on the ground using the method presented in Section 6.4. Then, detected individuals are tracked within each camera independently, given a global optimization framework (solved with Linear Programming) similar to [13] (see Section 6.5). The resulting tracklets are matched across cameras to obtain the long-term tracks over the full area by modeling social affinities; see Section 6.6.1. Fig. 6.2 illustrates the distributed processing pipe.

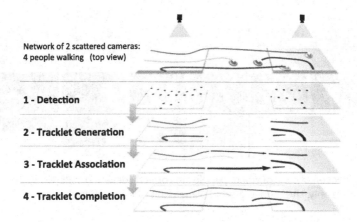

Figure 6.2 Overview of the system: each camera extracts independently tracklets with high confidence. Then, the tracklets are linked across cameras. For completeness, we illustrate the fourth step, namely, tracklet completion, which will be presented in Chapter 9. The goal of this step is to predict the detailed trajectories of pedestrians.

6.4 HUMAN DETECTION IN 3D

6.4.1 Method

The first step in our system involves locating 3D position of people on the ground within the field-of-view of each single camera. The scene geometry needs to be estimated across thermal and optical cameras. Most previous works locating people on the ground (in 3D) use calibration data to map image coordinates to the real-world [7,8,10,6]. A set of 3D coordinates or the cameras extrinsic parameters are needed to estimate the homography matrices. However, when dealing with large-scale setups, it is difficult to obtain calibration data for all cameras. Some works exist to automatically estimate the scene geometry using vanishing lines [35] and additional human poses [36]. However, such approaches do not work on thermal images where binary silhouettes are only observed (see Fig. 6.3).

We propose to solve the single-view scene geometry and 3D localization problem using only extracted binary human silhouettes. As a result, it works across any camera modality (i.e., thermal or optical). We address the problem as a dictionary-based inverse problem where both the dictionary and the occupancy vector are unknown. We use the same sparsity driven formulation as in [6]. An inverse problem is formulated to deduce people location points (i.e., occupancy vector x), given a sparsity constraint on the ground occupancy grid. Let y be our observation vector (i.e., the binary foreground silhouettes), and D the dictionary of atoms approximating the binary foreground silhouettes of a single person at all locations (see Fig. 6.4). We present an algorithm to find the ground occupancy vector x satisfying the following equation when the dictionary

Thermal imaging   Optical imaging

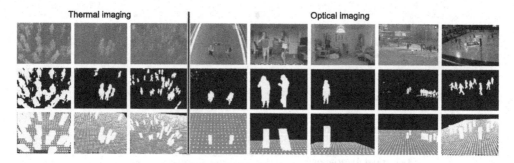

Figure 6.3 A collection of camera where people are correctly detected in 3D given the observed foreground silhouettes only. The learned ground plane points are also plotted.

D is unknown as opposed to previous work:

$$x, D = \underset{x \in \{0,1\}, D_i \in D^\varepsilon}{\arg\min} \; \|y - D_i x\|_2^2 \;\; \text{s.t.} \;\; \|x\|_0 < \varepsilon_p, \tag{6.1}$$

where D^ε is the space of all potential dictionaries and ε_p is the maximum number of people to be detected.

Algorithm 1 illustrates our "Detection while Learning" algorithm solving a relaxation of Eq. (6.1). We iterate over the space of dictionary $\forall D_i \in D^\varepsilon$. We solve Eq. (6.1) for a fixed D_i using the "Set Covering Object Occupancy Pursuit" (SCOOP) algorithm presented in [11]. The latter iteratively recovers one element of the support set. More precisely, at each iteration, it selects the atom a of the dictionary which contributes the most in the signal energy (the most correlated atom with the signal) and fits well the image. The output of SCOOP algorithm is an occupancy vector associated with a residual error on the data fidelity term ($\|y - D_i x\|_2^2$). We compute SCOOP over all dictionaries and select the one with the smallest residual errors. Note that the dictionaries D_i are not random matrices. They lie on a manifold. The space of solutions D^ε is much smaller than the dimension of the matrix. It is related to the extrinsic parameters of the cameras, i.e., its height to the ground, and its orientation (see Fig. 6.4). A coarse-to-fine sampling step is used to select potential dictionaries approximating a gradient descent.

6.4.2 Evaluation

First, qualitative results are available in Fig. 6.3. It illustrates the performance of the proposed SCOOP-learning algorithm in locating people in 3D as well as the scene geometry. Various heights and viewing angles are illustrated. Then, quantitative results are shared in Fig. 6.5. It presents the accuracy of the presented algorithm to estimate the camera parameters with respect the estimated heights and angles of various cameras setups. We can see that the camera parameters are correctly estimated once enough

---

Algorithm 1: SCOOP-learning: Detecting while Learning.

Input: signal γ, regularization parameter w;

Output: occupancy vector x and dictionary D.

1. **Init**

$E_{s+1} = E_s = 0, s_{min} = 0;$

2. $\forall D_i \in D^c$

– **SCOOP algorithm with a fixed D_i**

– Initialization

$\widehat{S} \Leftarrow \{\}, r \Leftarrow \gamma, \widehat{\gamma} \Leftarrow 0, e_{s+1} = \left| \text{supp}(\gamma) \right|, e_s = \left| \text{supp}(\gamma) \right|;$

– Matching pursuit-like process

while $(e_{s+1} - e_s \leq 0)$ **do**

$$j \Leftarrow \underset{j' \in \mathcal{U}}{\arg\min} \left\{ w \frac{\left| \text{supp}(r) \backslash \text{supp}(d_{j'}) \right|}{\left| \text{supp}(r) \right|} + (1 - w) \frac{\left| \text{supp}(d_{j'}) \backslash \text{supp}(r) \right|}{\left| \text{supp}(d_{j'}) \right|} \right\}$$

– Updates

Recovered support, $\widehat{S} \Leftarrow \widehat{S} \cup \{j\};$

Recovered $\widehat{\gamma}$, $\text{supp}(\widehat{\gamma}) \Leftarrow \text{supp}(\widehat{\gamma}) \cup \text{supp}(d_j);$

Remainder, $\text{supp}(r) \Leftarrow \text{supp}(r) \backslash \text{supp}(d_j);$

Error, $e_s \Leftarrow e_{s+1};$

Error, $e_{s+1} \Leftarrow \left| \text{supp}(\gamma \oplus \widehat{\gamma}) \right|;$

end

– **Updates**

$E_s = E_{s+1};$

$E_{s+1} = e_i;$

if $E_{s+1} \leq E_s$ **then**

$\qquad D = D_i;$

$\qquad x = \widehat{S}.$

end

---

people are present in the scene. Indeed, the more silhouettes are used to select the dictionary, the less error-prone the estimations are, since many possible solutions exist with a single silhouette as opposed to several ones. Note that the error on the 3D estimation is on average less than 50 cm once 3 people are present in the field of view, although the camera parameters still have about 15% errors in them.

Once the camera parameters are found, i.e., the dictionary is known, we can evaluate the performance of human detection in 3D, in both low and high density crowds. We refer the readers to [6] and [11] for a detailed analysis on this topic.

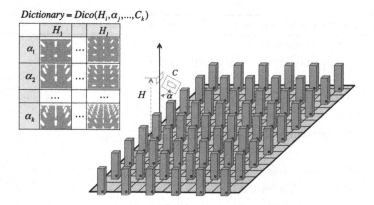

Figure 6.4 Illustration of the dictionary space. We approximate the ideal silhouettes of people with cuboids on the ground. A dictionary is the collection of atoms representing the observation of the cuboid in the image plane of the camera with a given height and viewing angle.

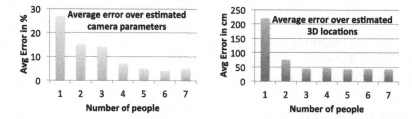

Figure 6.5 Performance of SCOOP-learning algorithm to estimate the scene geometry. The left graph illustrates the measured error over the estimated camera parameters with respect to the number of people observed in the sequence of images. The right graph illustrates the error in the final 3D localization of people on the ground.

6.5 TRACKLET GENERATION

Once humans are located in the field-of-view of each camera, we need to track them in time (second step in Fig. 6.2). As a reminder, we refer to tracklets as the short trajectories of humans captured by a single camera. We want to find the set of tracklets X, where each tracklet $x \in X$ is represented as an ordered set of detections (L_x), representing the detected coordinates of humans (using the method described in the previous section). Similarly, $L_x = (l_x^{(1)}, \dots, l_x^{(n)})$ is an ordered set of intermediate detections which are linked to form the tracklets. These detections are ordered by the time of initiation. The problem can be written as a maximum a posteriori estimation problem similar to [19,20]

$$X^* = \max_X P(L|X)P(X), \tag{6.2}$$

where $P(L|X)$ is the probability of the detections in L being true positive detection. The probability $P(L|X)$ is

$$P(L|X) \propto \prod_{x \in X} \prod_{l \in L_x} \frac{P_{tp}(l)}{P_{fp}(l)}, \tag{6.3}$$

where $P_{tp}(l)$ and $P_{fp}(l)$ are probabilities of the detection being a true positive, and false positive, respectively.

Next, similar to [20], we assume a Markov-chain model connecting every intermediate detection $l_x^{(i)}$ in the tracklet X, to the subsequent detection $l_x^{(i+1)}$ with a probability given by $P(l_x^{(i+1)}|l_x^{(i)})$. The tracklet probability $P(X)$ is

$$P(X) = \prod_{x \in X} P(x), \tag{6.4}$$

$$P(x) = \prod_{i=1}^{n} P\left(l_x^{(i)}|l_x^{(i-1)}\right),$$

where $n = |L_x|$ is the number of intermediate detections in the tracklet.

The MAP problem from Eq. (6.2) can now be formulated as a linear integer program:

$$\min_{f} \quad C(f) \tag{6.5}$$

$$C(f) = \sum_{x_i \in X} \alpha_i f_i + \sum_{x_i, x_j \in X} \beta_{ij} f_{ij}$$

$$\text{s.t} \quad f_i, f_{ij} \in \{0, 1\}$$

$$\text{and} \quad f_i = \sum_{j} f_{ij},$$

where f_i is the flow variable indicating whether the corresponding detection is a true positive, and f_{ij} indicates if the corresponding detections are linked together. The variable β_{ij} denotes the transition cost given by $\log P\left(l_i|l_j\right)$ for the detection $l_i, l_j \in L$. The local cost α_i is the log-likelihood of an intermediate detection being a true positive. In our case, we suppose that all detections have the same likelihood.

We note that the optimization problem in Eq. (6.5) is equivalent to the flow optimization problem widely discussed in [20,19]. Such problems can be solved through k-shortest paths algorithm [20,12].

The main challenge in solving the tracklet generation step is to define the transition cost β_{ij}. For any two detections, it can be split into several components as shown below:

$$\beta_{ij} = \beta_{ij}^{appearance} + \beta_{ij}^{motion}, \tag{6.6}$$

where $\beta^{appearance}$ is the cost to ensure similar appearance and β^{motion} is the cost to ensure motion smoothness in the connected detections. The choice behind these similarity metrics is still an open research problem. Xiang et al. [16] have shown that learning these metrics from training data outperforms hand-designed features such as color histogram or Kalman filters. We refer the readers to [16] for a detailed evaluation of their method on the public Multi-Object Tracking (MOT) challenge [37].

6.6 TRACKLET ASSOCIATION

The third step of our intelligent system is to connect tracklets across cameras (see Fig. 6.2). This task becomes even more challenging when cameras are scattered and distant by several dozens of meters. Previous techniques based on appearance and motion similarities are not sufficient since the camera viewpoints might be very different, leading to strong appearance changes, and the linear motion assumption is not valid anymore on long distances. In this section, we present a descriptor that models social interactions to reason on the data association step. We show how to use the same graph-based framework presented in previous section to solve the tracklet association step, although additional constraints need to be modeled.

6.6.1 Social Affinity Map – SAM

When walking in crowded environments, humans often have social affinities that remain stable over time.

Definition 1. We define "social affinity" as the motion affinity of neighboring individuals.

Social affinities can be consciously formed by friends, relatives or coworkers. However, in crowded environments, subconscious affinities exist. For example, the "*leader-follower*" phenomenon [38] represents a spontaneous formation of lanes in dense flows, as a result of fast pedestrians, passing slower ones. More formally, the leader–follower pattern captures the behavior of a pedestrian (a follower) who adjusts his/her motion to follow a leader to enable smooth travel. We propose to learn the various social affinities which bind people in a crowded scene through a feature called Social Affinity Map (SAM).

6.6.2 The SAM Feature

We observed that in public settings, social forces are mostly determined by the proximity of people to each other as noted in previous works [27]. Since people are more easily influenced by others in their vicinity, we develop a social affinity feature which captures the spatial position of the tracklet's neighbors. As shown in Fig. 6.6, we achieve this by radially binning the position of neighboring tracklets.

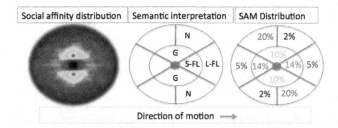

Figure 6.6 The left-hand side shows the heatmap of the relative positions of all neighboring pedestrians across all tracklets. The middle column represents the SAM with our semantic description where "G" is the group affinity (such as couples, friends), "S-FL" is the short distance follow–leader behavior, "L-FL" is the long distance FL behavior, and "N" can be seen as the comfortable distance to maintain while walking in the same direction. The right-hand side represents the distribution of presented behavior.

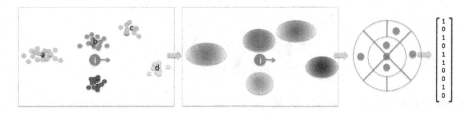

Figure 6.7 Illustration of a Social Affinity Map extraction (top view). The relative positions of neighboring individuals are clustered into a radial histogram. The latter is 1 bit quantized.

We further learn the spatial binning by first clustering the relative position of surrounding individuals over all captured trajectories. We considered relative positions within a limit of 3 m to avoid outliers. The distribution of the relative positions across millions of trajectories is visualized in Fig. 6.6. We obtain 10 bins as a result of this clustering, as shown in the figure. The percentage of relative positions pooled into this bins is also shown in the figure. It is interesting to point out that the most used bin is the one on the far right side (labeled "N" in Fig. 6.6). It can be interpreted as the comfortable pattern to walk with respect to other individuals as opposed to the left-hand side.

Given a new tracklet, we perform vector quantization (VQ) coding to obtain the SAM feature. We fit a Gaussian Mixture Model to the relative position of its surrounding tracklets. The inferred GMM values within the previously learned spatial bins are discretized to obtain a binary radial histogram, which represents the SAM feature vector. The complete process is illustrated in Fig. 6.7. Hamming distance is used to compare SAM across tracklets. Note that binary quantization has little impact on the efficacy of the feature, and is only used to speed up the comparison method.

Our SAM feature can differentiate between various configurations of social affinities such as "*couple walking*", or the "*leader–follower*" behavior. Fig. 6.8 illustrates the 8

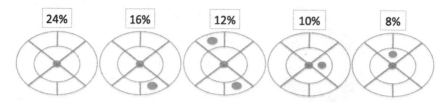

Figure 6.8 Illustration of the 8 most observed social affinities learned from the data. The above percentage represents the frequency of occurrence of the corresponding SAM.

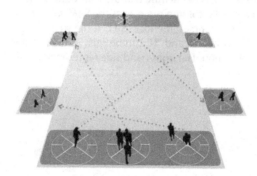

Figure 6.9 Predicting the behavior of pedestrians given Social Affinity Maps (SAMs) with few cameras. Orange regions represent the monitoring areas of cameras. We illustrate the extreme case when cameras are only placed at entrance or exit zones, referred to as OD cameras. (For interpretation of the references to color in this figure legend, the reader is referred to the web version of this chapter.)

most observed SAM over millions of trajectories. It is worth pointing out that 76% of individuals belong to a group, hence an SAM provides valuable information in crowded settings, motivating the use of these cues in forecasting the mobility of pedestrians.

6.6.3 Tracklet Association Method

Often there is a sparse network of cameras monitoring the transit of people in a public setting like a railway terminal. The terminal has a set of entry points referred to as the *origin*, and exit points referred to as the *destination*. One key motivation behind tracking humans in the terminal is to identify the Origin and Destination (OD) of every person entering and exiting the camera network. We achieve this by identifying the trajectories which connect the tracklets starting at the origin to the tracklets ending at the destination. The number of intermediate tracklets linked to obtain these trajectories decreases with the sparsity of the camera network. Fig. 6.9 illustrates an extreme case with only origin and destination tracklets.

We have a set of origin tracklets O and an equal number of destination tracklets D. Each tracklet in O is captured at one of the many entrances into the area, and a destination tracklet in D is captured at an exit. We also have a set of intermediate tracklets X

obtained by our sparse camera network. We want to find the set of trajectories T, where each trajectory $t \in T$ is represented as an ordered set of tracklets, (o_t, X_t, d_t), with $o_t \in O$ and $d_t \in D$ representing the origin and destination tracklets of the trajectory. Similarly, $X_t = (x_t^{(1)}, \ldots, x_t^{(n)})$ is an ordered set of intermediate tracklets which are linked to form the trajectory. These tracklets are ordered by the time of initiation. The problem can be written as a maximum a posteriori estimation problem similar to that in Section 6.5:

$$T^* = \max_{T} P(X|T)P(T), \tag{6.7}$$

where $P(X|T)$ is the probability of the tracklets in X being true positive tracklets. The probability $P(X|T)$ is

$$P(X|T) \propto \prod_{t \in T} \prod_{x \in X_t} \frac{P_{tp}(x)}{P_{fp}(x)}, \tag{6.8}$$

where $P_{tp}(x)$ and $P_{fp}(x)$ are probabilities of the tracklet being a true positive and false positive, respectively.

We define $P_{OD}(o, d)$ as the OD-prior term which states the probability of a person entering at the origin corresponding to o and exiting at the destination corresponding to d. Such prior is often neglected and assumed to be uniform. However, in many applications, it is a strong prior, such as avoiding forbidden paths in airports.

Next, similar to Section 6.5, we assume a Markov-chain model connecting every intermediate track $x_t^{(i)}$ in the trajectory T, to the subsequent track $x_t^{(i+1)}$ with a probability given by $P(x_t^{(i+1)}|x_t^{(i)})$. The trajectory probability $P(T)$ is

$$P(T) = \prod_{t \in T} P(t), \tag{6.9}$$

$$P(t) = P_{OD}(o_t, d_t)P\left(x_t^{(1)}|o_t\right)$$

$$\times \prod_{i=2}^{n} P\left(x_t^{(i)}|x_t^{(i-1)}\right) P\left(d_t|x_t^{(n)}\right),$$

where $n = |X_t|$ is the number of intermediate tracklets in the trajectory.

The MAP problem from Eq. (6.7) can now be formulated as a linear integer program in a manner similar to [20]:

$$\min_{f} \quad C(f) \tag{6.10}$$

$$C(f) = \sum_{x_i \in X} \alpha_i f_i + \sum_{x_i, x_j \in X} \beta_{ij} f_{ij}$$

$$+ \sum_{\substack{x_i \in X, \\ o \in O}} \beta_{oi} f_{oi} + \sum_{\substack{x_i \in X, \\ d \in D}} \beta_{id} f_{id} + \sum_{\substack{o \in O, \\ d \in D}} \gamma_{od} f_{od}$$

$$\text{s.t} \quad f_i, f_{ij}, f_{od} \in \{0, 1\}$$

$$\text{and} \quad f_i = \sum_j f_{ij} + \sum_d f_{id} = \sum_i f_{ji} + \sum_o f_{oi},$$

$$\sum_{od} f_{od} = |O| = |D|,$$

$$\sum_d f_{od} = \sum_i f_{oi},$$

$$\sum_o f_{od} = \sum_i f_{id} \quad \forall \, x_i, x_j \in X, \, o \in O, \, d \in D,$$

where f_i is the flow variable indicating whether the corresponding tracklet is a true positive, and f_{ij} indicates if the corresponding tracklets are linked together. The variable β_{ij} denotes the transition cost given by $\log P(x_i|x_j)$ for the tracks $x_i, x_j \in X$. The log-likelihoods β_{oi}, β_{id} are also defined similarly, for the origin track o and destination track d. The local cost α_i is the log-likelihood of an intermediate track being a true positive. Finally, the OD-prior cost is represented as $\gamma_{od} = \log P_{OD}(o, d)$.

We note that the optimization problem in Eq. (6.10) is equivalent to the flow optimization problem in Eq. (6.5) in the absence of the OD-prior term. The addition of the OD-prior term leads to loops in the network-flow problem, and can no longer be solved exactly through shortest path algorithms. Hence, we adopt a heuristic approach to solve Eq. (6.10), as discussed in Section 6.6.4.

The local cost α_i is proportional to the length of a tracklet. This helps us to remove short tracklets that might represent false positives. The transition cost β_{ij} for any two tracklets is split into two components as shown below

$$\beta_{ij} = \beta_{ij}^{SAM} + \beta_{ij}^M, \tag{6.11}$$

where β^{SAM} is the social-affinity cost and β^M is a cost to ensure smoothness in the connected tracklets.

Social affinity cost. In our model, we wish to ensure that tracklets moving in similar social groups have a stronger likelihood of being linked to each other. This affinity forms an important component in large scale tracking scenarios like ours, where the appearance of an individual is not very discriminative. The SAM features introduced in Section 6.6.1 are used to measure the social affinity distance between tracklets moving in groups, namely,

$$\beta_{ij}^{SAM} = \mathbf{H}(sam_i, sam_j), \tag{6.12}$$

where $\mathbf{H}(\cdot)$ denotes the Hamming distance between two binary vectors, and sam_i, sam_j denote the SAM feature vector of the two tracks.

Motion similarity. Another cue, β^M, which is used to ensure smoothness in trajectory motion is obtained by measuring the distance between the motion patterns of two tracklets similar to [25,23].

The OD-prior cost is the log-likelihood of the prior probability of transiting from an origin point to the destination. In most surveillance settings, we can use prior knowledge on the geography of the terminal, as well as rough estimates of the passenger freight to obtain an OD prior. In addition, the OD prior can be used to enforce constraints such that passengers entering a certain entry point would not return to the same location from a parallel entrance. In our experiments in later sections, the OD prior is obtained by a short survey in the location.

6.6.4 Optimization

As stated before, the optimization in Eq. (6.10) cannot be trivially solved through existing shortest path algorithms [20] as in the case of traditional tracking. Hence, we adopt a heuristic approach as explained below.

Greedy optimization with OD-prior. We first run a greedy algorithm to identify the low-cost solutions in the graph:

1. Find the shortest path which links an origin tracklet to the destination tracklet in Eq. (6.10).
2. Remove the tracklets which are part of the trajectory obtained in the previous step and repeat.

The greedy algorithm provides an approximate solution to the problem and is computationally efficient. However, it does not solve the global optimization problem. We use a simple heuristic explained below to obtain a better solution.

Optimization with OD reweighted cost. The solution of the greedy algorithm helps us identify the paths which agree with the OD prior. Hence, the transition flow variables set by this algorithm provide a rough estimate of the pairwise affinity between tracklets in the presence of OD prior. We use this intuition to add an additional cost which penalizes the link between tracklets which were not originally connected by the greedy algorithm. While adding this cost, we remove the original OD-prior cost γ_{od}, thus resulting in a network-flow problem which can be solved by the k-shortest path approach. The modified cost $\tilde{C}$ is given by

$$\tilde{C}(f) = \sum_{x_i \in X} \alpha_i f_i + \sum_{x_i, x_j \in X} \tilde{\beta}_{ij} f_{ij} \qquad (6.13)$$

$$+ \sum_{\substack{x_i \in X, \\ o \in O}} \tilde{\beta}_{oi} f_{oi} + \sum_{\substack{x_i \in X, \\ d \in D}} \tilde{\beta}_{id} f_{id},$$

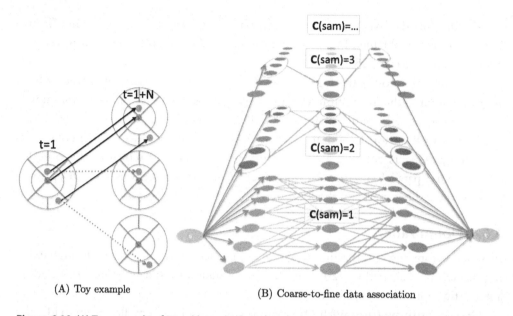

(A) Toy example

(B) Coarse-to-fine data association

Figure 6.10 (A) Toy example of 3 tracklets which could be wrongly linked. Dashed red arrows illustrate wrong assignments that are likely to occur without a coarse-to-fine data association. (B) Coarse-to-fine data association given SAM cardinality. Each subgraph corresponds to the tracklet association problem over tracklet groups of specific cardinalities, denoted by $C(sam)$ representing the sum of the elements of the SAM feature. The flow variables obtained by solving these subproblems are used to defined additional transition costs used in the final optimization. (For interpretation of the references to color in this figure legend, the reader is referred to the web version of this chapter.)

where $\tilde{\beta}$ is the OD-reweighted cost defined as

$$\tilde{\beta}_{ij} = \beta_{ij} + \lambda \mathbf{1}(f_{ij}^{greedy} = 1), \tag{6.14}$$

f_{ij}^{greedy} is the solution obtained from the greedy algorithm, and λ is a parameter indicating the strength of the OD-prior cost. The transition cost is reweighted for all pairs of tracklets including the origin and destination tracklets.

6.6.5 Coarse-to-Fine Data Association

The model presented in Section 6.6.3 uses a social affinity cost to ensure that tracklets with similar grouping cues are connected. However, it does not account for the fact that people belonging to groups of different cardinalities (number of people in a group) can still share the same SAM feature. An example is shown in Fig. 6.10A, where two tracklets belonging to groups of different cardinalities are wrongly connected (indicated in red) due to similar SAM. However, we want to encourage tracklets from groups of similar sizes to be connected together (black arrows). We account for this by proposing a coarse-to-fine data association method.

We cluster tracklets cooccurring at the same time into different groups based on the social separation. The cardinality of a tracklet denoted by $\mathbf{C}(x_i)$ is the number of people belonging to the group corresponding to the tracklet x_i. We can imagine that if the clustering is perfect and people moved in the same configuration across the entire camera network, it would suffice to link the tracklet groups instead of the tracklets. This would also solve the problem of tracklets being linked across groups of different cardinalities. However, in practical setting, the grouping is not perfect, and people break away from groups. Hence, we link the groups of same cardinality and use the links obtained from this group tracking to define additional transition costs. The complete method is explained in the supplementary document. The method is briefly visualized in Fig. 6.10B.

6.7 EXPERIMENTS

6.7.1 Large-Scale Evaluation

The data collection campaign helps us conduct various experiments in real life setting with a large and dynamic crowd. In this section, we present a set of experiments to address the tracking problem in scattered camera network. We select a subset of cameras in our network and measure the performance of our algorithm to track mobility with only these cameras.

Measurement. In this section, we evaluate the correct estimation of the origin and destination of a person entering the camera network. We have limited the monitoring to 14 origins and destinations leading to 196 possible OD paths for a trajectory. We have clustered the cameras into two groups: cameras belonging to OD locations (i.e., capturing the beginning or ending of long-term tracks), and cameras in-between these locations. We compute the OD error rate as the percentage of wrong predictions out of the total number of people covered by the camera network.

Ground truth. Since Big Data is collected, it is not realistic to label the millions of trajectories. We hence use as labels the output of our detection and tracking algorithm. To reach high level of tracking accuracy, we have installed a dense network of cameras to reduce the blind spots as much as possible and link tracklets that are only a few centimeters away from each other. The trajectories computed from this dense network are used as a baseline (our labels). While the trajectories (and OD) computed from the dense network is not the perfect ground-truth, in practice they are easier and less expensive to obtain than manually annotating trajectories at our scale. The goal of our forecasting algorithm is to reach the same performance as the dense network of cameras while using a sparse network.

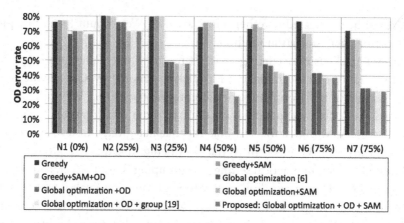

Figure 6.11 Performance of OD forecasting with different number of in-between cameras. The percentage of in-between cameras are shown in brackets. Seven network configurations are evaluated (referred to as N1 to N7).

6.7.2 OD Forecasting

Fig. 6.11 presents the resulting OD error rates for 7 sparse networks of cameras. The evaluation is carried out at several levels of network sparsity, from 0% to 75% of in-between cameras. For instance, networks N4 and N5 use only half of the cameras available in the corridor (see Fig. 6.1). The cameras are selected to heuristically minimize the average distance between them at any given sparsity. At a given sparsity, we also evaluate different camera configurations such as N4 and N5 for 50% sparsity. On average, tracklets from network N1 to N3 are several dozens of meters away from each other, and tracklets from networks N4 to N7 are a dozen of meters away from each other. To validate our algorithm, we evaluate the performance of greedy optimization methods against the proposed global one. We measure the impact of using SAM as an additional feature, as well as the impact of modeling the OD prior with coarse-to-fine tracking.

As expected, the global optimization methods always outperform the greedy methods with or without OD prior. The performance improvement is more than doubled in the global optimization method. The SAM feature and use of OD-reweighted cost (use of OD prior) are both seen to have a positive impact while using global optimization. This justifies our decision to heuristically model the effect of OD prior during optimization.

We also compare with the algorithms from [12] and [13]. Our final full model, i.e., "Global optimization + OD + SAM", outperforms these methods when observations are limited to the corridor. Note that the camera placement has an impact on the forecasting. Although the same number of cameras are used by networks N2 and N3, or N4 and N5, the forecasting accuracy differs for these networks. If an in-between camera

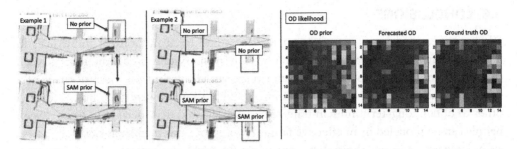

Figure 6.12 Qualitative results on the linked tracklets within the sparse network 1 where 50% of the in-between cameras within the corridor are not used. Only tracklets selected by the method are shown. The lines illustrate the linked tracklets. On the right side, we illustrate the OD prior as a heatmap, as well as show the forecast and the ground truth. We can see that although the prior is different, the final result is still similar to the ground truth.

is strategically placed to capture frequent route choices, it reduces the uncertainty in the linking strategy. This leads to different performance for networks with same number of cameras as shown in Fig. 6.11.

We evaluate the extreme setup when there are no in-between cameras (labeled N1), i.e., we only have cameras at entrance and exit zone (OD cameras). In such a setup, tracklets are up to 100 m away from each other. Fig. 6.11 presents the resulting drop in performance. The gap between greedy and global optimization is much smaller. In addition, the SAM feature and OD prior do not have a significant impact on such extreme case. These results motivate our future work to handle such an extreme case.

Fig. 6.12 illustrates some qualitative results demonstrating the power of SAM. We also plot the OD prior, forecasted OD with a sparse network of cameras using only half as many cameras as the dense network (ground truth).

Impact of SAM. We illustrate the tracklet linking achieved by our full method and compare it with a global optimization method which does not use SAM in Fig. 6.12. As expected, we see that in the absence of SAM, tracklets traveling in similar group configurations are not connected together, leading to erroneous results. On the other hand, SAM helps disambiguate between tracklet choices which are similar to each other, except for the group configuration.

Impact of OD prior. In Fig. 6.12 we present the final OD matrices estimated by our full model, and compare them with the OD prior and the ground truth OD (from dense camera network). Clearly, the prior only provides weak cues about the true OD, but helps by down-weighting paths which are highly unfavorable, like blocked corridors. The OD matrix forecasted by our method is close to the ground truth OD matrix obtained from a dense camera network.

6.8 CONCLUSIONS

We have presented an efficient system to detect and track millions of individuals in real-world crowded environments. The first step in the system used a dictionary-based sparsity promoting method to detect and track people within the field-of-view of a single camera. These short "tracklets" from multiple cameras were then linked to each other to obtain long-term human trajectories. We showed that social affinities between people can be modeled in an effective fashion to improve this tracklet association. These affinities were captured through a new powerful SAM descriptor, which empowers tractable global optimization of the tracklet association problem. We also deployed a large network of cameras to enable large-scale analysis of real-world crowd motion. Several hundred thousands of trajectories were collected per day leading to more than 100 million trajectories to date. It helps in the development of new motion priors to predict human behavior in crowded scenes. In the next chapter, we will show that it is possible to not just track but also *predict* long-term human behaviors from these millions of trajectories.

REFERENCES

[1] Felzenszwalb PF, Girshick RB, McAllester D. Cascade object detection with deformable part models. In: CVPR. IEEE; 2010.

[2] Girshick R, Donahue J, Darrell T, Malik J. Rich feature hierarchies for accurate object detection and semantic segmentation. In: CVPR. IEEE; 2014.

[3] Tuzel O, Porikli F, Meer P. Pedestrian detection via classification on Riemannian manifolds. In: PAMI. IEEE; 2008.

[4] Benenson R, Mathias M, Timofte R, Van Gool L. Pedestrian detection at 100 frames per second. In: CVPR. IEEE; 2012.

[5] Enzweiler M, Gavrila DM. Monocular pedestrian detection: survey and experiments. In: PAMI. 2009.

[6] Alahi A, Jacques L, Boursier Y, Vandergheynst P. Sparsity driven people localization with a heterogeneous network of cameras. J Math Imaging Vis 2011.

[7] Delannay D, Danhier N, Vleeschouwer CD. Detection and recognition of sports(wo)man from multiple views. In: Proc. ACM/IEEE international conference on distributed smart cameras. 2009.

[8] Eshel R, Moses Y. Homography based multiple camera detection and tracking of people in a dense crowd. In: Proc. IEEE international conference on computer vision and pattern recognition. 2008. p. 1–8.

[9] Khan SM, Shah M. Tracking multiple occluding people by localizing on multiple scene planes. In: PAMI. 2009.

[10] Fleuret F, Berclaz J, Lengagne R, Fua P. Multicamera people tracking with a probabilistic occupancy map. In: PAMI. 2008.

[11] Golbabaee M, Alahi A, Vandergheynst P. Scoop: a real-time sparsity driven people localization algorithm. J Math Imaging Vis 2014;48(1):160–75.

[12] Berclaz J, Fleuret F, Turetken E, Fua P. Multiple object tracking using k-shortest paths optimization. In: PAMI. 2011.

[13] Leal-Taixe L, Pons-Moll G, Rosenhahn B. Everybody needs somebody: modeling social and grouping behavior on a linear programming multiple people tracker. In: ICCV workshops. 2011.

[14] Khan Z, Balch T, Dellaert F. MCMC-based particle filtering for tracking a variable number of interacting targets. PAMI 2005;27(11):1805–19.

[15] Nillius P, Sullivan J, Carlsson S. Multi-target tracking-linking identities using Bayesian network inference. In: CVPR. IEEE; 2006. p. 2187–94.

[16] Xiang Y, Alahi A, Savarese S. Learning to track: online multi-object tracking by decision making. In: Proceedings of the IEEE international conference on computer vision. 2015. p. 4705–13.

[17] Kuo C, Huang C, Nevatia R. Inter-camera association of multi-target tracks by on-line learned appearance affinity models. In: ECCV. 2010.

[18] Ess A, Leibe B, Schindler K, Van Gool L. A mobile vision system for robust multi-person tracking. In: CVPR. IEEE; 2008.

[19] Zhang L, Li Y, Nevatia R. Global data association for multi-object tracking using network flows. In: CVPR. 2008.

[20] Pirsiavash H, Ramanan D, Fowlkes CC. Globally-optimal greedy algorithms for tracking a variable number of objects. In: CVPR. 2011.

[21] Andriluka M, Roth S, Schiele B. People-tracking-by-detection and people-detection-by-tracking. In: CVPR. 2008.

[22] Javed O, Rasheed Z, Shafique K, Shah M. Tracking across multiple cameras with disjoint views. In: Proc. IEEE international conference on computer vision. Washington (DC): IEEE Computer Society; 2003. p. 952.

[23] Song B, Jeng T, Staudt E, Roy-Chowdhury A. A stochastic graph evolution framework for robust multi-target tracking. In: ECCV. 2010.

[24] Perera AA, Srinivas C, Hoogs A, Brooksby G, Hu W. Multi-object tracking through simultaneous long occlusions and split–merge conditions. In: CVPR. 2006.

[25] Yu Q, Medioni G, Cohen I. Multiple target tracking using spatio-temporal Markov chain Monte Carlo data association. In: CVPR. 2007. p. 1–8.

[26] Alahi A, Ramanathan V, Fei-Fei L. Socially-aware large-scale crowd forecasting. In: CVPR. 2014.

[27] Helbing D, Molnar P. Social force model for pedestrian dynamics. Phys Rev E 1995.

[28] Luber M, Stork J, Tipaldi G, Arras K. People tracking with human motion predictions from social forces. In: ICRA. 2010. p. 464–9.

[29] Pellegrini S, Ess A, Schindler K, Van Gool L. You'll never walk alone: modeling social behavior for multi-target tracking. In: ICCV. 2009.

[30] Qin Z, Shelton CR. Improving multi-target tracking via social grouping. In: CVPR. IEEE; 2012.

[31] Antonini G, Bierlaire M, Weber M. Discrete choice models of pedestrian walking behavior. Transp Res Part B.

[32] Pellegrini S, Ess A, Van Gool L. Improving data association by joint modeling of pedestrian trajectories and groupings. In: ECCV. 2010.

[33] Lan T, Sigal L, Mori G. Social roles in hierarchical models for human activity recognition. In: Computer vision and pattern recognition (CVPR). 2012.

[34] Yang B, Huang C, Nevatia R. Learning affinities and dependencies for multi-target tracking using a CRF model. In: CVPR. 2011.

[35] Hedau V, Hoiem D, Forsyth D. Recovering the spatial layout of cluttered rooms. In: ICCV. IEEE; 2009. p. 1849–56.

[36] Fouhey D, Delaitre V, Gupta A, Efros A, Laptev I, Sivic J. People watching: human actions as a cue for single view geometry. Int J Comput Vis 2014.

[37] Leal-Taixé L, Milan A, Reid I, Roth S, Schindler K. MOTChallenge 2015: towards a benchmark for multi-target tracking. Available from arXiv:1504.01942 [cs].

[38] Moussaïd M, Perozo N, Garnier S, Helbing D, Theraulaz G. The walking behaviour of pedestrian social groups and its impact on crowd dynamics. PLoS ONE 2010;5(4):e10047.

CHAPTER 7

Subject-Centric Group Feature for Person Reidentification

Li Wei, Shishir K. Shah
Quantitative Imaging Laboratory, Department of Computer Science, University of Houston, Houston, TX, USA

Contents

7.1 INTRODUCTION

Person reidentification is a fundamental task in a multicamera surveillance system to associate people across camera views at different locations and times [1]. With a growing network of cameras being used for security applications, manual reidentification that relies on a human operator is ineffective and lacking in reliability and scalability [2,3]. Therefore, an automatic solution to person reidentification has received increasing attention from the computer vision community. Person reidentification is a challenging task and relies predominantly on visual features, such as clothing and the accessories that people carry. The visual features are intrinsically weak for matching people [1] because different people may be dressed similarly, while the visual features of the same people may change significantly due to the changes in view angle, lighting and observed occlusions.

Group and Crowd Behavior for Computer Vision
DOI: 10.1016/B978-0-12-809276-7.00008-4

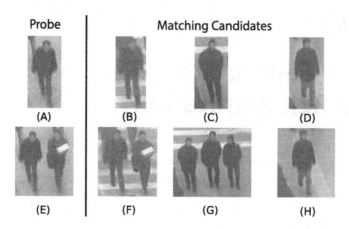

Figure 7.1 An example of group information assisting person reidentification. The first row is the persons' individual image, with (A) being the probe and (B)–(D) being candidates that match with (A). (E)–(H) are the group images probing the person and candidates.

Many recent approaches have focused on solving the reidentification problem by developing a feature representation of a person, using low-level appearance features, such as color [4], texture [5], or their combinations [6]. Once a suitable representation is obtained, a distance metric is used to measure the similarity/dissimilarity between samples. In this paper, we refer to this methodology as the "baseline method", on which we introduce the group information to improve the accuracy of reidentification.

The motivation of our approach is the observation that people often tend to walk alongside others or in a group. Such information can serve as context to reduce the ambiguity of person reidentification. If cameras are not geographically far apart, the same group structure could reappear in neighboring cameras. Although the visual feature of one person could be different between cameras, by taking the cotravelers' information (geometry and visual) in to consideration, we can reduce reidentification ambiguity significantly. An intuitive example is shown in Fig. 7.1, where (A) is the probe, and (B) to (D) are the matching candidates' images. Considering only the individual images of candidates, it is difficult to point out the image that is most similar to (A) since all persons are dressed in dark color coats and long pants. The situation would be better if we also look into persons' group context. From (E) we can observe the probe person walking with a cotraveler carrying a white object on the left side. With this information, we can tell that the first candidate has the highest possibility to match with (A), because in (F) we can observe there is a person carrying a white object walking on the left side of the first candidate, while in (G) we see that the candidate walks with two other persons, and in (H) we find that the candidate walks alone.

Motivated by this example, we introduce a subject centroid feature, named person–group feature, to describe the person's profile within their belonging group. By combining the person–group feature with other approaches that measures the similarity/dissimilarity between individuals, we can improve the accuracy of reidentification. The idea of matching people with group context has been explored by previous works, such as [7,8]. The novelty of our proposed feature is that it utilizes not only appearance but also the geometric attribute of groups for reidentification to improve matching accuracy. The proposed approach is unsupervised and can be applied to reidentification of subjects appearing in multiple videos. The advantage of keeping this approach unsupervised is that it is simple to implement and independent of the scene of the videos. We evaluate our approach on the NLPR_MCT [9] and PRID-Group [10] datasets using videos obtained from real scenarios and find an improvement in reidentification accuracy.

The main contributions of our work include:

- A framework that can improve the baseline reidentification result using people grouping information.
- A person–group feature that encodes the person's profiles within the group, including in-group position and cotravelers' baseline features. We also propose the metric for computing the distance between person–group features.
- A rich set of experiments to demonstrate that our approach improves the baseline results to achieve higher accuracy (around 90% matching rate at rank 5 for group members), and outperforms other reidentification methods that also utilize group information.

7.2 RELATED WORKS

With the ubiquitous growth in cameras, recent approaches for reidentification have addressed both the single-shot and the multi-shot challenges. Multiple shot person reidentification means that there are multiple images or video sequences that can be exploited for person reidentification. Compared to single shot reidentification, which mainly relies on appearance features [11,12,5,4] from a single image, multiple shot reidentification could rely on much boarder types of features, such as spatiotemporal features [13,14], accumulated appearance variability [15], etc. People gait information can also be extracted if video is provided, and reidentification can be solved using gait recognition [16,17]. The time index of frames that are associated with a person can also be used to learn a probability model for nonoverlapping camera tracking [18]. For a more comprehensive survey of reidentification approaches, please refer to recent surveys found in [19,20]. Our method, presented in this paper, can take advantage of state-of-the-art approaches and improve the reidentification results by incorporating group information.

Group information based reidentification. The group information has been explored to improve reidentification in recent approaches. The papers [7,8] contain the methods that are most similar to ours. Zheng et al. [7] proposed a method to associate groups of people in nonoverlapping camera views. Their method explores group information as a contextual cue for reducing the ambiguity in person reidentification

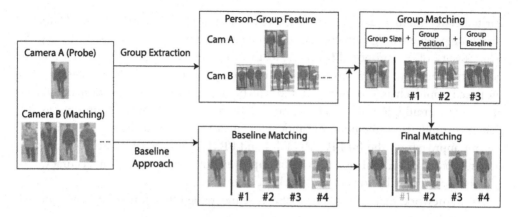

Figure 7.2 The overview of our approach.

if a person appears in the group. They propose a rotation-invariant descriptor named Center Rectangular Ring Ratio-Occurrence Descriptor (CRRRO) to handle the person position change and camera viewpoint change. This approach addresses single shot reidentification and cannot easily be extended for multiple shot scenarios. The inputs of this method are manually selected person–group images. This task in itself is time consuming because finding groups manually in large video datasets is quite tedious and requires expertise. Our approach detects people groups automatically by clustering the person trajectories, and we introduce a person–group feature that is also robust to person position and camera view point changes. Cai et al. [8] compute relative appearance context model of groups to mitigate ambiguities in individual appearance matching. Different to [7], Cai et al. use a relaxed definition of a group named neighboring set, which is a set of people that enter/exit at similar locations within a time frame. The groups under this definition have no social connection, therefore the assumption that the same set of people will reappear in a different camera view is weak. Cai et al. [8] also assume that appearance difference between a pair of persons is similar across cameras, however, this assumption is also weak because the person appearance would significantly change due to the background, illumination, and camera setting changes. In our approach, we use a group extraction method to detect groups that form social connections, and take advantage of state-of-the-art reidentification approach, to improve individual matching accuracy. We compare the results obtained using our approach against both of the above discussed methods [7,8].

7.3 METHODOLOGY

An overview of our method is illustrated in Fig. 7.2. Given a probe from Camera A and a set of matching candidates from Camera B, our method computes a baseline matching that measures the dissimilarity score between persons from the two cameras. The baseline approach is a method that estimates the dissimilarity score of persons using the individual information only. There are many features that can be used in the baseline approach, such as appearance features, spatiotemporal features, and so on. In our evaluation we use Symmetry-Driven Accumulation of Local Features (SDALF) [12] and Local Maximal Occurrence (LOMO) [21] to calculate our baseline score. For each person appearing in one camera, we calculate multiple baseline features of that person across all the frames showing that person. When we compute the dissimilarity score between persons from two cameras, we simply average the baseline feature distance of all possible feature pairs between them. The baseline approach results in a pair-wise score matrix, and it serves as an initial reidentification result.

Our method uses group information to improve the baseline score. First, we perform group extraction (Section 7.3.1) to extract groups from persons' trajectories. Then the person–group features (Section 7.3.2) are computed for each person. The person–group feature includes the in-group position of a person and the information of group members. In the group matching step, we evaluate the dissimilarity between groups by considering three aspects: group size, group position, and the baseline features of group members. The final matching is obtained by combining group matching and the baseline matching (Section 7.3.3).

7.3.1 Group Extraction

In this section, we present a group extraction approach by clustering the person's trajectories observed in a camera view. In this paper, the group is defined as a set of persons traveling together through the scene. In social science research, McPhail and Wohlstein [22] analyzed and summarized pedestrian behavior from a set of film records, and proposed an objective measure for people traveling together. The group members are determined by thresholds of difference in people's positions and velocities. Ge et al. [23] directly applied these thresholds to automatically detect small groups in a crowd automatically. However, we found that directly applying a threshold does not provide robust results when persons' positions and velocities are noisy because both are computed from person's on-the-ground trajectories, which is reconstructed from persons' tracking data. To improve the robustness of group extraction, we use a kernel function to compute the possibility of person grouping over frames. Next, we use affinity propagation to discover the clusters/groups of people.

Consider the trajectory of the person P_i in the scene as a set of sequence $L_i = \{(s_i^t, v_i^t)\}$, where s_i^t and v_i^t are the person's centroid (back-projected onto the ground using estimated homography) and velocity vector of P_i at frame t. Similar to [23], we compute the aggregated pairwise grouping possibility $W = [w_{ij}]$ over time as

Figure 7.3 Two examples of group extraction results. The images are video frames from two nonoverlapping cameras. The persons' bounding boxes and trajectories of 2 seconds are shown on the figures. In each figure, the persons belonging to the same group are marketed using the same color. (For interpretation of the references to color in this figure legend, the reader is referred to the web version of this chapter.)

$$w_{ij} = \sum_{t=0}^{\infty} \delta_{ij}^t \exp\left(-\frac{\|s_i^t - s_j^t\|^2}{2\tau_s^2} - \frac{\|v_i^t - v_j^t\|^2}{2\tau_v^2}\right) \Big/ \sum_{t=0}^{\infty} \delta_{ij}^t, \tag{7.1}$$

$$\delta_{ij}^t = \begin{cases} 1, & \text{if both } P_i \text{ and } P_j \text{ appear in the scene at frame } t, \\ 0, & \text{otherwise,} \end{cases}$$

where τ_s and τ_v are the thresholds of spatial and velocity difference.

To identify the groups, we use clustering method to find the groups with the great internal grouping possibility. As we already compute the group possibilities between trajectories in Eq. (7.1), we can use any clustering algorithm that takes pairwise distance/similarity as input, such as K-medoids or spectral clustering. However, both methods require the number of clusters as input, which is not easy to obtain in our problem. Therefore, we use Affinity Propagation (AP) [24] to discover both the group numbers and group members. Each person forms a data point, and the grouping possibility matrix W is used as the similarity matrix, which is the input to AP. The output of AP is a set of exemplars and corresponding clusters/groups. We denote these groups as $G = \{g_i\}$. We also use $G(P_i)$ to denote the group that P_i belongs to. Fig. 7.3 shows two examples of the group extraction results.

7.3.2 Person–Group Feature

In this section, we introduce the person–group feature, which describes two things about a subject within a group: who are the people that the subject is traveling with, and how they travel with that person. For the first part, we collect the subject's cotravelers' baseline feature, and reutilize the baseline score to evaluate similarity of cotravelers. For

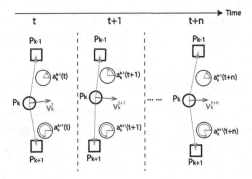

Figure 7.4 In-group position signature. The circle denotes the subject and the rectangle denotes the cotraveler. Red arrows point to the subject's moving direction. (For interpretation of the references to color in this figure legend, the reader is referred to the web version of this chapter.)

the second part, we propose an in-group position signature to encode position of subject within group. We compute the local positions of cotravelers with respect to the subject's moving direction through time, the in-group position signature is a set of cotravelers' positions. The distance measure between in-group position signatures can be computed by solving the integer programming problem inspired by Earth Mover Distance [25].

7.3.2.1 In-Group Position Signature

Assume we want to construct the in-group position signature of P_i, where P_i belongs to group $G(P_i)$. First, for each $P_j \in G(P_i)$ and $P_j \neq P_i$, we have to compute the angles between P_j and the moving direction, from perspective of P_i through all frames. We denote by (s_i^t, v_i^t) the P_i's position and velocity at frame t. Then the angle between P_j and moving direction is computed as

$$\alpha_i^j(t) = \begin{cases} \Delta, & \vec{\Gamma} \cdot \vec{Z} \geq 0, \\ 2\pi - \Delta, & \text{otherwise}, \end{cases} \qquad (7.2)$$

$$\Delta = \cos^{-1} \frac{(s_j^t - s_i^t) \cdot v_i^t}{|s_j^t - s_i^t||v_i^t|},$$

$$\vec{\Gamma} = v_i^t \times (s_j^t - s_i^t),$$

$$\vec{Z} = (0, 0, 1).$$

We collect $\alpha_i^j(t)$ through all frames, which is fitted by a Gaussian distribution, and we denote this distribution as $\alpha_i^j = (\mu_i^j, \sigma_i^j)$, where μ_i^j is the mean angle and σ_i^j is the angle deviation. An illustration of an in-group position signature is shown in Fig. 7.4.

As we collect the distributions for all group members in $G(P_i)$ except for P_i, it forms a distribution set that is represented as $H_i = \{\alpha_i^j | P_j \in G(P_i), P_i \neq P_j\}$, which is the in-group position signature of P_i. We denote P_j's cotravelers baseline features as $B_i = \{\beta_i^j | P_j \in G(P_i), P_i \neq P_j\}$. Hence, we represent the person–group feature of P_i as $PG_i = (H_i, B_i)$.

7.3.2.2 Metric of Person–Group Feature

Given person–group features, the distance measure between features is based on a linear combination of three terms: group size score, in-group position score, and group baseline score. Let PG_i and PG_j denote the person–group feature of P_i and P_j. Their distance takes the form

$$D(PG_i, PG_j) = D_g(G(P_i), G(P_j)) + D_p(H_i, H_j) + D_b(B_i, B_j). \tag{7.3}$$

The first term D_g is the group size score, which returns the size difference of groups that include P_i and P_j. The group size score is computed by

$$D_g(G(P_i), G(P_j)) = \big| |G(P_i)| - |G(P_j)| \big| \tag{7.4}$$

where $|G|$ is the group size (number of group members) of group G.

The second term D_p is the in-group position score, which evaluates the difference between in-group position signatures. As we know, $H_i = \{\alpha_i^j | P_j \in G(P_i), P_i \neq P_j\}$ is a set of distributions that encode the cotraveler's location around P_i. H_i is a distribution in metric space. The problem of computing distance between H_i and H_j becomes one of computing the distance between two distributions. There are many metrics that define distance between distributions. We found that the intuition behind Earth Mover Distance (EMD) [25] fits our problem best. EMD computes the distance between distributions in space by computing minimum cost of turning one distribution to another, where costs are assumed to be amount of weights moved, times the distance by which it is moved in space. The minimum cost can be solved as a linear programming problem. In our problem, we define the distance between in-group position signature as the minimum amount of deformations that transfer one feature to another. However, unlike the original EMD algorithm, the person can only be transformed as a complete part, therefore integer programming is required to solve the minimum deformation in our problem.

Let $H_s = \{\alpha_s^1, \ldots, \alpha_s^m\}$ be the in-group position signature of P_s, $H_t = \{\alpha_t^1, \ldots, \alpha_t^n\}$ be the in-group position signature of P_t. As we mentioned above, all possible angle distribution belongs to a metric space M. The distance function of M is simply defined as the distance between the distributions' mean angle

$$Dis(\alpha_s^m, \alpha_t^n) = \begin{cases} \dfrac{\Theta}{\pi}, & \Theta \leq \pi, \\ 2 - \dfrac{\Theta}{\pi}, & \text{otherwise,} \end{cases}$$

$$\Theta = |\mu_s^m - \mu_t^n|.$$

Let $D = [d_{ij}]$ be the difference between the ith element in H_s and the jth element in H_t. We try to find a flow $F = [f_{ij}]$, where f_{ij} is a binary variable, with $f_{ij} = 1$ when the ith element of H_s is moved to the same location of the jth element in H_t after the deformation. This optimization can be formulated as a binary integer programming problem

$$F = \arg\min_{F} \sum_{i=1}^{m} \sum_{j=1}^{n} f_{ij} d_{ij} \qquad (7.5)$$

subject to the following constrains:

$$f_{ij} \in \{0, 1\}, \ 0 \leq i \leq m, 0 \leq j \leq n,$$

$$\sum_{i=1}^{m} f_{ij} \leq 1, \ 1 \leq j \leq n,$$

$$\sum_{j=1}^{n} f_{ij} \leq 1, \ 1 \leq i \leq m,$$

$$\sum_{i=1}^{m} \sum_{j=1}^{n} f_{ij} = \min(m, n).$$

After we solve the above optimization problem, the in-group position signature distance is calculated using

$$D_p(H_s, H_t) = \frac{\sum_{i=1}^{m} \sum_{j=1}^{n} f_{ij} d_{ij}}{\sum_{i=1}^{m} \sum_{j=1}^{n} f_{ij}}. \qquad (7.6)$$

The final term D_b is the group baseline score. It computes the aggregated differences of cotravelers' baseline features, under the condition that the cotraveler's correspondence is known by solving Eq. (7.5). Let $R = [r_{ij}]$ be the pairwise baseline score matrix, where r_{ij} denotes the baseline score between the ith element in B_s and the jth element in B_t. The group baseline score takes the form

$$D_b(B_s, B_t) = \frac{\sum_{i=1}^{m} \sum_{j=1}^{n} f_{ij} r_{ij}}{\sum_{i=1}^{m} \sum_{j=1}^{n} f_{ij}}. \qquad (7.7)$$

When a person is traveling alone, the person–group feature is empty. In this case the distance to an empty person–group feature D_p and D_b is set to zero, and only group size score, G_g, contributes to the person–group feature difference.

7.3.3 Person Reidentification with Person–Group Feature

In Section 7.3.2, we introduced the person–group feature and defined the distance function between features. We argue that by combining the metric of person–group and baseline features, we can improve the performance of person reidentification.

A simple way to combine two distance measurements is by linearly adding them as

$$D(P_i, P_j) = D(B_i, B_j) + D(PG_i, PG_j) \tag{7.8}$$

where B_i is the baseline feature of P_i, and $D(B_i, B_j)$ means the baseline score of persons P_i and P_j.

7.4 RESULTS

To evaluate our approach, we test our method on the NLPR_MCT [9] and PRID-Group [10] dataset. Other reidentification datasets (e.g., CAVIAR, VIPeR, ETHZ) either contain single person's images only, or do not have group information provided in the dataset, therefore they are not suitable for evaluation of our approach. Datasets 1 and 2 of NLPR_MCT are used for evaluation. For both datasets, there are three synchronous videos (resolution is 320×240 at 20 fps) from three nonoverlapping cameras. We use the videos produced by two outdoor cameras for evaluation. The numbers of people in each dataset are presented in Table 7.1. The dataset provides the ground truth annotation, which includes the bounding box tracking for each person. To better evaluate our approach, we create PRID-Group as an additional dataset to evaluate our subject centric group feature. PRID dataset [10] contains two synchronous videos (resolution is 720×576 at 25 fps) of two nonoverlapping street views. It provides tracking information of 200 individuals from two videos for reidentification task evaluation. In the original dataset, most of individuals walk alone. A small subset of individuals walk in groups, but most group members are not recorded by the dataset due to occlusion among cowalkers. Therefore, the original dataset is not suitable for group based reidentification evaluation. In order to record the persons traveling in groups, we find groups of person by observing the persons location and interaction in video, then manually annotate 38 individuals, which forms 16 groups appearing in both camera views. The tracking information of each individual is computed using [26]. The challenge of person reidentification of PRID-Group dataset is threefold: first, the videos from two cameras contain a large viewpoint change; second, there is a stark difference in illumination, background, and camera characteristics between the two videos; and third, occlusion among group members is frequent in the first video. Some example group images are shown in Fig. 7.11.

In both NLPR_MCT and PRID-Group datasets, the persons' X–Y plane locations are computed by back-projecting the mid-bottom of bounding boxes, and the homography is estimated interactively off-line. The group information of NLPR_MCT dataset

Table 7.1 Numbers of people in evaluation datasets

	Camera 1	Camera 2	Common
NLPR_MCT Dataset 1	76	78	72
NLPR_MCT Dataset 2	115	111	105
PRID–Group	38	38	38

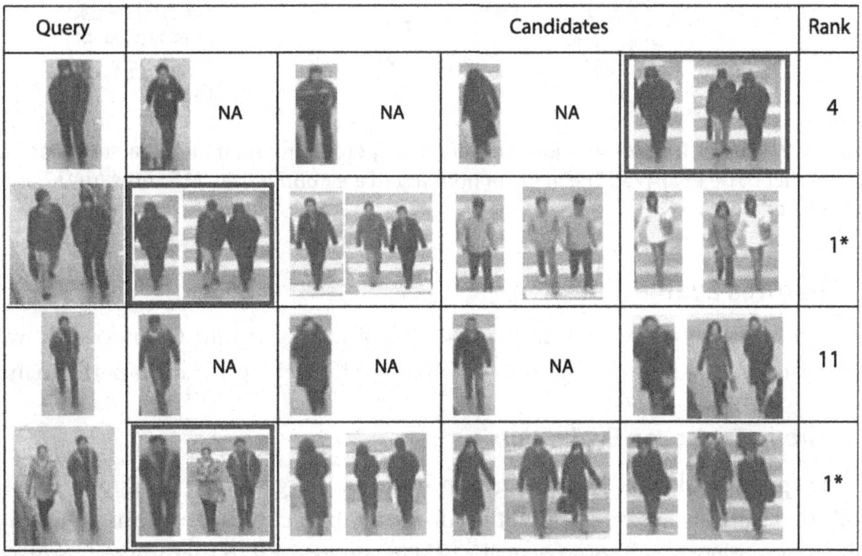

Figure 7.5 Two examples of reidentification results. For each query, the image of query person and the group that person belongs to are shown. We display the matching results of baseline approach [12] and our approach. The top four candidates are shown; we display the image of candidate and the group that candidate belongs to in each grid. The ground truth matching is labeled by blue boxes, where the rank is also given at right. The ranks with star symbols are the results obtained using our approach, otherwise the ranks are computed by the baseline approach. (For interpretation of the references to color in this figure legend, the reader is referred to the web version of this chapter.)

is extracted using the proposed algorithm in Section 7.3.1. In Dataset 1, both Camera 1 and Camera 2 have 18 persons traveling with cotravelers and forming 8 groups (with size greater than 1). In Dataset 2, 35 and 31 persons travel with cotravelers, and they form 16 and 15 groups viewed by Camera 1 and Camera 2, respectively.

When one person in a camera view is given, we compute his/her person–group feature, compute the distance to all the persons in another camera view, and sort the persons in an ascending order based on the distance value. The rank score is the order of ground truth person in the sorted persons' list. Some examples of query and candidates person–group images are illustrated in Fig. 7.5. The results obtained using [12] are also provided.

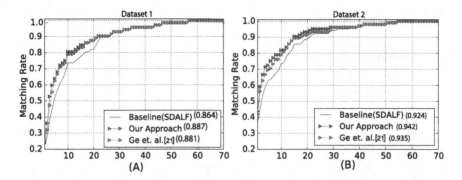

Figure 7.6 CMC curves for person reidentification using group information extracted using our approach and that of Ge et al. [23]. The value of normalized area under the CMC curve (uAUC) is shown within the parentheses.

7.4.1 Features Evaluation

In this section, we evaluate each individual step of the algorithm to discover how each step and feature contribute to the final accuracy of the reidentification performance.

7.4.1.1 Group Extraction Evaluation

Since our approach depends on the group information given by the group extraction method, we want to discover how different group extraction algorithms affect the reidentification results. We choose Ge et al.'s [23] group extraction method for comparison. The results are illustrated in Fig. 7.6. As seen, using group information extracted by either method leads to an improvement in accuracy compared to the baseline approach. In Dataset 1 our approach provides similar accuracy as that of [23], while in Dataset 2 our approach is slightly better. The reason is that Dataset 1 has less crowded scenes, and the group extraction task is relatively easier. However, the scenes are more crowded in Dataset 2 and the performance of [23] is affected by directly using the threshold and noisy trajectories. Our group extraction algorithm handles noise better by computing the grouping probability using a kernel function, which leads to more accurate group information and benefits the reidentification task.

7.4.1.2 Group Features Evaluation

As shown in Eq. (7.3), the score between person–group features contains three terms: group size (GS), in-group position (GP), and group baseline (GB). We evaluate the contribution of the three parts by showing how well the person reidentification performs when only individual parts are added into the final score. We use SDALF as the baseline and show the CMC curve of GS, $GS + GP$, and $GS + GB$. The reason that we combine GP and GB with GS is that GP and GB are only meaningful to compare the groups of the same size, and we have to combine them with GS in order to calculate

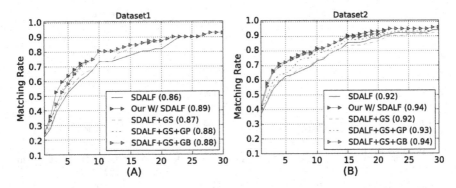

Figure 7.7 CMC evaluation of group size (GS), in-group position (GP) and group baseline (GB) of person–group feature.

Table 7.2 Matching rates comparison of SDALF baseline score combined with group size (GS), in-group position (GP), and group baseline (GB) of person–group feature

Dataset	Rank	SDALF	SDALF + GS	SDALF + GS + GP	SDALF + GS + GB	Our w/ SDALF
1	1	0.22	0.24	0.24	0.25	0.25
	5	0.52	0.54	0.60	0.58	0.64
	10	0.74	0.74	0.80	0.80	0.80
2	1	0.37	0.40	0.40	0.41	0.44
	5	0.62	0.61	0.64	0.72	0.72
	10	0.73	0.74	0.79	0.81	0.82

any score between groups of different size. The results are illustrated in Fig. 7.7, and the rate comparison is given in Table 7.2. As seen in both Datasets 1 and 2, the *GS* score alone provides very little improvement. This is due to the fact that *GS* only considers the size of the group and does not take any individual information into account. We also observe that both *GS* + *GP* and *GS* + *GB* perform better than *GS* as they consider the location and appearance information of the group members. The performance of *GS* + *GB* is slightly better than *GS* + *GP*, the reason being that *GB* is calculated based on the in-group position mapping of *GP* and it leads to less ambiguity by considering the person's appearance in *GB*. In both datasets, our approach, which uses all three terms, shows the overall best performance.

7.4.2 Comparison with Baseline Approaches

To test the performance of our method with respect to different baseline methods, we conduct experiments using the Symmetry-Driven Accumulation of Local Features (SDALF) [12] and Local Maximal Occurrence (LOMO) [21]. SDALF requires background subtraction, which is obtained using ViBe [27]. We measure the performance

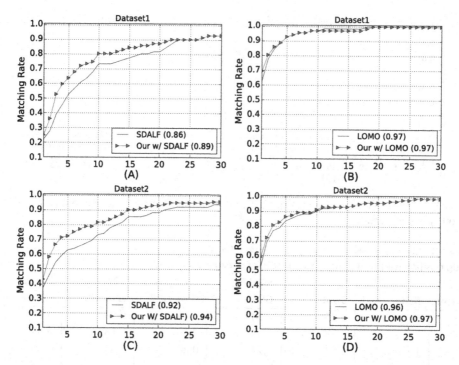

Figure 7.8 Comparison of CMC using baseline methods SDALF and LOMO on two datasets. The value of uAUC is shown within the parentheses.

Table 7.3 Matching rates comparison between our approach and baseline methods (SDALF and LOMO)

Dataset	Rank	SDALF	Our w/ SDALF	LOMO	Our w/ LOMO
1	1	0.22	0.25	0.62	0.69
	5	0.52	0.64	0.93	0.93
	10	0.74	0.80	0.97	0.97
2	1	0.37	0.44	0.52	0.59
	5	0.62	0.72	0.83	0.87
	10	0.73	0.82	0.90	0.91

using Cumulative Match Curve (CMC) [28]. The results are shown in Fig. 7.8. The matching rates comparisons between our approach and baseline methods at rank 1, 5, and 10 are given in Table 7.3. Our approach outperforms the baseline method, while the improvements are more significant with Dataset 2 than with Dataset 1. This can be attributed to the fact that there are more persons traveling with cotravelers in Dataset 2. We also observe that the improvement of SDALF is more significant than of LOMO. The reason is that our approach mainly boosts the reidentification accuracy of those who travel within groups. The accuracy improvement of the overall dataset is bounded by the maximal possible accuracy improvement of group members. The reidentification of group members IDs using SDALF as the baseline feature is shown in Fig. 7.9, B and F. The baseline approach does not perform well on group members IDs and leaves room for improvement by using group information. The improvements are 11%, 43%, and 27% at rank 1, 5, and 10, respectively, in Dataset 1, and 23%, 33%, and 29% at rank 1, 5, and 10, respectively, in Dataset 2. The reidentification of group members IDs using LOMO as the baseline feature is shown in Fig. 7.9, D and H. As seen, LOMO performs much better than SDALF and already reaches high rank 1 accuracy. By introducing group information, it is not able to gain as much improvement as SDALF, especially at high ranks. By using LOMO as baseline, the improvements are 21%, 0%, and 0% at rank 1, 5, and 10, respectively, in Dataset 1, and 23%, 10%, and 3% at rank 1, 5, and 10, respectively, in Dataset 2.

7.4.3 Comparison with Group-Based Approaches

We also compare our approach to [7] and [8], both of which use group information as context to improve the accuracy of individual reidentification. The first approach [7] extracts Center Rectangular Ring Ratio-Occurrence (CRRRO) descriptor as the group context feature from a manually selected static group image. Although our dataset consists of videos, we generate group images by cropping the video frames that include all group members and computing CRRRO representations of the groups across multiple frames. When we compute the CRRRO score of two persons, we use the average score of all possible CRRRO feature pairs between the groups that the two persons belong to. The distance between CRRRO features are linearly combined with other appearance-based distance as the final score. The second work uses Relative Appearance Context (RAC) feature as group context, which measures the appearance difference of the person to the nearby people. The distance of appearance feature is also linearly combined with relative appearance context distance as the final distance value. To make sure the comparison is fair, in both comparison methods we use both SDALF and LOMO to represent the individual appearance feature. We use parameters as suggested by respective authors in all our experiments.

Results are shown in Fig. 7.9. The matching rates at rank 1, 5, and 10 are given in Table 7.4. As seen, there is an overall improvement in reidentification accuracy. To further evaluate the impact of the person–group feature, we specifically restrict the dataset to those IDs that are found in a group. Results obtained on this restricted dataset are as shown in Fig. 7.9, B and F, for SDALF baseline, and Fig. 7.9, D and H, for LOMO baseline. As can be seen from the results, our method provides the best performance in both datasets. By looking into the CMC for all persons (Fig. 7.9, A, C, E, and G), we can observe that the accuracy is boosted through our approach. In general, the accuracy is slightly better than compared approaches. However, as seen through the CMC of

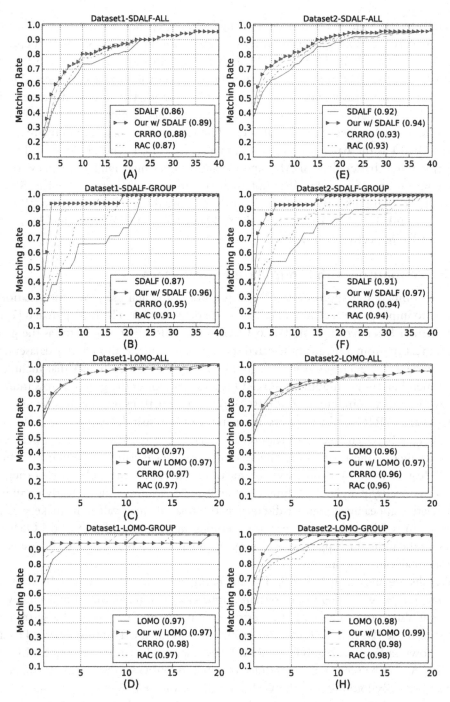

Figure 7.9 Comparison of the CMC of person reidentification using our approach, CRRRO descriptor, and RAC feature. The value of uAUC is shown within the parentheses.

Table 7.4 Comparison of NLPR_MCT and PRID-Group dataset matching rates across methods that use group information for reidentification

| Dataset | Rank | Baseline = SDALF | | | | Baseline = LOMO | | | |
		Baseline	Our w/ Baseline	CRRRO	RAC	Baseline	Our w/ Baseline	CRRRO	RAC
1-ALL	1	0.22	0.25	0.25	0.21	0.62	0.69	0.67	0.64
	5	0.52	0.64	0.62	0.53	0.93	0.93	0.93	0.93
	10	0.74	0.80	0.80	0.78	0.97	0.97	0.97	0.97
1-Group	1	0.28	0.39	0.38	0.22	0.67	0.89	0.83	0.72
	5	0.51	0.94	0.88	0.51	0.94	0.94	0.94	0.94
	10	0.67	0.94	0.94	0.83	0.94	0.94	1.00	0.94
2-ALL	1	0.37	0.44	0.42	0.38	0.52	0.59	0.58	0.54
	5	0.62	0.72	0.70	0.63	0.83	0.87	0.82	0.83
	10	0.73	0.82	0.79	0.77	0.90	0.91	0.90	0.90
2-Group	1	0.19	0.42	0.39	0.23	0.48	0.71	0.68	0.55
	5	0.54	0.87	0.77	0.58	0.87	0.97	0.90	0.84
	10	0.64	0.93	0.84	0.77	0.97	1.00	0.94	0.97
PRID-Group	1	0.11	0.16	0.16	0.13	0.38	0.59	0.57	0.46
	5	0.32	0.74	0.66	0.45	0.49	0.95	0.81	0.76
	10	0.39	0.87	0.76	0.63	0.78	1.00	0.89	0.86

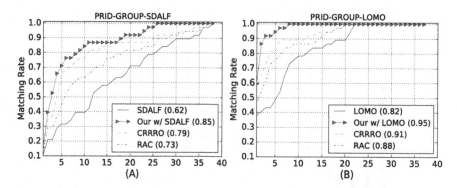

Figure 7.10 CMC of PRID-Group person reidentification using our approach and compared approaches.

group persons (Fig. 7.9, B, D, F, and H), our method is able to reach accuracy of around 90% at rank 5 using SDALF baseline, which is significantly better then the baseline method and compared approaches; and is able to reach around 95% using LOMO baseline, which is also better than compared approaches (Fig. 7.10).

To further evaluate the proposed approach, we conduct reidentification on PRID-Group dataset, which contains only people traveling within groups. We also evaluate the performance using both SDALF and LOMO as baseline features. Example probes and top four matching results from the PRID-Group data with SDALF baseline are shown in Fig. 7.11. The matching rates at rank 1, 5, and 10 are shown in Table 7.4. The results show that SDALF reaches 39% matching rate at rank 10. However, by additionally including group information, all group based reidentification approaches show an improvement in matching accuracy in this experiment. This time RAC reaches 13%, 45%, and 63% at rank 1, 5, and 10, respectively. CRRRO reaches 16%, 66%, and 76% at rank 1, 5, and 10, respectively. Our approach has the best overall performance compared to CRRRO and RAC, and reaches 16%, 74%, and 87% at rank 1, 5, and 10, respectively. LOMO baseline approach does much better than SDALF baseline and reaches 38%, 49%, and 78% at rank 1, 5, and 10, respectively; and our approach has the best overall performance, reaching 46%, 76%, and 86% at rank 1, 5, and 10, respectively.

7.5 CONCLUSION

In this chapter, we addressed the problem of person reidentification using subject-centric group features. We proposed person–group feature that encodes the geometry and visual information of groups. The distance between person–group features were computed by solving an integer programming problem. The final distance was a linear combination of person–group feature distance and a baseline distance obtained by considering a feature of an individual. We demonstrated that our proposed method could

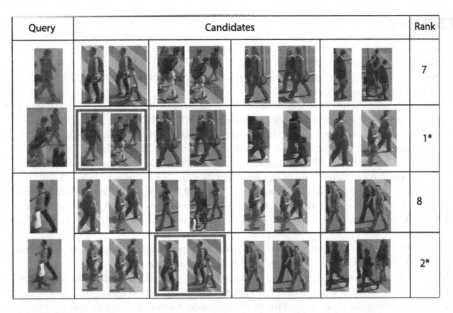

Figure 7.11 Two examples of reidentification results with SDALF baseline in PRID-Group dataset. The ground truth matching is labeled by blue boxes, where the rank is also given on the right. The ranks with star symbols are the results obtained using our approach, otherwise the ranks are computed by the baseline approach. (For interpretation of the references to color in this figure legend, the reader is referred to the web version of this chapter.)

always improve the accuracy of a baseline approach, and outperform the state-of-the-art group information based reidentification approaches.

One limitation of the proposed algorithm is the assumption that the human position within group does not change across the camera views. This is a valid assumption when the camera views are close to each other, but limits the application of this method on the camera networks that contain cameras far apart. Another limitation is that the method assumes a reasonable crowd density, which contains a number of groups, where robust group extraction algorithm can succeed. We are looking to overcome above limitations in our future works by introducing a dynamic group model and a nonrigid mapping of groups to allow changes in the position of members within a group. We also plan to explore the group/crowd behaviors to further reduce the ambiguity of person reidentification.

ACKNOWLEDGMENTS

This work was supported in part by the U.S. Department of Justice 2009-MU-MU-K004. Any opinions, findings, conclusions, or recommendations expressed in this paper are those of the authors and do not necessarily reflect the views of our sponsors.

REFERENCES

[1] Gong Shaogang, Cristani Marco, Loy Chen Change, Hospedales Timothy M. The re-identification challenge. In: Person re-identification. Springer; 2014. p. 1–20.

[2] Keval Hina. CCTV control room collaboration and communication: does it work? In: Proceedings of human centred technology workshop. 2006. p. 11–2.

[3] Williams David. Effective CCTV and the challenge of constructing legitimate suspicion using remote visual images. J Invest Psychol Offender Profiling 2007;4(2):97–107.

[4] Yang Yang, Yang Jimei, Yan Junjie, Liao Shengcai, Yi Dong, Li Stan Z. Salient color names for person re-identification. In: Computer vision – ECCV 2014. Springer; 2014. p. 536–51.

[5] Zhao Rui, Ouyang Wanli, Wang Xiaogang. Person re-identification by salience matching. In: 2013 IEEE international conference on computer vision (ICCV). IEEE; 2013. p. 2528–35.

[6] Zhao Rui, Ouyang Wanli, Wang Xiaogang. Learning mid-level filters for person re-identification. In: 2014 IEEE conference on computer vision and pattern recognition (CVPR). IEEE; 2014. p. 144–51.

[7] Zheng Wei-Shi, Gong Shaogang, Xiang Tao. Associating groups of people. In: Proc. BMVC. ISBN 1-901725-39-1, 2009. p. 23.1–11.

[8] Cai Yinghao, Medioni Gerard. Exploring context information for inter-camera multiple target tracking. In: IEEE winter conference on applications of computer vision. 2014. p. 761–8.

[9] Multi-camera object tracking challenge. Available from http://mct.idealtest.org/Datasets.html, 2014.

[10] Hirzer Martin, Beleznai Csaba, Roth Peter M, Bischof Horst. Person re-identification by descriptive and discriminative classification. In: Proc. Scandinavian conference on image analysis (SCIA). Springer; 2011.

[11] Prosser Bryan, Zheng Wei-Shi, Gong Shaogang, Xiang Tao. Person re-identification by support vector ranking. In: Proc. BMVC. ISBN 1-901725-40-5, 2010. p. 21.1–11.

[12] Farenzena Michela, Bazzani Loris, Perina Alessandro, Murino Vittorio, Cristani Marco. Person re-identification by symmetry-driven accumulation of local features. In: 2010 IEEE conference on computer vision and pattern recognition (CVPR). IEEE; 2010. p. 2360–7.

[13] Wang Taiqing, Gong Shaogang, Zhu Xiatian, Wang Shengjin. Person re-identification by video ranking. In: Computer vision – ECCV 2014. Springer; 2014. p. 688–703.

[14] Gheissari Niloofar, Sebastian Thomas B, Hartley Richard. Person reidentification using spatiotemporal appearance. In: 2006 IEEE Computer Society conference on computer vision and pattern recognition, vol. 2. IEEE; 2006. p. 1528–35.

[15] Hamdoun Omar, Moutarde Fabien, Stanciulescu Bogdan, Steux Bruno. Person re-identification in multi-camera system by signature based on interest point descriptors collected on short video sequences. In: Second ACM/IEEE international conference on distributed smart cameras. IEEE; 2008. p. 1–6.

[16] Han Ju, Bhanu Bir. Individual recognition using gait energy image. IEEE Trans Pattern Anal Mach Intell 2006;28(2):316–22.

[17] Martín-Félez Raúl, Xiang Tao. Gait recognition by ranking. In: Computer vision – ECCV 2012. Springer; 2012. p. 328–41.

[18] Javed Omar, Rasheed Zeeshan, Shafique Khurram, Shah Mubarak. Tracking across multiple cameras with disjoint views. In: Ninth IEEE international conference on computer vision, proceedings. IEEE; 2003. p. 952–7.

[19] Vezzani Roberto, Baltieri Davide, Cucchiara Rita. People reidentification in surveillance and forensics: a survey. ACM Comput Surv 2013;46(2):29.

[20] Gong Shaogang, Cristani Marco, Yan Shuicheng, Loy Chen Change. Person re-identification. ISBN 978-1-4471-6296-4, 2014.

[21] Liao Shengcai, Hu Yang, Zhu Xiangyu, Li Stan Z. Person re-identification by local maximal occurrence representation and metric learning. In: Proceedings of the IEEE conference on computer vision and pattern recognition. 2015. p. 2197–206.

[22] McPhail Clark, Wohlstein Ronald T. Using film to analyze pedestrian behavior. Sociol Methods Res 1982;10(3):347–75.

[23] Ge Weina, Collins Robert T, Ruback R Barry. Vision-based analysis of small groups in pedestrian crowds. IEEE Trans Pattern Anal Mach Intell 2012;34(5):1003–16.

[24] Frey Brendan J, Dueck Delbert. Clustering by passing messages between data points. Science 2007;315(5814):972–6.

[25] Rubner Yossi, Tomasi Carlo, Guibas LJ. The Earth Mover's distance as a metric for image retrieval. Int J Comput Vis 2000:1–20.

[26] Nebehay Georg, Pflugfelder Roman. Clustering of static-adaptive correspondences for deformable object tracking. In: Computer vision and pattern recognition. IEEE; 2015.

[27] Barnich Olivier, Van Droogenbroeck Marc. ViBe: a universal background subtraction algorithm for video sequences. IEEE Trans Image Process 2011;20(6):1709–24.

[28] Grother Patrick, Phillips P Jonathon. Models of large population recognition performance. In: Proceedings of the 2004 IEEE Computer Society conference on computer vision and pattern recognition. IEEE; 2004. p. II-68.

Group and Crowd Behavior Modeling

CHAPTER 8

From Groups to Leaders and Back
Exploring Mutual Predictability Between Social Groups and Their Leaders

Francesco Solera, Simone Calderara, Rita Cucchiara
Department of Engineering, University of Modena and Reggio Emilia, Modena, Italy

Contents

8.1 INTRODUCTION

Understanding crowd dynamics has engaged many scientists in the past century from different heterogeneous points of view, ranging from collective psychology to system theory, involving sociology and computer vision as the main analytic tools. Crowd phenomena are complex and alluring modern elephant men (Reicher [1]), because their logic still escapes formal rules and contemporarily exposes fascinating challenges. Eventually, the ambition is always to precisely characterize people behavior in a crowd, to

Group and Crowd Behavior for Computer Vision
DOI: 10.1016/B978-0-12-809276-7.00010-2

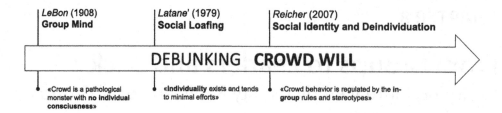

Figure 8.1 Evolution of crowd theories from mass phenomena to groups formation.

predict and prevent potentially dangerous situations by means of either synthetic simulation models or real time visual analysis. In his pioneering work on crowd behavior, Gustave Le Bon [2] defined crowds as hidden and inherent threats to society. In his writings Le Bon asserted that as members of a crowd, people tend to display a loss of self-awareness and an increased inclination toward violence. Far from this approach, the modern elaborated *Social Identity Model* [1] proposes a social-normative conception of collective behavior based on members spontaneous transition from an individual identity to a common and shared one among small subset of people, also known as *groups*.

In accordance with recent theories, empirical observations [3] recognize groups as the basic elements which the crowd is composed of, leading to an intermediate level of abstraction that is placed between the crowd as a flow of people and its interpretation as a collection of individuals; see Fig. 8.1. Identifying groups is consequently a mandatory step in order to grasp the complex social dynamics ruling collective behaviors in crowds. Nevertheless, automatic group detection in video streams is definitely less studied than pedestrian analysis or crowd flow motion estimation. One of the greater challenges resides in the lack of a single, agreed, computational definition of a group, formal definitions of the mechanisms which govern them and insights on the relations arising among people during social gatherings. Conversely, there seems to be agreement on the fact that not all the members of a group (and of a crowd, more generally) undergo the same level of identity shift [2,4,5]. People who define the norms and the values which then become shared among all the other members are recognized as *leaders*, thus identifying leaders is crucial in crowd management, emergency planning and sociological analysis.

The purpose of this chapter is twofold:

- We aim to provide a learning framework that can be useful for the visual recognition of both groups and leaders. These problems are neatly casted into the same structured learning framework [6,7], and solved efficiently using Structural SVMs; see Section 8.3 and thereinafter.
- We aim to discuss, starting from empirical evidence and original experiments, the roles of leaders in forming and structuring a group and the mutual influence groups and leaders have in their automatic visual detection; see Section 8.6.

We hope such a discussion will bring benefit to the computer vision community by raising awareness of the social mechanisms underpinning crowds and groups, and by providing straightforward solutions to leaders and groups detection.

8.2 MODELING AND OBSERVING GROUPS AND THEIR LEADERS IN LITERATURE

In this section we briefly survey past and recent sociological theories that have tried to explain crowds' behavior and their structure, as well as computational models employed by the computer vision community to automatically analyze crowd and related events.

8.2.1 Sociological Perspective

Most of the research work has tried to tackle the crowd as an exclusively collective phenomenon, where individuality and social groups do not exist. This recalls the primitive *Popular Mind Theory* [2], where the crowd was defined as a "pathological monster with no individual consciousness." Accordingly, crowds have been modeled by means of physical models (e.g., hydrodynamics [8]), neglecting the existence of single individual purposes and goals. Conversely, many other studies have been inspired by the 1970s *Social Loafing Theory* [9], which stated that individuality was a strong requirement for the pursuit of personal goals. Helbing's *Social Force Model* [10], which asserts that anyone's movements toward his/her goals are influenced by the surrounding pedestrians, has also been the main building block for many crowd modeling and analysis works. Recently, studies on pedestrians attending events have underlined that most people tend to move in groups, and social relations influence the way people behave in crowds [11,3]; see Fig. 8.1. These empirical observations are supported by Reicher in the recent *Social Identity Model of Deindividuation Effects* [12], which assumes that crowd behavior is regulated by the social rules and behaviors groups choose to adopt. Actual field observations of temporal gatherings by McPhail [13] indicate that members are rarely violent, leaders provide direction through verbal and reasonable interventions, and composing groups do not act in a capricious, unpredictable fashion. These groups form, change, and disband, and the internal structures and processes of the crowd and its groups are more similar than different.

If crowds can be understood by taking into account group processes, then their leaders' actions can be understood by taking into account leadership processes. Le Bon's leadership model [2] resembles the need of the crowd to place trust in someone's ability to provide orientation and contribute to its overall stability. This leader neither is a founding figure nor is permanently established; see Fig. 8.2A. Instead, the crowd formation is an emergent process that uses the leader as a stabilizer. While for Le Bon the leader is an elected anonymous figure, in Freud's *Group Psychology* [4] the leader comes to play a constitutive role and every member of the crowd identifies the leader as their

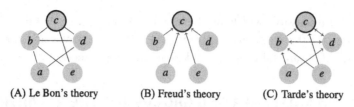

Figure 8.2 Different interpretation of relationships between a leader and other members in different crowd psychology theories.

"I"-ideal. With respect to Le Bon self-organizing and emergent notion of crowds, Freud provides a highly centralized model of the social community, appreciable in Fig. 8.2B. The relation between the leader and the crowd is thus radically asymmetric as the members are submitted under the leader; relations between crowd members are secondary. Eventually, and to even a greater extent than Le Bon, Tarde [5] emphasizes the self-referential emergence of crowd phenomena but characterizing the leader as the "spark" behind the organizational patterns. Unlike Freud, Tarde was not interested in the leader's foundational role. Instead, he analyzes how the leader contributes to the flow of imitation in a society and builds its theory assuming that every process of imitation begins with asymmetry, as highlighted in Fig. 8.2C. Nevertheless, as individuals initiative converge with leadership, the leader's identity may change without altering the crowd and groups stability [13].

Due to the lack of accepted theories on leader emergence in groups and following McPhail's on-the-field observations that crowd and groups share similar and hierarchical formation and structuring processes, in the rest of this work we will borrow sociological theories about crowd leaders and apply them to group leaders.

8.2.2 Computational Approaches

The computational modeling of pedestrian dynamics in crowd situations represents a relatively recent, but already established field of research with many different applications. Most models employ basic proxemic notions [14], like the tendency of pedestrians to preserve a personal space around them whenever possible. However, recent approaches take a more comprehensive viewpoint on proxemics, which also comprises the preference of an individual to stay close to other members that belong to the same group [15,16]. Sociological concepts such as *F-formations* by Kendon [17] have been exploited as a foundation for an interesting line of research on group detection [18,19]. F-formations can be interpreted as specific positional and orientational patterns that people assume in order to be considered engaged in a social interaction. Nevertheless, the theory holds only for stationary groups and is not defined for moving groups, a case which cannot be ignored in crowd analysis. Differently, motion paths are considered as the main feature by the most recent group detection approaches. There are three approaches: In *group-based* approaches, groups are considered as atomic entities in the scene and no higher level information can be extracted neatly, typically due to high noise or high complexity of crowded scenes [20–22]. Up to now these models were confined to the detection problem and not used to further infer on group paths and identities in time compared to the *individual–group joint* approaches that also include individual tracking while tracking groups at a coarser level [23,24]. Finally, *individual-based* approaches build up on single pedestrians trajectories. This kind of approach is subjected to the challenge of tracking individuals even in high density crowds and can be applied with significant limitations in real life scenarios [25–28].

While visual group detection has gained momentum in the computer vision community, leader identification is still an emerging topic. Pioneering works tackled the problem of leadership identification as finding the group member that contributes the most to the group proxemic formation. This line of work builds on the intuition that leaders cover the central position inside the group [29,30]. Despite being effective there is still a lack of empirical evidence that the leader's spatial centrality holds independently from the crowd type and its density and, as a matter of fact, this is typically neglected in sociological literature. More complex approaches adhered to Tarde's referential model of leadership exploiting either Bayesian inference on causal graph or ranking techniques in the feature space to establish the leader inside a group, [31–33]. Referential models are up to now the most effective models of leadership. Unluckily, most of the models solve the problem considering one group at a time, and not exploring the mutual relations between members and leaders of different groups. Conversely, in crowd psychology these intergroup relations have been supposed in recent theories.

8.3 TECHNICAL PRELIMINARIES AND STRUCTURED OUTPUT PREDICTION

Many inference tasks related to crowd analysis are structured prediction tasks. Structured output prediction is the inference task related to a set of random variables whose outcome is interdependent in complex, but observable ways. Canonical examples of such structured objects may include matrices, sequences or graphs, with applications to image segmentation, natural language processing or bio-informatics, among others. Structured prediction neatly applies to crowds because individual behavior is oftentimes interlinked with the behavior of other people as well.[1]

This section introduces Structural Support Vector Machines (SSVM) [34], a discriminative method for complex and structured output prediction. On top of the SSVM

---

[1] Here and in the rest of this work we will use the word *behavior* without implying any further sociological claim. Eventually, to what extent behavior is really captured is up to the features extracted from the videos and employed in the presented methods.

learning framework, in Section 8.4 we develop algorithms to detect groups and their leaders in social gatherings and crowded environments.

8.3.1 Problem Statement

Let us consider the input $\mathbf{x} \in \mathcal{X}$ to be some representation of the crowd, possibly describing all unary, pairwise, and higher order interaction terms. As an example, imagine $\mathbf{x}$ as describing crowd members' position, mutual distance, engagement, and so on. We want to learn a mapping from input $\mathbf{x}$ to output variable $\mathbf{y} \in \mathcal{Y}(\mathbf{x})$, possibly describing each member's social group (Section 8.4.1) or its role (Section 8.4.2), based on a set of samples $\mathcal{D} = \{(\mathbf{x}_i, \mathbf{y}_i)\}_{i=1,\dots,n}$ drawn from a fixed but unknown distribution. Depending on the task and on the input $\mathbf{x}$, the output space $\mathcal{Y}(\mathbf{x})$ will have different sizes and characterization.

A discriminant score function $F : \mathcal{X} \times \mathcal{Y} \to \mathbb{R}$ is defined over the joint input–output space such that $F(\mathbf{x}, \mathbf{y})$ can be interpreted as measuring the compatibility of output proposal $\mathbf{y}$ given a specific input $\mathbf{x}$. Now, the prediction function $f : \mathcal{X} \to \mathcal{Y}$ can be defined as

$$f(\mathbf{x}) = \arg \max_{\mathbf{y} \in \mathcal{Y}(\mathbf{x})} F(\mathbf{x}, \mathbf{y}) \tag{8.1}$$

where the maximizer over the label space $\mathcal{Y}(\mathbf{x})$ is the predicted label, i.e., the solution of the inference problem. For simplicity, we choose to restrict the space of F to linear functions over some combined feature representation $\Psi(\mathbf{x}, \mathbf{y})$ subject to a $\mathbf{w}$-parametrization so that $F(\mathbf{x}, \mathbf{y}) = \mathbf{w}^T \Psi(\mathbf{x}, \mathbf{y})$.

The problem of learning in structured and interdependent output spaces can been formulated as a maximum-margin problem. We adopt the n-slack, margin-rescaling formulation:

$$\min_{\mathbf{w}, \xi} \quad \frac{1}{2} \|\mathbf{w}\|^2 + \frac{C}{n} \sum_{i=1}^{n} \xi_i$$
$$\text{s.t.} \quad \forall i : \xi_i \geq 0, \tag{8.2}$$
$$\forall i, \forall \mathbf{y} \in \mathcal{Y}(\mathbf{x}_i) \backslash \mathbf{y}_i : \mathbf{w}^T \delta \Psi_i(\mathbf{y}) \geq \Delta(\mathbf{y}, \mathbf{y}_i) - \xi_i,$$

where $\delta \Psi_i(\mathbf{y}) \overset{\text{def}}{=} \Psi(\mathbf{x}_i, \mathbf{y}_i) - \Psi(\mathbf{x}_i, \mathbf{y})$, ξ_i are the slack variables introduced in order to accommodate for margin violations, $\Delta(\mathbf{y}_i, \mathbf{y})$ is the loss function measuring distance between two outputs, and C is the regularization trade-off. Intuitively, we want to maximize the margin and jointly guarantee that for a given input, every possible output result is considered worst than the correct one by at least a margin of $\Delta(\mathbf{y}_i, \mathbf{y}) - \xi_i$, where $\Delta(\mathbf{y}_i, \mathbf{y})$ is larger when the two predictions are known to be more different.

Note that both the feature map $\Psi(\mathbf{x}, \mathbf{y})$ and the loss function cannot be defined out of the context of the problem, as it is a problem itself that specifies (i) given a particular input, the nature of the desired solution; and (ii) how to account for differences in output objects. As a result, SSVM is more of a framework than an off-the-shelf algorithm. Section 8.4 will introduce the feature map and the loss function for the tasks of group detection and leader identification.

8.3.2 Stochastic Optimization

The quadratic program (QP) (8.2) introduces a constraint for every possible wrong prediction of the n examples in the training set $\mathcal{D}$, more precisely $\sum_{i=1}^{n}(|\mathcal{Y}(\mathbf{x}_i)| - 1)$. Unfortunately, the number of possible solutions typically involved in combinatorial objects, such as graphs, scales exponentially (or worse) with the size of the input, making the optimization intractable.

In order to deal with this high number of constraints, many approximation schemes have been proposed, where cutting plane algorithms or subgradient methods (e.g., [35, 36]) are among the most common. In particular, if for each example we rearrange all the constraints of QP (8.2) and focus on satisfying just the one requiring the highest ξ_i, we can define the structured hinge-loss as the highest classification penalty for a specific example:

$$\widetilde{H}(\mathbf{x}_i) \overset{\text{def}}{=} \max_{\mathbf{y} \in \mathcal{Y}} \Delta(\mathbf{y}_i, \mathbf{y}) - \mathbf{w}^T \delta \Psi_i(\mathbf{y}). \tag{8.3}$$

The computation of the structured hinge-loss for each element i of the training set amounts to finding the most "violating" output $\mathbf{y}^*$ for a given input $\mathbf{x}_i$ and its correct associated output $\mathbf{y}_i$. Eq. (8.3) suggests the violation resides in having simultaneously a high loss and a high compatibility, which is a contradiction by definition. We are now left with an unconstrained, nonsmooth version of QP (8.2),

$$\min_{\mathbf{w}} \quad \frac{1}{2}\|\mathbf{w}\|^2 + \frac{C}{n}\sum_{i=1}^{n}\max\{0, \widetilde{H}(\mathbf{x}_i)\}. \tag{8.4}$$

By disposing of a maximization oracle, i.e., a solver for Eq. (8.3), and a computed solution $\mathbf{y}^*$, subgradient methods can easily be applied to QP (8.4), yielding $\partial_{\mathbf{w}}\widetilde{H}(\mathbf{x}_i) = -\delta\Psi_i(\mathbf{y}^*)$.

To exploit the domain separability of the constraints and limit the number of oracle calls needed to converge to the optimal solution, we choose to adopt a Block-Coordinate version of the Frank–Wolfe algorithm (BCFW) [37], delineated in Algorithm 1.

The algorithm works by minimizing the objective function of Eq. (8.4), but restricted to a single random example at each iteration. By calling the max oracle upon the selected training sample (line 4), we obtain a new suboptimal parameter set $\mathbf{w}_s$ by simple derivation (line 5). The best update is then found through a closed-form line

---

Algorithm 1 Block-coordinate Frank–Wolfe algorithm.

1: Let $\mathbf{w}^{(0)}, \mathbf{w}_i^{(0)} := \mathbf{0}$ and $l^{(0)}, l_i^{(0)} := 0$

2: **for** $k \leftarrow 0$ **to** n_{it} **do**

3:     Pick i at random in $\{1, \ldots, n\}$

4:     Solve $\mathbf{y}^* \leftarrow \arg\max_{\mathbf{y} \in \mathcal{Y}} \Delta(\mathbf{y}_i, \mathbf{y}) - \mathbf{w}^T \delta \Psi_i(\mathbf{y})$

5:     Let $\mathbf{w}_s \leftarrow \frac{C}{n} \delta \Psi_i(\mathbf{y}^*)$ and $l_s := \frac{C}{n} \Delta(\mathbf{y}_i, \mathbf{y}^*)$

6:     Let $\gamma \leftarrow \frac{(\mathbf{w}_i^{(k)} - \mathbf{w}_s)^T \mathbf{w}^{(k)} + \frac{C}{n}(l_s - l_i^{(k)})}{\|\mathbf{w}_i^{(k)} - \mathbf{w}_s\|^2}$ and clip to $[0, 1]$

7:     Update $\mathbf{w}_i^{(k+1)} \leftarrow (1 - \gamma)\mathbf{w}_i^{(k)} + \gamma \mathbf{w}_s$ and $l_i^{(k+1)} := (1 - \gamma)l_i^{(k)} + \gamma l_s$

8:     Update $\mathbf{w}^{(k+1)} \leftarrow \mathbf{w}^{(k)} + \mathbf{w}_i^{(k+1)} - \mathbf{w}_i^{(k)}$ and $l^{(k+1)} := l^{(k)} + l_i^{(k+1)} - l_i^{(k)}$

9: **end for**

---

search (line 6), greatly reducing convergence time compared to other subgradient or cutting plane methods.

8.4 THE TOOLS OF THE TRADE IN SOCIAL AND STRUCTURED CROWD ANALYSIS

In this section we report and summarize authors' approaches to group detection [6] and leader identification [7] in crowds. Previous work has shown that the concept of *group* and *leader* cannot be uniquely specified, but varies according to the crowdedness, the environment, the cultural habits, and other factors that are difficult to detect and encode. As a consequence, learning seems an adequate paradigm to tackle social related computer vision tasks. At the same time, the methods presented in the remainder of this section have demonstrated good generalization ability, implying they can be applied off-the-shelf to scenarios similar to the ones used for training; otherwise a short training stage might be required.

Learning is accomplished through Structural SVM, introduced in Section 8.3. In Sections 8.4.1 and 8.4.2, we delineate the feature map and loss function for the tasks of group detection and leader identification. Eventually, these methods will be the two key ingredients employed in the investigation on the relationships between a leader behavior and the behavior of the rest of the group.

8.4.1 Socially Constrained Structural Learning for Groups Detection in Crowd

As previously described, modern crowd theories agree that collective behavior is the result of the underlying interactions among small groups of individuals and individuals (even if singletons are rarer in some kind of crowds than others). This is why detecting groups of socially connected pedestrians (social groups) already is a central topic in computer vision aided crowd analysis. Here we propose a solution for visually detecting groups in low/medium density crowds under the hypothesis that the *groups* can be visually discerned and people walking paths can be tracked up to some extent. In order to design a computational model, we rely on Turner's definition of groups as *two or more people interacting to reach a common goal and perceiving a shared membership, based on both physical and social identity* [1], and on a set of extracted features grounded on this definition.

8.4.1.1 Task Formulation

We cast the group detection task as a clustering problem. Consider a set of pedestrians $\mathcal{M} = \{a, b, \ldots\}$ and let $\mathcal{Y}(\mathcal{M})$ be the set of all possible ways to partition $\mathcal{M}$. Defining y as a subset of pedestrians (also referred to as a group or cluster) in $\mathcal{M}$, a generic set of subsets $\mathbf{y} = \{y_1, y_2, \ldots\}$ is a valid solution in $\mathcal{Y}(\mathcal{M})$ if the partitioning axioms are satisfied, i.e., $\forall a \in \mathcal{M}, \exists! y \in \mathcal{Y}(\mathcal{M}) : a \in y$ and $\cup_{y \in \mathcal{Y}(\mathcal{M})} y = \mathcal{M}$. There are two trivial partitioning solutions: one where all the pedestrians are clustered as a single group ($|\mathbf{y}| = 1$) and one where each pedestrian belongs to different clusters ($|\mathbf{y}| = |\mathcal{M}|$). In the remained of this section we call *singletons* those pedestrians who form clusters by themselves, i.e., $|y| = 1$.

We propose to solve the crowd partitioning problem by employing the *Correlation Clustering* (CC) [38]. The CC algorithm takes as input an affinity matrix W where, if $W^{ab} > 0$ ($W^{ab} < 0$), elements a and b belong to the same (different) cluster with certainty $|W^{ab}|$. The algorithm returns the partition $\mathbf{y}$ of a set of elements $\mathcal{M} = \{a, b, \ldots\}$ so that the sum of the affinities between item pairs in the same clusters y is maximized:

$$\text{CC} = \arg\max_{\mathbf{y} \in \mathcal{Y}(\mathcal{M})} \sum_{y \in \mathbf{y}} \sum_{a \neq b \in y} W_{\mathbf{d}}^{ab}. \tag{8.5}$$

The pairwise elements affinity in W is **w**-parametrized as a weighted linear combination of a bounded dissimilarity measure and its complement:

$$\begin{aligned} W_{\mathbf{d}}^{ab} &= \boldsymbol{\alpha}^T (\mathbf{1} - \mathbf{d}(a, b)) - \boldsymbol{\beta}^T \mathbf{d}(a, b) \\ &= \underbrace{-(\boldsymbol{\alpha} + \boldsymbol{\beta})^T \mathbf{d}(a, b)}_{\text{from distance to correlation}} + \underbrace{\boldsymbol{\alpha}^T \mathbf{1}}_{\text{threshold}} . \end{aligned} \tag{8.6}$$

To be consistent with Turner's definition of groups, we design the pairwise distance between pedestrian a and b, $\mathbf{d}(a, b)$, as a vector of pairwise distances built upon different aspects that concur to unveil the presence of groups. In detail, we measure the physical relation between pedestrians d_{ph}, their mutual influences in motion pattern by causality and trajectories shape analysis, d_{ca} and d_{sh}, and their simultaneous convergence to peculiar zones in the scene d_{he}, obtaining $\mathbf{d}(a, b) = [d_{ph}, d_{sh}, d_{ca}, d_{he}]$. For a deeper presentation and discussion about the employed features, please refer to author's previous work [6].

8.4.1.2 SSVM Adaptation to Group Detection

By tuning $\mathbf{w} = [\boldsymbol{\alpha}, \boldsymbol{\beta}]$ parameters in Eq. (8.6), we can evaluate many different groupings. In order to efficiently learn those parameters according to different peculiarities groups exhibit in different scenarios, we now introduce the feature map and a loss function specifically designed for accurately measuring the compatibility among possible crowd partitions.

Following the definition of correlation clustering in Eq. (8.5) and its parametrization introduced in Eq. (8.6), the compatibility of an input–output pair of Eq. (8.1), and thus the feature map $\Psi(\mathbf{x}, \mathbf{y})$, is directly described as

$$F(\mathbf{x}, \mathbf{y}; \mathbf{w}) = \mathbf{w}^T \Psi(\mathbf{x}, \mathbf{y}) = \mathbf{w}^T \sum_{y \in \mathbf{y}} \sum_{a \neq b \in y} [\mathbf{1} - \mathbf{d}(a, b), -\mathbf{d}(a, b)]. \qquad (8.7)$$

Inference and Max Oracle

Solving Correlation Clustering exactly is known to be NP-hard, and the problem is also hard to approximate [39]. To deal with this complexity, we adopt a standard greedy procedure [40] where, initially, all pedestrians have their own separate cluster and then, iteratively, the two clusters with the highest correlation are merged. The procedure stops when the best merge would decrease the overall correlation. A similar procedure can be devised for the loss augmented problem of Eq. (8.3) where, at each iteration, the two clusters to be merged are chosen according to both the correlation gain and the score of the loss function. Of course, by following a greedy procedure, there is no guarantee of selecting the most violated constraint. Interestingly enough, Lacoste-Julien et al. [37] show that all convergence results known for exact maximizer of the loss augmented problem also hold for approximate maximizers by allowing the algorithm to iterate longer. For further details, please refer to their work.

Loss Function

Due to efficiency constraints, the loss function $\Delta(\mathbf{y}_i, \mathbf{y})$ is usually required to decompose with respect to the output. Nevertheless, the iterative nature of aforementioned procedure allows for a more complex and problem-tailored choice. The proposed loss function is based on the *MITRE score* [41]. This score is founded on the understanding that connected components are sufficient to describe groups, inducing a linear amount of positive links (a tree that connects people belonging to the same group) and negative links (a tree that connects groups' connected components). Nevertheless, by working on links, the MITRE score fails to evaluate errors proportionally to the size of the crowd as the number of singletons varies (since singletons have no links). To further square the loss function to the group detection problem, for each pedestrian we add a fake counterpart to which only singletons are connected. Through this shrewdness we can now take into consideration singletons as well when computing the discrepancy between two solutions.

More formally, consider two clustering solutions, $\mathbf{y}_i$ and $\mathbf{y}$, and representatives of their respective spanning forests, Q and R. The connected components of Q and R are identified respectively by the set of trees $Q_1, Q_2, \ldots$ and $R_1, R_2, \ldots$. Note that if the number of elements in Q_j is $|Q_j|$, then only $c(Q_j) \stackrel{\text{def}}{=} |Q_j| - 1$ links are needed in order to create a spanning tree. Let us define $\pi_R(Q_j)$ as the partition of a tree Q_j with respect to the forest R, that is, the set of subtrees obtained by considering only the membership relations in Q_j also found in R. Besides, if R partitions Q_j in $|\pi_R(Q_j)|$ subtrees then $v(Q_j) \stackrel{\text{def}}{=} |\pi_R(Q_j)| - 1$ links are sufficient to restore the original tree. It follows that the recall error for Q_j can be computed as the number of missing links divided by the minimum number of links needed to create that spanning tree. Accounting for all trees Q_j the global recall measure of $\mathbf{y}_i$ is

$$\mathcal{R}(\mathbf{y}_i) = 1 - \frac{\sum_j v(Q_j)}{\sum_j c(Q_j)} = \frac{\sum_j |Q_j| - |\pi_R(Q_j)|}{\sum_j |Q_j| - 1}. \tag{8.8}$$

The precision of $\mathbf{y}_i$ (recall of $\mathbf{y}$) can be computed by exchanging Q and R. Given the definition of precision, recall and employing the standard F-score F_1, the loss is defined as

$$\Delta(\mathbf{y}_i, \mathbf{y}) = 1 - \frac{2\mathcal{R}(\mathbf{y}_i)\mathcal{R}(\mathbf{y})}{\mathcal{R}(\mathbf{y}_i) + \mathcal{R}(\mathbf{y})}. \tag{8.9}$$

8.4.2 Learning to Identify Group Leaders in Crowd

Once groups have been discovered, the crowd structure can be further investigated by discerning its leaders. To underline the key role of leaders in crowd analysis, we quote a famous case reported in Gustave Le Bon's *The Crowd* [2]:

> *During the last strike of the Parisian omnibus employees the arrest of the two leaders directing it was at once sufficient to bring it to an end.*

By finding and disconnecting the leaders from the rest of the crowd, efficient containment can be accomplished. At the same time, an influential voice of nonviolence in a crowd can lead to a mass sit-in, and a strong leader can take control of an emergency situation, initiate movement, and guide suitable crowd behavior avoiding panic [42]. Either way, leaders are key subjects to pay attention to when dealing with otherwise unmanageable crowds. Yet, not every crowd qualifies to have a unique influential leader. Many crowds are social gatherings and can be thought of as disconnected groups sharing the same location because of their similar goals, like families in a shopping mall. In such cases, each group has its own leader and it makes sense to restrict the analysis to one group at a time by considering it a *small crowd on its own*.

In this section, we investigate a computational model for the individuation of group leaders in crowded scenes – and of crowd leaders in the case of one group only. We deal with the lack of a formal definition of leadership by learning, in a supervised fashion, a metric space based exclusively on people spatiotemporal information and their partitioning into social groups.

8.4.2.1 Task Formulation

Leader identification amounts to finding the higher scored member for each group, given some leadership scoring criteria. Learning in such a setting might be difficult because evaluating results just by looking at the top scored element makes the objective nonsmooth with respect to the input. Nevertheless, it is easy to see that leader identification is lower bounded by leader ranking, where we care about predicting the complete leadership hierarchy. Of course, correctly predicting the leadership hierarchy will also predict the correct leader.

More formally, given as input a social group of size q, $\mathbf{x} = \{a, b, \ldots\} \in \mathcal{M}$, we want to predict its correct order $\mathbf{y}$ as a permutation of the first q natural numbers, among all possible orderings of that set $\mathcal{Y}(\mathbf{x})$. The leader of $\mathbf{x}$ is the jth member if $\mathbf{y}(j) = 1$. It is easy to see that the best ordering satisfies

$$\mathrm{PR} = \arg \max_{\mathbf{y} \in \mathcal{Y}(\mathbf{x})} -R(\mathbf{x})^T \mathbf{y}$$

$$= \arg \max_{\mathbf{y} \in \mathcal{Y}(\mathbf{x})} - \left((\mathbf{I} - dM)^{-1} \frac{1-d}{|\mathbf{x}|} \mathbf{1} \right)^T \mathbf{y}, \tag{8.10}$$

where R is a leadership scoring criterion, and – specifically to our case – the Page-Rank [43]. When the damping factor d is in $(0, 1)$ and $M = (D^{-1}G)^T$ is a column stochastic matrix, the PageRank has a unique solution. D is the outdegrees diagonal matrix of the graph $G = (\mathbf{x}, f)$. With a slight abuse of notation, G is also the graph matrix such that $G_{a,b} = f(a, b)$ where nodes $(a, b) \in \mathbf{x}$ are the group members and edges $f(a, b)$ encode the probability of b being a leader for a through the generic feature f. We choose the PageRank algorithm because of its ability to take advantage of the referential and asymmetric structure of the crowd model [5] to assess the importance of each member.

To let different social aspects intervene in the final ranking, we learn how to combine different contributions from different features. Unluckily, the nonlinear nature of PageRank does not allow us to learn how to combine features as edges of the same graph. Instead, we need to compute different PageRanks, each of which works on one graph with edges defined by a single feature, and then aggregate the ranks together. Formally, we generalize $R(\mathbf{x})$ to a linear combination of many PageRanks, each of which is computed on a different feature:

$$R(\mathbf{x}) = \sum_f \mathbf{w}^f \underbrace{\left((\mathbf{I} - dM^f)^{-1} \frac{1-d}{|\mathbf{x}|} \mathbf{1} \right)}_{R^f(\mathbf{x}): \text{ PageRank on feature } f}. \tag{8.11}$$

The features employed in Eq. (8.11) are pairwise time-lagged features leveraging on mutual position, relative speed, and DTW. Eventually, member centrality and group size are also considered. Details about the definition of these features can be found in authors' original work [7].

8.4.2.2 SSVM Adaptation to Leader Identification

During training each feature is used to produce a separate ranking of the members of the considered group and Structural SVM is employed to combine different features contribution. In this section we introduce the feature map and the loss function required for this learning to take place.

In order to define the feature map $\Psi(\mathbf{x}, \mathbf{y})$, that is, the compatibility of an input–output pair, let $R^*(\mathbf{x}) = [R^{f_1}, R^{f_2}, \ldots]$ be the column-wise concatenation of different ranks computed on the whole feature set. According to Eq. (8.10) and leveraging on the introduced parameterization, we specify Eq. (8.1) as

$$F(\mathbf{x}, \mathbf{y}; \mathbf{w}) = \mathbf{w}^T \Psi(\mathbf{x}, \mathbf{y}) = -\mathbf{w}^T R^*(\mathbf{x})^T \mathbf{y}. \tag{8.12}$$

Inference and Max Oracle

At test time, for each group, the algorithm returns a ranking of the members, among which the highest will be predicted as the leader. By looking at Eq. (8.10), it is easy to see that the ranking $\mathbf{y}$ maximizing the objective is the one representing the descending sort of $R(\mathbf{x})$. As a result, inference turns out to be extremely quick even for large groups. To extend the same inference procedure to the computation of the maximization oracle of Eq. (8.3), we need a loss that decomposes with respect to the output elements. Such a loss will be presented in the following paragraph.

Loss Function

The loss function $\Delta(\mathbf{y}, \mathbf{y}_i)$ should evaluate discrepancies between two predictions. Less obviously, the loss function is also a good place to store and employ prior knowledge. This is because SSVM basically learns how to mimic the loss function by looking merely at inputs (instead of outputs). In our case, the loss function should consider (i) the overall ranking and, even more importantly, (ii) the leader position. One way to handle these requirements is taking a sum of squared differences of predicted positions, i.e., $\Delta(\mathbf{y}, \mathbf{y}_i) = \|\mathbf{y}_i - \mathbf{y}\|^2$. This loss strongly penalizes members whose rank position is predicted further from the true one. With such a loss, all errors are taken into account, but the most costly ones involve members at the boundary of the ranking (first and last positions) as they can be moved further from their original positions. Thereby, it becomes crucial to be able to correctly predict these positions, while allowing some level of mistakes in the central positions of the hierarchy, where even humans might find the ranking task difficult.

Moreover, the proposed loss is linear w.r.t. the maximization argument $\mathbf{y}$, and a search for the most violating constraint can be accomplished as follows:

$$
\begin{aligned}
\mathbf{y}_i^* &= \arg \max_{\mathbf{y} \in \mathcal{Y}(\mathbf{x}_i)} \|\mathbf{y}_i - \mathbf{y}\|^2 - \mathbf{w}^T R^*(\mathbf{x})^T \mathbf{y} \\
&= \arg \max_{\mathbf{y} \in \mathcal{Y}(\mathbf{x}_i)} -(2\mathbf{y}_i + R^*(\mathbf{x})\mathbf{w})^T \mathbf{y},
\end{aligned}
\tag{8.13}
$$

by noting that $\|\mathbf{y}_i\|^2 = \|\mathbf{y}\|^2$ does not depend on the particular choice of $\mathbf{y}$. Through this shrewdness, the maximization oracle can be efficiently computed as in Eq. (8.12).

8.5 RESULTS ON VISUAL LOCALIZATION OF GROUPS AND LEADERS

For the group detection task, we selected two publicly available datasets, namely the *BIWI Walking Pedestrians* dataset [44] and the *Crowds-By-Examples (CBE)* dataset [45]. The former dataset records two low crowded scenes, outside a university and at a bus stop (eth and hotel). The *CBE* dataset records a medium density crowd outside another university (student003, briefly stu003) providing different challenges: the density of the pedestrians is significantly higher and multiple entry and exit points are present. While *BIWI* and *CBE* are standard datasets in crowd analysis, we also use the more recent *Vittorio Emanuele II Gallery (GVEII)* dataset [46].

Quantitative results of the SSVM approach, introduced in Section 8.4.1.2, are given in Table 8.1, while visual results are depicted in Fig. 8.3. To better appreciate the results of Structural SVM supervised clustering, we report accuracy in terms of both G-MITRE Δ_{GM} and the more intuitive pairwise loss Δ_{PW}^+ [47], which accounts only for positive (intra-group) relations. The slightly lower performances on the stu003 sequence are due to the higher complexity of the scene and higher density as well.

Leaders have been tested on a subset of the previous datasets, namely stu003 and GVEII. Sociologist manually provided leadership ground truth annotation. Visual examples of the datasets and the achieved results are shown in Fig. 8.3. Both stu003 and GVEII present mildly dense but highly group-structured crowds, characterized by the high variability of groups' size and motion patterns. In all scenarios approximately 65% of groups are pairs, but triplets and larger groups are present as well. For testing purposes, we employed the ground truth trajectories and group annotation as the input of the SSVM based algorithm of Section 8.4.2. The training of the SSVM is performed

Table 8.1 (Left) Quantitative results for visual group detection in terms of precision $\mathcal{P}$ and recall $\mathcal{R}$ computed according to Δ_{GM} and Δ_{PW}^+, respectively. (Right) Quantitative accuracy in leader identification

		hotel	eth	stu003	GVEII
Δ_{GM}	$\mathcal{P}$	97.3	91.8	81.7	73.1
	$\mathcal{R}$	97.7	94.2	82.5	74.3
Δ_{PW}^+	$\mathcal{P}$	89.1	91.1	82.3	70.9
	$\mathcal{R}$	91.9	83.4	74.1	71.1

	SSVM	SVM
GVEII groups $117 - 75 - 11$	83.2	67.4
stu003 groups $87 - 20 - 8$	82.3	78.3

(A) eth

(B) hotel

(C) stu003

(D) GVEII

Figure 8.3 Visual results on the employed datasets. Groups are identified by the color of the shape containing their members (first row) and leaders are marked with dots of pertinent colors (bottom row). (For interpretation of the references to color in this figure legend, the reader is referred to the web version of this chapter.)

independently on every video sequence on the first 20% of the groups. Leader identification accuracy results are reported in terms of binary classification, see Table 8.1. We also report a binary SVM baseline, where the leader in a group is the member, among the properly labeled ones, with the highest distance from the margin.

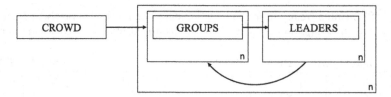

Figure 8.4 Scheme summarizing our different approaches in plate notation. Starting from the crowd, our group detections algorithm retrieves n clusters (i.e., groups) simultaneously. Then for each group, and separately from the others, we identify its leader. The last task, recover a group from its leader, cannot be accomplished separately for each leader, but requires taking all the leaders and finding all the groups simultaneously.

8.6 THE PREDICTIVE POWER OF LEADERS IN SOCIAL GROUPS

Until now we have taken a zoomed-in look at the crowd, going from the whole set of pedestrians to groups and from groups to leaders. In this section we start tackling the reverse problem or, more formally, whether it is possible to recover groups just by knowing their leaders. The final aim of this investigation is a better understanding of the relationship between groups and their leaders, measured as the group detection performance improvement when leaders are used as prior knowledge. Eventually, under the proper investigative tools, we hope to shed some light onto these computationally unexplored questions.

8.6.1 Experimental Settings

Going from leaders to groups is a complex task, mainly due to (i) the noninvertible nature of PageRank, and (ii) the requirement that groups should be recovered simultaneously and not separately, as in the case of leaders. Considering leaders independently would create inconsistent partitions, dependent on the order followed to visit each leader. As a consequence, it is important to approach the problem by considering all leaders as input and by output, with global constraints, all respective groups. This requirement, together with the iterative nature of PageRank, makes going from groups to leader a noninvertible path. Fig. 8.4 briefly captures the different approaches and the local or global level at which they operate.

Building on the fact that the leaders-to-groups task can be seen as a group detection problem with prior knowledge on the leaders, we propose to computationally tackle it through clustering with seeds. As already stated in Section 8.4.1, clustering is indeed a proper choice for group detection. By following the same reasoning, we employ both the same features used for groups detection and the same − already learned − metric space. However, Correlation Clustering cannot be adopted because it doesn't support constraining on cluster seeds. In the rest of this section we experiment and validate the k-medoids and the nearest neighbor (NN) clustering algorithms, where leaders

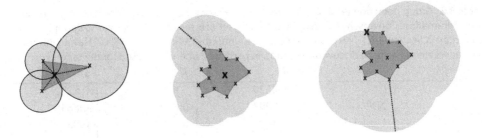

Figure 8.5 *k*-Medoids requirements to guarantee a correct clustering. See text for details.

can play a role in the initialization step of the clustering. Moreover, as a guarantee for experiments to be sound, we only consider groups with at least three members so that leaders can be properly identified.

8.6.2 Leader Centrality in Feature Space

We start by questioning whether a simple *k*-medoids approach, with cluster medoids initialized on leaders, is able to recover the groups which generated those leaders. The *k*-medoids partitions members trying to minimize the distance between points labeled to be in a cluster and the medoid point of that cluster. In the case of distances (Euclidean or proxemics) on the ground plane, sociologists agree on that the leader doesn't necessarily have to be in the center of its group. We rephrase this question by forcing *k*-medoids to use our learned distance in the features space from Section 8.4.1.

8.6.2.1 Group Recovery Guarantees

Finding the optimal partition of a multivariate set of data with more than two clusters is known to be NP-hard. In practice, *k*-medoids is solved fast through Lloyds' algorithm, which iteratively alternates between estimating clusters and their medoids. Under strong hypothesis on the data, applying this iterative algorithm always yields the correct results. Fig. 8.5 depicts this hypothesis, better explored below.

Let us initialize a clustering seed for each leader. Since we assume to know the leader, we end up having a leader for each unknown group. Now, define *safe-zone* for each member as the region (in feature space) required by the cluster to avoid containing other initializing seed. This space is a hypersphere in feature space and its radius is determined by the distance of that member from the seed belonging to his group. Moreover, define *safe-zone* for a group the overlap of all safe-zones from all members of that group. It is important to note that these safe-zones strongly depend on the initialization seeds.

In this setting the correctness guarantee is data-dependent and not seed-dependent. It is obtained by creating for every group the *worst case safe-zone*, that is, a group safe-

Table 8.2 Group detection results. Under Nearest Neighbor (NN) and k-medoids, leaders are used as prior information. Conversely, Correlation Clustering reports results of the crowd-to-groups approach presented in Section 8.4.1. Precision and Recall are computed as in Section 8.4.1.2. Having information on leaders always helps in discovering groups, in particular when NN is used for the task

		$\mathcal{P}$	$\mathcal{R}$	$1 - \Delta_{GM}$
stu003	CC	0.8329	0.8227	0.8272
	NN	0.8997	0.8708	0.8848
	k-medoids	0.8434	0.82112	0.8319
GVEII	CC	0.7752	0.7932	0.7834
	NN	0.9601	0.9528	0.9563
	k-medoids	0.8825	0.8690	0.8755

zone where members contribute through the distances from their furthest member. Note that the worst case safe-zone does not depend from the cluster seed but from members mutual arrangement. Consequently, the correct cluster can be retrieved given any member, not necessarily the leader. Moreover, if the leader is also the medoid of its group, the clustering terminates in one iteration.

8.6.2.2 Validation and Results

Of course, it rarely happens that we can assure the worst case safe-zone to be empty. In such cases, it is important to study how the group safe-zone changes with respect to the initialization seed, i.e., the leader in our setting. As shown in Fig. 8.5, a central initialization yields a more homogeneous and averagely smaller group safe-zone. Oppositely, having an initialization on the border of the cluster makes the safe-zone much smaller in some parts as well as much larger in others. Ideally, a central initialization is to be preferred.

In Table 8.2 we report results from group detection experiments when leaders were used as initializing seeds for NN and k-medoids. To better understand these results, we correlate them with the concept of leader *centrality*, defined as the one-complement normalized distance between the leader and the group centroid. Intuitively, a more central leader should favor a better group recovery. More formally, centrality is computed as follows:

$$c(\mathbf{x}_l, \mathbf{x}_c) = 1 - \frac{\|\mathbf{x}_l - \mathbf{x}_c\|}{\max_m \|\mathbf{x}_m - \mathbf{x}_c\|}, \tag{8.14}$$

with $\mathbf{x}_l$ and $\mathbf{x}_c$ being the leader and centroid position, respectively, and $\mathbf{x}_m$ the position of any other member of that group. Fig. 8.6, which depicts group detection correctness, leader centrality and safe-zone violation, confirms that: (i) as long as no safe-zone violation occurs, the leader can always recover its group, and (ii) as the safe-zone is violated leader centrality in feature space starts to play a discriminant role. Interestingly,

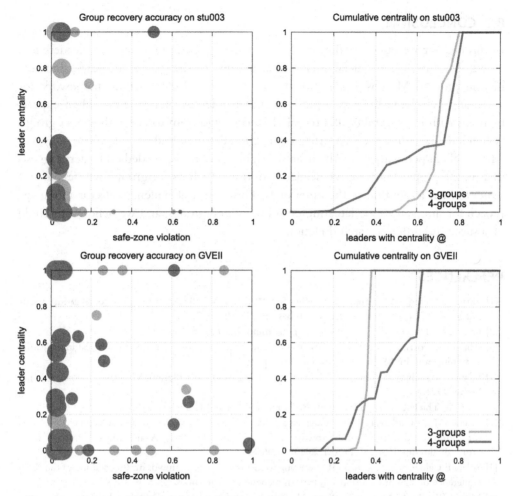

Figure 8.6 (Left) The recall ability of *k*-medoids is tested on `stu003` and `GVEII` (larger dots mean higher recall). If the safe-zone is not violated, leader centrality doesn't influence the recall. As the violation increases, leader centrality plays an important role. (Right) Cumulative distribution of leader with centrality up to a specific value. These figures are obtained by summing points on the scatter plot on the left from bottom to top, always increasing the centrality threshold up to which leaders are considered. Faster growing plots indicate more frequent leader centrality.

from Table 8.2, we observe that having prior information on the leader always brings improvements to group detection performance – particularly in the case of fixed seeds (1 iteration *k*-medoids, equivalent to NN). While *k*-medoids can shift the clusters' center, moving *leaders* from their original groups to higher density zones of the crowd, nearest neighbor preserves leader locality, thereby reaching higher scores. This experimental evidence confirms that knowledge about leaders can significantly boost group detection accuracy, suggesting these problems should be jointly approached.

8.7 CONCLUSION

In this chapter we discussed the problem of visually detecting groups and leaders in a crowd. We briefly introduced sociological insights to the group and leadership formation process. Moreover, we provided a single structural learning framework for automatically solving these two problems independently. Nevertheless, when experimenting with the capability of a trivial clustering algorithm to correctly detect groups (when initialized on leaders), a strong connection between leaders and groups emerged. The excellent performance of the nearest neighbor clustering seeded on leaders suggests that, in the proper feature space, leaders' positioning allows for group members to stay closer to their leader than to the other ones. This empirical evidence calls for a joint approach for simultaneously detecting both leaders and groups, and we believe this could be a successful future research direction.

REFERENCES

[1] Turner JC, Hogg MA, Oakes PJ, Reicher SD, Wetherell MS. Rediscovering the social group: a self-categorization theory. Cambridge (MA): Basil Blackwell; 1987.

[2] Le Bon G. The crowd: a study of the popular mind. Macmillan; 1896.

[3] Bandini S, Gorrini A, Manenti L, Vizzari G. Crowd and pedestrian dynamics: empirical investigation and simulation. In: Measuring behavior. 2012. p. 308–11.

[4] Freud S. Massenpsychologie und Ich-Analyse. Die Zukunft einer Illusion. Vienna: Psychoanalytischer Verlag; 1921.

[5] Tarde G. The laws of imitation. New York: Henry Holt and Company; 1903.

[6] Solera F, Calderara S, Cucchiara R. Socially constrained structural learning for groups detection in crowd. IEEE Trans Pattern Anal Mach Intell 2015. http://dx.doi.org/10.1109/TPAMI.2015.2470658.

[7] Solera F, Calderara S, Cucchiara R. Learning to identify leaders in crowd. In: Proceedings of the IEEE conference on computer vision and pattern recognition workshops. 2015. p. 43–8.

[8] Moore BE, Ali S, Mehran R, Shah M. Visual crowd surveillance through a hydrodynamics lens. Commun ACM 2011;54:64–73.

[9] Ingham AG, Levinger G, Graves J, Peckham V. The Ringelmann effect: studies of group size and group performance. J Exp Soc Psychol 1974;10:371–84.

[10] Helbing D, Molnár P. Social force model for pedestrian dynamics. Phys Rev E 1995;51:4282–6.

[11] Moussaid M, Perozo N, Garnier S, Helbing D, Theraulaz G. The walking behaviour of pedestrian social groups and its impact on crowd dynamics. PLoS ONE 2010;5.

[12] Reicher SD, Spears R, Postmes T. A social identity model of deindividuation phenomena. Eur Rev Soc Psychol 1995;6:161–98.

[13] McPhail C. The myth of the madding crowd. Transaction Publishers; 1991.

[14] Manenti L, Manzoni S, Vizzari G, Ohtsuka K, Shimura K. An agent-based proxemic model for pedestrian and group dynamics: motivations and first experiments. In: Multi-agent-based simulation XII. Lect Notes Comput Sci. Berlin, Heidelberg: Springer; 2012. p. 74–89.

[15] Calderara S, Cucchiara R. Understanding dyadic interactions applying proxemic theory on video surveillance trajectories. In: Proc. IEEE int'l conference on computer vision and pattern recognition workshops (CVPRW). 2012. p. 20–7.

[16] Cristani M, Paggetti G, Vinciarelli A, Bazzani L, Menegaz G, Murino V. Towards computational proxemics: inferring social relations from interpersonal distances. In: Proc. IEEE int'l conference on social computing. 2011. p. 290–7.

[17] Kendon A. Conducting interaction: patterns of behavior in focused encounters. Cambridge University Press; 1990.

[18] Cristani M, Bazzani L, Paggetti G, Fossati A, Tosato D, Del Bue A, Menegaz G, Murino V. Social interaction discovery by statistical analysis of F-formations. In: Proc. British machine vision conference (BMVC). 2011. p. 1–12.

[19] Setti F, Russell C, Bassetti C, Cristani M. F-formation detection: individuating free-standing conversational groups in images. PLoS ONE 2015;10(5):e0123783.

[20] Wang YD, Wuand JK, Kassim AA, Huang WM. Tracking a variable number of human groups in video using probability hypothesis density. In: Proc. int'l conference on pattern recognition (ICPR). 2006. p. 1127–30.

[21] Feldmann M, Fränken D, Koch W. Tracking of extended objects and group targets using random matrices. IEEE Trans Signal Process 2011;59:1409–20.

[22] Lin WC, Liu Y. A lattice-based MRF model for dynamic near-regular texture tracking. IEEE Trans Pattern Anal Mach Intell 2007;29:777–92.

[23] Pang SK, Li J, Godsill S. Detection and tracking of coordinated groups. IEEE Trans Aerosp Electron Syst 2011;47:472–502.

[24] Bazzani L, Zanotto M, Cristani M, Murino V. Joint individual–group modeling for tracking. IEEE Trans Pattern Anal Mach Intell 2015;37(4):746–59.

[25] Rodriguez M, Laptev I, Sivic J, Audibert JY. Density-aware person detection and tracking in crowds. In: Proc. int'l conference on computer vision (ICCV). 2011. p. 2423–30.

[26] Pellegrini S, Ess A, Van Gool L. Improving data association by joint modeling of pedestrian trajectories and groupings. In: Proc. European conference on computer vision (ECCV). 2010. p. 452–65.

[27] Yamaguchi K, Berg A, Ortiz L, Berg T. Who are you with and where are you going? In: Proc. IEEE int'l conference on computer vision and pattern recognition (CVPR). 2011. p. 1345–52.

[28] Chang MC, Krahnstoever N, Ge W. Probabilistic group-level motion analysis and scenario recognition. In: Proc. int'l conference on computer vision (ICCV). 2011. p. 747–54.

[29] Andersson M, Gudmundsson J, Laube P, Wolle T. Reporting leaders and followers among trajectories of moving point objects. GeoInformatica 2007;12(4).

[30] Yu T, Lim SN, Patwardhan K, Krahnstoever N. Monitoring, recognizing and discovering social networks. In: Conference on computer vision and pattern recognition. 2009. p. 1462–9.

[31] Carmi A, Mihaylova L, Septier F, Pang SK, Gurfil P, Godsill S. MCMC-based tracking and identification of leaders in groups. In: International conference on computer vision workshops. 2011. p. 112–9.

[32] Kjargaard M, Blunck H, Wustenberg M, Gronbask K, Wirz M, Roggen D, Troster G. Time-lag method for detecting following and leadership behavior of pedestrians from mobile sensing data. In: International conference on pervasive computing and communications. 2013. p. 56–64.

[33] Sanchez-Cortes D, Aran O, Mast M, Gatica-Perez D. A nonverbal behavior approach to identify emergent leaders in small groups. IEEE Trans Multimed 2012;14(3):816–32.

[34] Tsochantaridis I, Joachims T, Hofmann T, Altun Y. Large margin methods for structured and interdependent output variables. J Mach Learn Res 2005:1453–84.

[35] Joachims T, Finley T, Yu CNJ. Cutting-plane training of structural SVMs. Mach Learn 2009;77(1):27–59.

[36] Shalev-Shwartz S, Zhang T. Accelerated proximal stochastic dual coordinate ascent for regularized loss minimization. Math Program 2016;155(1–2):105–45.

[37] Lacoste-Julien S, Jaggi M, Schmidt M, Pletscher P. Block-coordinate Frank–Wolfe optimization for structural SVMs. Available from arXiv:1207.4747, 2012.

[38] Bansal N, Blum A, Chawla S. Correlation clustering. Mach Learn 2004;56(1–3):89–113.

[39] Tan J. A note on the inapproximability of correlation clustering. Inf Process Lett 2008. http://dx.doi.org/10.1016/j.ipl.2008.06.004.

[40] Finley T, Joachims T. Supervised clustering with support vector machines. In: Proceedings of the 22nd international conference on machine learning. ACM; 2005. p. 217–24.

[41] Vilain M, Burger J, Aberdeen J, Connolly D, Hirschman L. A model-theoretic coreference scoring scheme. In: Proceedings of the 6th conference on message understanding. Association for Computational Linguistics; 1995. p. 45–52.

[42] Challenger W, Clegg W, Robinson A. Understanding crowd behaviours: guidance and lessons identified. UK Cabinet Office; 2009.

[43] Page L, Brin S, Motwani R, Winograd T. The PageRank citation ranking: bringing order to the web. Technical report. Stanford InfoLab; 1999.

[44] Pellegrini S, Ess A, Schindler K, Van Gool L. You'll never walk alone: modeling social behavior for multi-target tracking. In: Proc. IEEE int'l conference on computer vision and pattern recognition (CVPR). 2009. p. 261–8.

[45] Lerner A, Chrysanthou Y, Lischinski D. Crowds by example. Comput Graph Forum 2007;26:655–64.

[46] Bandini S, Gorrini A, Vizzari G. Towards an integrated approach to crowd analysis and crowd synthesis: a case study and first results. Pattern Recognit Lett 2014;44:16–29.

[47] Zanotto M, Bazzani L, Cristani M, Murino V. Online Bayesian non-parametrics for social group detection. In: Proc. British machine vision conference (BMVC). 2012. p. 111.1–12.

CHAPTER 9

Learning to Predict Human Behavior in Crowded Scenes

Alexandre Alahi, Vignesh Ramanathan, Kratarth Goel, Alexandre Robicquet, Amir A. Sadeghian, Li Fei-Fei, Silvio Savarese

Stanford University, Stanford, CA, USA

Contents

9.1 INTRODUCTION

Humans are much more predictable in their transit patterns than we expect. In the presence of sufficient observations, it has been shown that our mobility is highly predictable even at a city-scale level [1]. The location of a person at any given time can be predicted with an average accuracy of 93% supposing 3 km^2 of uncertainty. How about at finer resolutions such as in shopping malls, in airports, or within train terminals for safety or resource optimization? What are the relevant cues to best predict human behavior within a margin of a few centimeters?

Recently, Kitani et al. [2] showed that scene semantics provide strong cues for forecasting pedestrians' trajectories. Helbing et al. [3,4] also showed that our mobility is influenced by our neighbors, either consciously, e.g., by relatives or friends, or even unconsciously, e.g., by following an individual to facilitate navigation. More broadly, when humans walk in a crowded public space such as a train terminal, mall, or city center, they obey a large number of (unwritten) common sense rules and comply with social conventions. For instance, as they consider where to move next, they respect personal space and yield right-of-way. The ability to model these "rules" and use them to understand and predict human motion in complex real world environments is extremely valuable for a wide range of applications from the deployment of socially-aware robots [5] to the design of intelligent tracking systems [6] in smart environments.

In this chapter, we present two families of methods to forecast human trajectories in crowded environments. The first one is based on the popular Social Forces model [3] where the causalities behind human navigation is hand-designed by a set of functions that have been carefully chosen based on our understanding of physics underlying social behavior. The second method is a fully data-driven approach based on Recurrent Neural Networks [7] that does not impose any hand-designed functions or explicit mobility based constraints.

The causality behind human mobility is an interplay between both observable and non-observable cues (e.g., intentions). Humans have the innate ability to "read" one another. When they need to avoid each other, there is an implicit cooperation on where to move next. They have the ability to get along well with each other by preserving a personal distance. These capabilities are often referred to as *Social Intelligence* [8]. Any forecasting method needs to infer the same behaviors to develop socially-aware intelligent systems. This requires understanding the complex and often subtle interactions that take place between people in crowded spaces.

In the reminder of this chapter, after presenting relevant works in forecasting human behavior (while sharing more details on the popular Social Forces model [3]), we present a novel characterization of humans that describes the "*social sensitivity*" at which two humans interact. It captures both the preferred distance an individual wants to preserve with respect to her surrounding and the necessity to avoid collision. Low values for the social sensitivity feature implies that individual motion is not affected by other interacting neighbors. High values for the social sensitivity feature mean that individual navigation is highly dependent on the position of other people. This characterization allows us to define the "*navigation style*" humans follow while interacting with their surrounding. We obtain different classes of navigation styles by clustering trajectory samples in the *social sensitivity space* (see Fig. 9.2 for examples). This allows increasing the flexibility in characterizing various modalities of interactions, for instance, some

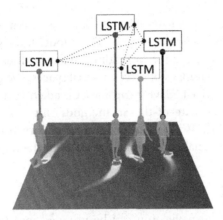

Figure 9.1 The goal of this chapter is to predict the motion dynamics in crowded scenes. This is, however, a challenging task as the motion of each person is typically affected by their neighbors. After presenting relevant methods to solve the forecasting task, we describe in Section 9.4.1 a new model which we call "Social" LSTM (Social-LSTM) which can jointly predict the paths of all the people in a scene by taking into account the common sense rules and social conventions that humans typically utilize as they navigate in shared environments. The predicted distribution of their future trajectories is shown in the heatmap.

pedestrians who are rushed may appear more "aggressive" whereas others might exhibit a milder behavior because they are just enjoying their walk. Navigation style classes are used to select the appropriate set of parameters for the Social Forces model to improve prediction of human trajectories.

The ability to model social sensitivity is a key step toward learning common sense conventions based on social etiquette for enhancing forecasting tasks. However, this approach still depends on hand-crafted functions to model "interactions" for specific settings rather than inferring them in a data-driven fashion. This results in favoring models that capture simple interactions (e.g., repulsion/attraction) and might fail to generalize for more complex crowded settings. It also focuses on modeling interactions among people in close proximity to each other (to avoid immediate collisions). It does not anticipate interactions that could occur in the more distant future. Consequently, we end the chapter by presenting a data-driven architecture for predicting human trajectories in the future. Inspired by the success of Long Short-Term Memory networks (LSTM) for different sequence prediction tasks such as handwriting [7] and speech [9] generation, we extend them for human trajectory prediction as well. While LSTMs have the ability to learn and reproduce long sequences, they do not capture dependencies between multiple correlated sequences. We address this issue through a novel architecture which connects the LSTMs corresponding to nearby sequences (see Fig. 9.1). In particular, we introduce a "Social" pooling layer which allows the LSTMs of spatially proximal sequences to share their hidden-states with each other. This architecture, which we refer to as the "Social-LSTM", can automatically learn typical interactions that take place among trajectories which coincide in time. This model leverages existing human trajectory datasets without the need for any additional annotations to learn common sense rules and conventions that humans observe in social spaces. We conclude the chapter by demonstrating that the Social-LSTM is capable of predicting trajectories of pedestrians much more accurately than state-of-the-art methods on two publicly available datasets: ETH [10], and UCY [11]. We also analyze the trajectory patterns generated by our model to understand the social constraints learned from the trajectory datasets.

9.2 RELATED WORK

Methods to forecast human navigation can be grouped into two categories: the ones modeling human–human interactions, and the ones focusing on human–space interactions. We briefly present an overview of past works for both approaches. We also discuss relevant Recurrent Neural Network (RNN) models for sequence prediction tasks.

9.2.1 Human–Human Interactions

Pioneering work from Helbing and Molnar [3] presented a pedestrian motion model with attractive and repulsive forces referred to as the *Social Force* model. This has been shown to achieve competitive results even on modern pedestrian datasets [11,10]. This method was later extended to robotics [5] and activity understanding [6,12–17].

Similar approaches have been used to model human–human interactions with strong priors for the model. Treuille et al. [18] use continuum dynamics, Antonini et al. [19] propose a Discrete Choice framework, and Wang et al. [20] and Tay et al. [21] use Gaussian processes. Such functions have also been used to study stationary groups [22, 23]. These works target smooth motion paths and do not handle the problems associated with discretization.

Another line of work uses well-engineered features and attributes to improve tracking and forecasting. Alahi et al. [24] present a social affinity feature by learning from human trajectories in crowd their relative positions, while Yu et al. [22] propose the use of human attributes to improve forecasting in dense crowds. They also use an agent-based model similar to [25]. Rodriguez et al. [26] analyze videos with high-density crowds to track and count people.

Most of these models provide hand-crafted energy potentials based on relative distances and rules for specific scenes. In contrast, we propose a method to learn human–human interactions in a more generic data-driven fashion.

9.2.2 Activity Forecasting

Activity forecasting models try to predict the motion and/or action to be carried out by people in a video. A large body of work learns motion patterns through clustering trajectories [27–30]. More approaches can be found in [31–36]. Kitani et al. in [37] use *Inverse Reinforcement Learning* to predict human paths in static scenes. They infer walkable paths in a scene by modeling human–space interactions. Walker et al. in [38] predict the behavior of generic agents (e.g., a vehicle) in a visual scene given a large collection of videos. Ziebart et al. [39,40] present a planning based approach.

Turek et al. [41,42] use a similar idea to identify the functional map of a scene. Other approaches like [43–46] show the use of scene semantics to predict goals and paths for human navigation. Scene semantics has also been used to predict multiple object dynamics [47,46,33,48]. These works are mostly restricted to static scene information usage to predict human motion or activity. In our work, we focus on modeling dynamic crowd interactions for path prediction.

More recent works have also attempted to predict future human actions. In particular, Ryoo et al. [49–54] forecast actions in streaming videos. More relevant to our work is the idea of using an RNN model to predict future events in videos [55–59]. Along similar lines, we predict future trajectories in scenes.

9.2.3 RNN Models for Sequence Prediction

Recently Recurrent Neural Networks (RNN) and their variants, including Long Short-Term Memory (LSTM) [60] and Gated Recurrent Units [61], have proven to be very successful for sequence prediction tasks: speech recognition [9,62,63], caption generation [64–68], machine translation [69], image/video classification [70–73], human dynamics [74], to name a few. RNN models have also proven to be effective for tasks with densely connected data such as semantic segmentation [75], scene parsing [76], and even as an alternative to Convolutional Neural Networks [77]. These works show that RNN models are capable of learning the dependencies between spatially correlated data such as image pixels. This motivates us to extend the sequence generation model from Graves et al. [7] to our setting. In particular, Graves et al. [7] predict isolated handwriting sequences, while in our work we jointly predict multiple correlated sequences corresponding to human trajectories.

9.3 FORECASTING WITH SOCIAL FORCES MODEL

We first present the popular Social Forces model [12] to forecast human trajectory. In this section, we introduce the basic theory behind the model and how to adapt it to multi-class settings. The model is also our inspiration for our *social sensitivity* feature described in Section 9.3.2.

9.3.1 Basic Theory

The Social Forces model is commonly used to predict trajectories of pedestrians in a crowded environment. In this model pedestrians are viewed as decision making agents who consider a multitude of personal, social and environmental factors to decide where to go next. Each agent makes a decision on the velocity $\mathbf{v}_i^{(t+\Delta t)}$. At each time step t, the object i is defined by a state variable $s_i^{(t)} = \{\mathbf{p}_i^{(t)}, \mathbf{v}_i^{(t)}, u_i^{(t)}, \mathbf{g}_i^{(t)}, A_i^{(t)}\}$, where $\mathbf{p}_i^{(t)}$ is the position, $\mathbf{v}_i^{(t)}$ the velocity, $u_i^{(t)}$ the preferred speed (according to the class and the past velocities), $\mathbf{g}_i^{(t)}$ the chosen destination (or goal), and $A_i^{(t)}$ is the set of objects in the same social group (including i). Similar to [12], the energy function, E_Θ, associated to every single agent is defined as

$$E_\Theta(\mathbf{v}; s_i, \mathbf{s}_{-\mathbf{i}}) = \lambda_0 E_{damping}(\mathbf{v}; s_i) \tag{9.1}$$

$$+ \lambda_1 E_{speed}(\mathbf{v}; s_i) \tag{9.2}$$

$$+ \lambda_2 E_{direction}(\mathbf{v}; s_i) \tag{9.3}$$

$$+ \lambda_3 E_{attraction}(\mathbf{v}; s_i, \mathbf{s}_{A_i}) \tag{9.4}$$

$$+ \lambda_4 E_{group}(\mathbf{v}; s_i, \mathbf{s}_{A_i}) \tag{9.5}$$

$$+ E_{collision}(\mathbf{v}; s_i, \mathbf{s}_{-\mathbf{i}} | \sigma_d, \sigma_w, \beta) \tag{9.6}$$

where $\Theta = \{\lambda_0, \lambda_1, \lambda_2, \lambda_3, \lambda_4, \sigma_d, \sigma_w, \beta\}$ is the model parameter set, $\mathbf{s}_{A_i}$ is the set of state variables of the agent in i's social group A_i. $\mathbf{s}_{-i}$ denotes the set of states of agents excluding agent i. The parameters λ_i are then weights to balance the importance of each of the energies $(E.)$. More details on the definition of each energy can be found in [12]. In our work, we use the collision energy to define our social sensitivity feature in Section 9.3.2. Consequently, we will describe the parameters $\{\sigma_d, \sigma_w, \beta\}$ in Section 9.3.2.

Previous works [12,13,5] only use one set of parameters for the whole crowd. This approximation implies that everyone maintains the same safety distance or grants the exact same weight to each energy function. We can easily see that someone in a hurry would be more likely to bump into or navigate close to others in order to navigate faster, granting more weight to his/her damping energy in order to go as straight as possible to his/her destination.

9.3.2 Modeling Social Sensitivity

We claim that modeling human trajectory with a single navigation style is not suitable for capturing a variety of social behaviors that targets exhibit when interacting in complex scenes. We believe that conditioning such models on *navigation style* (i.e., the way targets avoid each other) is a better idea and propose a characterization (feature) which we call *social sensitivity*. Given this characterization, we hence assign a navigation style to each target to better forecast its trajectory and improve tracking.

9.3.2.1 Social Sensitivity Feature

Inspired by the Social Forces model (SF) [12], we model targets' interactions with an energy potential E_{ss}. A high potential means that the target is highly sensitive to others. We define E_{ss} as follows:

At each time step t, the target i is defined by a state variable $s_i^{(t)} = \{\mathbf{p}_i^{(t)}, \mathbf{v}_i^{(t)}\}$, where $\mathbf{p}_i^{(t)}$ is the position, and $\mathbf{v}_i^{(t)}$ the velocity. The energy potential encoding the social sensitivity is computed as follows:

$$E_{ss}(\mathbf{v_i^{(t)}}; s_i, \mathbf{s}_{-i} | \sigma_d, \sigma_w, \beta) = \sum_{j \neq i} w(s_i, s_j) \exp\left(-\frac{d^2(\mathbf{v}, s_i, s_j)}{2\sigma_d^2}\right), \tag{9.7}$$

with $w(s_i, s_j)$ defined as

$$w(s_i, s_j) = \exp\left(-\frac{|\Delta\mathbf{p}_{ij}|}{2\sigma_\omega}\right) \cdot \left(\frac{1}{2}\left(1 - \frac{\Delta\mathbf{p}_{ij}}{|\Delta\mathbf{p}_{ij}|}\frac{\mathbf{v}_i}{|\mathbf{v}_i|}\right)\right)^\beta \tag{9.8}$$

and

$$d^2(\mathbf{v}, s_i, s_j) = \left|\Delta\mathbf{p}_{ij} - \frac{\Delta\mathbf{p}_{ij}(\mathbf{v} - \mathbf{v}_j)}{|\mathbf{v} - \mathbf{v}_j|^2}(\mathbf{v} - \mathbf{v}_j)\right|. \tag{9.9}$$

The energy E_{ss} is modeled as a product of Gaussians where the variances $\sigma_{w,d}$ represent the distances at which other targets will influence each other. For instance, if two targets i, j are close to each other ($\Delta\mathbf{p}_{ij}$ is small), E_{ss} will be large when $\sigma_{w,d}$ are small.

We define the parameter $\Theta_{ss} = \{\sigma_d, \sigma_w, \beta\}$ as the social sensitivity feature and interpret its dimension as follows:

- σ_d is the preferred distance a target maintains to avoid collision,
- σ_w is the distance at which a target reacts to prevent a collision (distance at which (s)he starts deviating from its linear trajectory),
- and β controls the peakiness of the weighting function.

In other words, the parameters $\{\sigma_d, \sigma_w, \beta\}$ aim at describing how targets avoid each other, i.e., their social sensitivity. We now present how we infer the parameters Θ_{ss} at training and testing time.

9.3.2.2 Training

At training time, since we observe all targets' velocities, V^{train}, we could learn a unique set of parameters, i.e., a single value for social sensitivity, that minimizes the energy potential as follows (similarly to what previous methods do [12,13,15,16]):

$$\{\sigma_d, \sigma_w, \beta\} = \underset{\{\sigma_d, \sigma_w, \beta\}}{\operatorname{argmin}} \left(\sum_{i=1}^{T-1} E_{ss}(v_i^{train}, s_i, s_{-i} | \sigma_d, \sigma_w, \beta)\right), \tag{9.10}$$

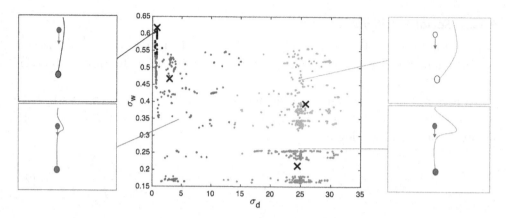

Figure 9.2 Illustration of the social sensitivity space where we have illustrated how targets avoid each other with four navigation styles (from a top view). Each point in the middle plot is a target. The x-axis shows the preferred distance σ_d a target keeps with its surrounding targets, and the y-axis provides the distance σ_w at which a target reacts to prevent a collision. Each color code represents a cluster (a navigation style). Even if our approach can handle an arbitrary number of classes, we only use 4 clusters for illustration purposes. In this plot, the green cluster represents targets with a mild behavior, willing to avoid other targets as much as possible and considering them from afar. The red cluster describes targets with a more aggressive behavior and with a very small safety distance. We illustrate on the sides of the plot examples of how targets follow different strategies in avoiding each other as different navigation styles are used. (For interpretation of the references to color in this figure legend, the reader is referred to the web version of this chapter.)

where T is the number of targets in the training data. This minimization is operated with an interior-point method and is set with the following constraint on σ_d: $\sigma_d > 0.1$ (it specifies that every target can't have a "vital space" smaller than 10 cm).

However, as mentioned previously, we claim that learning a unique set of parameters is not suitable when one needs to deal with complex multi–class target scenarios whereby targets can have different social sensitivity. To validate this claim, we visualize (Fig. 9.2) each target in a *social sensitivity space* where the x-axis contains the σ_d values and the y-axis shows the values of σ_w. This plot is generated using training images from our dataset (see Section 9.5 for more details). We did not plot the third parameter β since it does not change much across targets. Even though our approach can handle an arbitrary number of classes, we cluster the points into only four clusters for the ease of illustration. Each cluster corresponds to what we define as a "navigation style." A navigation style describes the sensitivity of a target to its surrounding. We illustrate on the sides of Fig. 9.2 how targets follow different strategies in avoiding each other as different navigation styles are used.

Thanks to the above analysis of the *social sensitivity space*, at training, we solve Eq. (9.10) for each target – without the summation over all targets – to get its social sensitivity feature. We then cluster the points with K-means clustering to have N clusters. Each cluster represents a navigation style.

9.3.2.3 Testing

At test time, we observe the targets until time t, and want to assign a navigation style.

In the presence of other targets, we solve Eq. (9.11) for each specific target i at time t:

$$\{\sigma_d(i), \sigma_w(i), \beta(i)\} = \underset{\{\sigma_d(i), \sigma_w(i), \beta(i)\}}{\text{argmin}} \left(E_{ss}(v_i^t, s_i, s_{-i} | \sigma_d(i), \sigma_w(i), \beta(i))\right). \qquad (9.11)$$

We obtain the social sensitivity feature $\Theta_{ss}(i) = \{\sigma_d(i), \sigma_w(i), \beta(i)\}$ for each target i. Given the clusters found at training, we assign each $\Theta_{ss}(i)$ to its corresponding cluster, i.e., navigation style.

In the absence of interactions, a target takes either a "neutral" navigation style (when entering a scene) or inherits the last inferred class from the previous interaction. The "neutral" navigation style is the most popular one (shown in green in Fig. 9.2). In Fig. 9.3, we show that when the target is surrounded by other targets, its class changes with respect to its social sensitivity.

9.3.3 Forecasting with Social Sensitivity

Thanks to our proposed social sensitivity feature, we have more flexibility in modeling target interactions to forecast future trajectories. In the remaining of this section, we present the details behind our forecasting model driven by social sensitivity.

Problem formulation. Given the observed trajectories of several targets at time t, we aim to forecast their future positions over the next N time frames (where N is in seconds).

We adapt the Social Forces model [12] from single class to multiple classes. Each target makes a decision on its velocity $\mathbf{v}_i^{(t+1)}$. The energy function, E_Θ, associated to every single target is defined as

$$E_\Theta(\mathbf{v^{t+1}}; s_i, s_{-i}) = \lambda_0(c)E_{damp}(\mathbf{v^{t+1}}; s_i) + \lambda_1(c)E_{speed}(\mathbf{v^{t+1}}; s_i)$$
$$+ \lambda_2(c)E_{dir}(\mathbf{v^{t+1}}; s_i) + \lambda_3(c)E_{att}(\mathbf{v^{t+1}}; s_i) + \lambda_4(c)E_{group}(\mathbf{v^{t+1}}; s_i, s_{A_i})$$
$$+ E_{ss}(\mathbf{v^{t+1}}; s_i, s_{-i} | \sigma_d(v^t), \sigma_w(v^t), \beta) \qquad (9.12)$$

where $\Theta = \{\lambda_0(c), \lambda_1(c), \lambda_2(c), \lambda_3(c), \lambda_4(c), \sigma_d(v^t), \sigma_w(v^t), \beta\}$ and c is the navigation class. More details on the definition of each of the energy terms can be found in [12].

In our work we propose to compute σ_d and σ_w directly from the observed velocity v^t using Eq. (9.11). Both distances, σ_d and σ_w, will then be used to identify the navigation class c. For each class c, the parameter Θ can be learned from training data by minimizing the energy in Eq. (9.12). We can visualize the impact of the navigation style on the

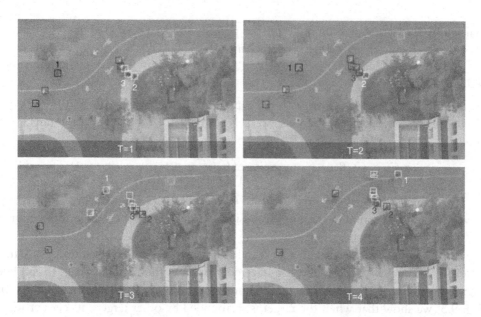

Figure 9.3 Illustration of the class assignment for each target. We follow the same color coding as in Fig. 9.2 to represent the different navigation styles. Note that for a given target, its class changes across time regardless of its physical class (i.e., whether it is a pedestrian, bike, etc.). When the target is surrounded by other targets, its class changes with respect to its social sensitivity. In this scene, first we can observe a cyclist (marked with label 1 in the images) belonging to a black cluster, i.e., being aggressive in his moves, then belonging to some milder clusters (purple and green). We also can see the evolution of a group of pedestrians (shown as labels 2, 3 in the images), initially "mild" (green at $T = 1$), who become red at time $T = 3$ when they accelerate to overtake another group. (For interpretation of the references to color in this figure legend, the reader is referred to the web version of this chapter.)

prediction. In Fig. 9.4 we show the predicted trajectories when several navigation styles are used to perform the forecasting. This shows the need to assign targets into specific classes.

9.4 FORECASTING WITH RECURRENT NEURAL NETWORK

Humans moving in crowded scenes adapt their motion based on the behavior of other people in their vicinity. For instance, a person could completely alter his/her path or stop momentarily to accommodate a group of people moving toward him. Such deviation in trajectory cannot be predicted by observing the person in isolation. Neither can it be predicted with simple "repulsion" or "attraction" functions (presented in the previous section).

This motivates us to build a model which can account for the behavior of other people within a large neighborhood, while predicting a person's path. In this section,

Learning to Predict Human Behavior in Crowded Scenes

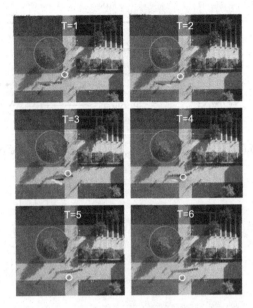

Figure 9.4 We show the predicted trajectory of a given target (red circle) in which 4 different navigation styles are used to perform the prediction. The corresponding predicted trajectories are overlaid over each other and shown with different color codes (the same as those used for depicting the clusters in Fig. 9.2). The ground truth is represented in blue. Predicted trajectories are shown for 6 subsequent frames indicated by $T = 1, \ldots, 6$ respectively. Interestingly, when the target is far away from other targets (no interactions are taking place) the predicted trajectories are very similar to each other (they almost overlap and show a linear trajectory). However, when the red target gets closers to other targets (e.g., those indicated in yellow), the predicted trajectories start showing different behaviors depending on the navigation style: a conservative navigation style activates trajectories' prediction that keep large distances to the yellow targets in order to avoid them (green trajectory) whereas an aggressive navigation style activates trajectories' prediction that are not too distant from the yellow targets (red trajectory). Notice that our approach is capable of automatically associating the target to one of the 4 clusters based on the characteristics in the "social sensitivity space" that have been observed until present. In this example, our approach selects the red trajectory which is the closest to the ground truth's predicted trajectory (in blue). (For interpretation of the references to color in this figure legend, the reader is referred to the web version of this chapter.)

we describe our pooling based LSTM model (Fig. 9.5) which jointly predicts the trajectories of all the people in a scene. We refer to this as the "Social" LSTM model.

Problem formulation. We assume that each scene is first preprocessed to obtain the spatial coordinates of the all people at different time-instants. Previous work follow this convention as well [5,24]. At any time-instant t, the ith person in the scene is represented by his/her xy-coordinates, (x_t^i, y_t^i). We observe the positions of all the people from time 1 to T_{obs}, and predict their positions for time instants T_{obs+1} to T_{pred}. This task can also be viewed as a sequence generation problem [7], where the input sequence corresponds

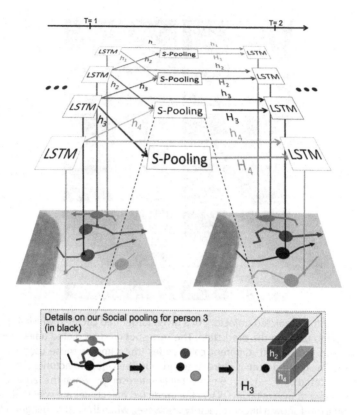

Figure 9.5 Overview of our Social-LSTM method. We use a separate LSTM network for each trajectory in a scene. The LSTMs are then connected to each other through a Social pooling (S-pooling) layer. Unlike the traditional LSTM, this pooling layer allows spatially proximal LSTMs to share information with each other. The variables in the figure are explained in Eq. (9.14). The bottom row shows the S-pooling for one person in the scene. The hidden-states of all LSTMs within a certain radius are pooled together and used as an input at the next time-step.

to the observed positions of a person and we are interested in generating an output sequence denoting his/her future positions at different time-instants.

9.4.1 Social LSTM

People have different motion patterns: they move with different velocity, acceleration, and have different gaits. We need a model which can understand and learn such person-specific motion properties from a limited set of initial observations corresponding to the person.

Long Short-Term Memory (LSTM) networks have been shown to successfully learn and generalize the properties of isolated sequences like handwriting [7] and speech [9]. Inspired by this, we develop an LSTM based model for our trajectory prediction

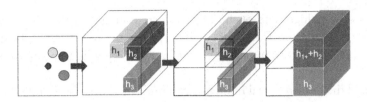

Figure 9.6 We show the Social pooling for the person represented by a black dot. We pool the hidden states of the neighbors (shown in yellow, blue, and orange) within a certain spatial distance. The pooling partially preserves the spatial information of neighbors as shown in the last two steps. (For interpretation of the references to color in this figure legend, the reader is referred to the web version of this chapter.)

problem as well. In particular, we have one LSTM for each person in a scene. This LSTM learns the state of the person and predicts his/her future positions as shown in Fig. 9.5. The LSTM weights are shared across all the sequences.

However, the naive use of one LSTM model per person does not capture the interaction of people in a neighborhood. The vanilla LSTM is agnostic to the behavior of other sequences. We address this limitation by connecting neighboring LSTMs through a new pooling strategy visualized in Figs. 9.5 and 9.6.

9.4.1.1 Social Pooling of Hidden States

Individuals adjust their paths by implicitly reasoning about the motion of neighboring people. These neighbors in-turn are influenced by others in their immediate surroundings and could alter their behavior over time. We expect the hidden states of an LSTM to capture these time varying motion-properties. In order to jointly reason across multiple people, we share the states between neighboring LSTMs. This introduces a new challenge: every person has a different number of neighbors and in very dense crowds [24], this number could be prohibitively high.

The "neighborhood" of a person changes dynamically, and the LSTM predicting future position should be able to process this time-varying "neighborhood" state.

Hence, we need a compact representation which combines the information from all neighboring states. We handle this by introducing "Social" pooling layers as shown in Fig. 9.5. At every time-step, the LSTM cell receives pooled hidden-state information from the LSTM cells of neighbors. While pooling the information, we try to preserve the spatial information through grid based pooling as explained below.

The hidden state h_t^i of the LSTM at time t captures the latent representation of the ith person in the scene at that instant. We share this representation with neighbors by building a "Social" hidden-state tensor H_t^i. Given a hidden-state dimension D, and neighborhood size N_o, we construct an $N_o \times N_o \times D$ tensor H_t^i for the ith trajectory:

$$H_t^i(m, n, :) = \sum_{j \in \mathcal{N}_i} \mathbf{1}_{mn}[x_t^j - x_t^i, y_t^j - y_t^i] h_{t-1}^j, \tag{9.13}$$

where h_{t-1}^j is the hidden state of the LSTM corresponding to the jth person at $t-1$, $\mathbf{1}_{mn}[x, y]$ is an indicator function to check if (x, y) is in the (m, n) cell of the grid, and $\mathcal{N}_i$ is the set of neighbors corresponding to person i. This pooling operation is visualized in Fig. 9.6.

We embed the pooled Social hidden-state tensor into a vector a_i^t and the coordinates into e_i^t. These embeddings are concatenated and used as the input to the LSTM cell of the corresponding trajectory at time t. This introduces the following recurrence:

$$
\begin{aligned}
e_t^i &= \phi(x_t^i, y_t^i; W_e), \\
a_i^t &= \phi(H_t^i; W_a), \\
h_i^t &= \text{LSTM}\left(h_i^{t-1}, e_i^t, a_i^t; W_l\right)
\end{aligned}
\tag{9.14}
$$

where $\phi(\cdot)$ is an embedding function with ReLU nonlinearity, W_e and W_a are embedding weights. The LSTM weights are denoted by W_l.

9.4.1.2 Position Estimation

The hidden-state at time t is used to predict the distribution of the trajectory position $(\hat{x}, \hat{y})_{t+1}^i$ at the next time-step $t+1$. Similar to Graves et al. [7], we assume a bivariate Gaussian distribution parametrized by the mean $\mu_{t+1}^i = (\mu_x, \mu_y)_{t+1}^i$, standard deviation $\sigma_{t+1}^i = (\sigma_x, \sigma_y)_{t+1}^i$, and correlation coefficient ρ_{t+1}^i. These parameters are predicted by a linear layer with a $5 \times D$ weight matrix W_p. The predicted coordinates $(\hat{x}_t^i, \hat{y}_t^i)$ at time t are given by

$$
(\hat{x}, \hat{y})_t^i \sim \mathcal{N}(\mu_t^i, \sigma_t^i, \rho_t^i).
\tag{9.15}
$$

The parameters of the LSTM model are learned by minimizing the negative log-likelihood loss (L^i for the ith trajectory):

$$
\begin{aligned}
[\mu_t^i, \sigma_t^i, \rho_t^i] &= W_p h_i^{t-1}, \\
L^i(W_e, W_l, W_p) &= -\sum_{t=T_{obs}+1}^{T_{pred}} \log\left(\mathbb{P}(x_t^i, y_t^i | \sigma_t^i, \mu_t^i, \rho_t^i)\right), \\
L(W_e, W_l, W_p) &= \sum_i L^i(W_e, W_l, W_p).
\end{aligned}
\tag{9.16}
$$

We train the model by minimizing this summation of loss (L) for all the trajectories in a training dataset. Note that our "Social" pooling layer does not introduce any additional parameters.

An important distinction from the traditional LSTM is that the hidden states of multiple LSTMs are coupled by our "Social" pooling layer and we jointly back-propagate through multiple LSTMs in a scene at every time-step. In other words, for a given snapshot in time, we evaluate the trajectory LSTMs along with the social pooling layer for all the trajectories in the scene. The sum of the loss from the predicted positions and true positions of these trajectories is jointly minimized through stochastic gradient descent.

9.4.1.3 Occupancy Map Pooling

The "Social" LSTM model can be used to pool any set of features from neighboring trajectories. As a simplification, we also experiment with a model which only pools the coordinates of the neighbors (referred to as O-LSTM in the experiments of Section 9.5). This is a reduction of the original model and does not require joint back-propagation across all trajectories during training. This model can still learn to reposition a trajectory to avoid immediate collision with neighbors. However, in the absence of more information from neighboring people, this model would be unable to smoothly change paths to avoid future collisions.

For a person i, we modify the definition of the tensor H_t^i, as an $N_o \times N_o$ matrix at time t centered at the person's position, and call it the occupancy map O_t^i. The positions of all the neighbors are pooled in this map. The (m, n)-element of the map is simply given by

$$O_t^i(m, n) = \sum_{j \in \mathcal{N}_i} \mathbf{1}_{mn}[x_t^j - x_t^i, y_t^j - y_t^i],$$ (9.17)

where $\mathbf{1}_{mn}[\cdot]$ is an indicator function as defined previously. This can also be viewed as a simplification of the social tensor in Eq. (9.13) where the hidden state vector is replaced by a constant value indicating the presence or absence of neighbors in the corresponding cell.

The vectorized occupancy map is used in place of H_t^i in Eq. (9.14) while learning this simpler model.

9.4.1.4 Inference for Path Prediction

During test time, we use the trained Social-LSTM models to predict the future position $(\hat{x}_t^i, \hat{y}_t^i)$ of the ith person. From time T_{obs+1} to T_{pred}, we use the predicted position $(\hat{x}_t^i, \hat{y}_t^i)$ from the previous Social-LSTM cell in place of the true coordinates (x_t^i, y_t^i) in Eq. (9.14). The predicted positions are also used to replace the actual coordinates while constructing the Social hidden-state tensor H_t^i in Eq. (9.13) or the occupancy map O_t^i in Eq. (9.17).

9.4.2 Implementation Details

We use an embedding dimension of 64 for the spatial coordinates before using them as input to the LSTM. We set the spatial pooling size N_o to be 32 and use an 8×8 sum pooling window size without overlaps. We used a fixed hidden state dimension of

128 for all the LSTM models. Additionally, we also use an embedding layer with ReLU (rectified Linear Units) nonlinearity on top of the pooled hidden-state features, before using them for calculating the hidden state tensor H_t^i. The hyperparameters were chosen based on cross-validation on a synthetic dataset. This synthetic was generated using a simulation that implemented the social forces model. This synthetic data contained trajectories for hundreds of scenes with an average crowd density of 30 per frame. We used a learning rate of 0.003 and RMS-prop [78] for training the model. The Social-LSTM model was trained on a single GPU with a Theano [79] implementation.

9.5 EXPERIMENTS

In this section, we present experiments on two publicly available human–trajectory datasets, ETH [10] and UCY [11]. The ETH dataset contains two scenes each with 750 different pedestrians and is split into two sets (*ETH* and *Hotel*). The UCY dataset contains two scenes with 786 people. This dataset has 3 components: *ZARA-01*, *ZARA-02*, and *UCY*. In total, we evaluate our model on 5 sets of data. These datasets represent real world crowded settings with thousands of nonlinear trajectories. As shown in [10], these datasets also cover challenging group behaviors such as couples walking together, groups crossing each other and groups forming and dispersing in some scenes.

We report the prediction error with three different metrics. Similar to Pellegrini et al. [10], we use:

1. *Average displacement error* – The mean square error (MSE) over all estimated points of a trajectory and the true points. This was introduced in Pellegrini et al. [10].
2. *Final displacement error* – The distance between the predicted final destination and the true final destination at end of the prediction period T_{pred}.
3. *Average nonlinear displacement error* – This is the MSE at the nonlinear regions of a trajectory. Since most errors in trajectory-prediction occur during nonlinear turns arising from human–human interactions, we explicitly evaluate the errors around these regions. We set a heuristic threshold on the norm of the second derivative to identify nonlinear regions.

In order to make full use of the datasets while training our models, we use a leave-one-out approach. We train and validate our model on 4 sets and test on the remaining set. We repeat this for all the 5 sets. We also use the same training and testing procedure for other baseline methods used for comparison.

During test time, we observe a trajectory for 3.2 s and predict their paths for the next 4.8 s. At a frame rate of 0.4, this corresponds to observing 8 frames and predicting for the next 12 frames. This is similar to the setting used by [10,11]. In Table 9.1, we compare the performance of the following methods:

- *Linear model (**Lin**).* We use an off-the-shelf Kalman filter to extrapolate trajectories with assumption of linear acceleration.

Table 9.1 Quantitative results of all the methods on all the datasets. We present the performance metrics as follows: First 6 rows are the average displacement error, row 7 to 12 are the average displacement error for nonlinear regions, and the final 6 rows are the final displacement error. All methods forecast trajectories for a fixed period of 4.8 s

Metric	Methods	Lin	LTA	SF [12]	IGP[a] [81]	SF-mc	LSTM	Our O-LSTM	Our Social-LSTM
Avg. displacement error	ETH [10]	0.80	0.54	0.41	**0.20**	0.41	0.60	0.49	0.50
	HOTEL [10]	0.39	0.38	0.25	0.24	0.24	0.15	**0.09**	0.11
	ZARA 1 [11]	0.47	0.37	0.40	0.39	0.35	0.43	**0.22**	**0.22**
	ZARA 2 [11]	0.45	0.40	0.40	0.41	0.39	0.51	0.28	**0.25**
	UCY [11]	0.57	0.51	0.48	0.61	0.45	0.52	0.35	**0.27**
	Average	0.53	0.44	0.39	0.37	0.37	0.44	0.28	**0.27**
Avg. nonlinear displacement error	ETH [10]	0.95	0.70	0.49	0.39	0.46	0.28	**0.24**	0.25
	HOTEL [10]	0.55	0.49	0.38	0.34	0.32	0.09	**0.06**	0.07
	ZARA 1 [11]	0.56	0.39	0.41	0.54	0.41	0.24	**0.13**	**0.13**
	ZARA 2 [11]	0.44	0.41	0.39	0.43	0.39	0.30	0.20	**0.16**
	UCY [11]	0.62	0.57	0.54	0.62	0.51	0.31	0.20	**0.16**
	Average	0.62	0.51	0.44	0.46	0.42	0.24	0.17	**0.15**
Final displacement error	ETH [10]	1.31	0.77	0.59	**0.43**	0.59	1.31	1.06	1.07
	HOTEL [10]	0.55	0.64	0.37	0.37	0.37	0.33	**0.20**	0.23
	ZARA 1 [11]	0.89	0.66	0.60	0.39	0.60	0.93	**0.46**	0.48
	ZARA 2 [11]	0.91	0.72	0.68	0.42	0.67	1.09	0.58	**0.50**
	UCY [11]	1.14	0.95	0.78	1.82	**0.76**	1.25	0.90	0.77
	Average	0.97	0.74	**0.60**	0.69	0.60	0.98	0.64	0.61

[a] Note that IGP uses the intended ground truth destination of a person during test time unlike other methods.

- *Collision avoidance (**LTA**).* We report the results of a simplified version of the Social Force [12] model which only uses the collision avoidance energy, commonly referred to as linear trajectory avoidance.
- *Social Force (**SF**).* We use the implementation of the Social Force model from [12] where several factors such as group affinity and predicted destinations have been modeled.
- *Iterative Gaussian Process (**IGP**).* We use the implementation of the IGP from [80]. Unlike the other baselines, IGP also uses additional information about the final destination of a person.
- *Our multiclass Social Force (**SF-mc**).* The approach presented in Section 9.3.3.
- *Our Vanilla LSTM (**LSTM**).* This is a simplified setting of our model where we remove the "Social" pooling layers and treat all the trajectories to be independent of each other.
- *Our LSTM with occupancy maps (**O-LSTM**).* We show the performance of a simplified version of our model (presented in Section 9.4.1). As a reminder, the model only pools the coordinates of the neighbors at every time-instance.
- *Our Social LSTM.* The approach presented in Section 9.4.1.

The naive linear model produces high prediction errors, which are more pronounced around nonlinear regions as seen from the average nonlinear displacement error. The vanilla LSTM outperforms this linear baseline since it can extrapolate nonlinear curves as shown in Graves et al. [7]. However, this simple LSTM is noticeably worse than the Social Force and IGP models which explicitly model human–human interactions. This shows the need to account for such interactions.

Our presented SF-mc performs the same as the single class Social Forces model in ETH dataset, and outperforms other methods in UCY datasets. This result can be justified by the fact that the UCY dataset is considerably more crowded, with more collisions, and therefore presenting different types of behaviors. Nonlinear behaviors such as people stopping and talking to each other, walking faster, or turning around each others are more common in UCY than in ETH. The SF-mc is able to infer these navigation patterns hence better predict the trajectories of pedestrians. We also report the performance of the IGP model for completeness. While IGP performs better on the less crowded dataset, it does not do well on the crowded ones. Notice that IGP uses the destination and time of arrival as additional inputs (which other methods don't use).

Our Social pooling based LSTM and O-LSTM outperform the heavily engineered Social Force and IGP models in almost all datasets. In particular, the error reduction is more significant in the case of the UCY datasets as compared to ETH. This can be explained by the different crowd densities in the two datasets: UCY contains more crowded regions with a total of 32K nonlinearities as opposed to the more sparsely populated ETH scenes with only 15K nonlinear regions.

In the more crowded UCY scenes, the deviation from linear paths is more dominated by human–human interactions. Hence, our model which captures neighborhood interactions achieves a higher gain in UCY datasets. The pedestrians' intention to reach a certain destination plays a more dominant role in the ETH datasets. Consequently, the IGP model which knows the true final destination during testing achieves lower errors in parts of this dataset.

In the case of ETH, we also observe that the occupancy and Social LSTM errors are at par with each other and in general better than the Social Force model. Again, our Social-LSTM outperforms O-LSTM in the more crowded UCY datasets. This shows the advantage of pooling the entire hidden state to capture complex interactions in dense crowds.

9.5.1 Analyzing the Predicted Paths

Our quantitative evaluation in Section 9.5 shows that the learned Social-LSTM model outperforms state-of-the-art methods on standard datasets. In this section, we try to gain more insights on the actual behavior of our model in different crowd settings. We qualitatively study the performance of our Social-LSTM method on social scenes where individuals interact with each others in a specific pattern.

We present an example scene occupied by four individuals in Fig. 9.7. We visualize the distribution of the paths predicted by our model at different time-instants. The first and third rows in Fig. 9.7 show the current position of each person as well as their true trajectory (solid line for the future path and dashed line for the past). The second and fourth rows show our Social-LSTM prediction for the next 12.4 s. In these scenes, we observe three people (2, 3, 4) walking close to each other and a fourth person (1) walking farther away from them.

Our model predicts a linear path for person (1) at all times. The distribution for person (1) is similar across time indicating that the speed of the person is constant.

We can observe more interesting patterns in the predicted trajectories for the 3-person group. In particular, our model makes intelligent route choices to yield for others and preempt future collisions. For instance, at time-steps 2, 4, and 5 our model predicts a deviation from the linear paths for person (3) and person (4), even before the start of the actual turn. At time-steps 3 and 4, we notice that the Social-LSTM predicts a "halt" for person (3) in order to yield for person (1). Interestingly, at time-step 4 the location of the halting point is updated to match the true turning-point in the path. At the next time-step, with more observations, the model is able to correctly predict the full turn anchored at that point.

In Fig. 9.8, we illustrate the prediction results of our Social-LSTM, the SF model [10], and the linear baseline on one of the ETH datasets. When people walk in a group or as, e.g., a couple, our model is able to jointly predict their trajectories. It is interesting to note that unlike Social Forces [12] we do not explicitly model group behavior.

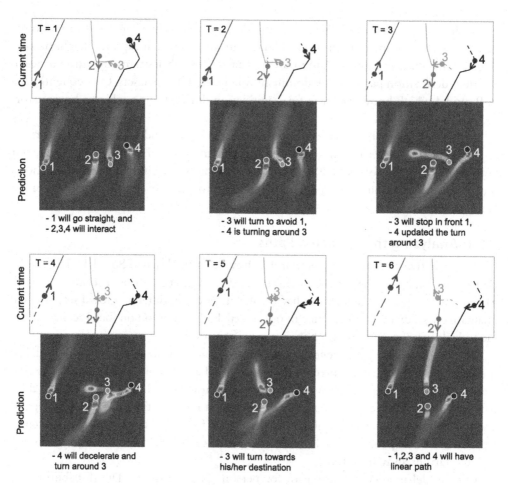

Figure 9.7 We visualize the probability distribution of the predicted paths for 4 people moving in a scene across 6 time steps. The subcaption describes what our model is predicting. At each time-step the solid lines in rows 1 and 3 represent the ground-truth future trajectories, the dashed lines refer to the observed positions till that time-step and the dots denote the position at that time-step. We notice that our model often correctly predicts the future paths in challenging settings with nonlinear motions. We analyze these figures in more details in Section 9.5.1. Note that T stands for time and the id 1 to id 4 denote person ids.

However, our model is better at predicting grouped trajectories in a holistic fashion. In the last row of Fig. 9.8, we show some failure cases, i.e., when our predictions are worse than previous works. We either predict a linear path (second column) or decelerate earlier (first and third column) than needed. Although the trajectories do not match the ground-truth in these cases, our Social-LSTM still outputs "plausible" trajectories, i.e., trajectories that humans could have taken. For instance, in the first and third columns, our model slows down to avoid a potential collision with the person ahead.

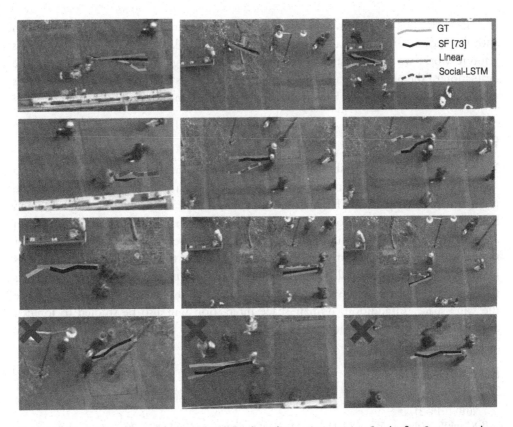

Figure 9.8 Illustration of our Social-LSTM method predicting trajectories. On the first 3 rows, we show examples where our model successfully predicts the trajectories with small errors (in terms of position and speed). We also show other methods such as Social Forces [12] and the linear method. The last row represents failure cases, e.g., person slowed down or took a linear path. Nevertheless, our Social-LSTM method predicts a plausible path. The results are shown on ETH dataset [10].

9.5.2 Discussions and Limitations

We are far from predicting all the nuances in human navigation. However, our experiments show encouraging results toward our claim that a data-driven approach has the potential to learn general rules on human navigation as well as nuances behind human behavior. Heuristic-based approaches that have been tried in the past can only capture the general rules of motion, but won't be adequate when it comes to capturing the characteristic and subtleties of human motion, which might at times be totally unexpected or random. We believe that a model that has observed human behavior for quite some time can come close to account for these irregularities in human motion, or if not that, then at least, react to this sudden anomaly in the most consistent way.

The current set of quantitative experiments assume that there is a single ground truth path to predict. Given the same social context, several plausible paths are possible. As a future work, we will investigate other metrics involving humans in the loop for the evaluation of the predicted paths. We can run experiments to study the number of paths generated by our forecasting model that are "plausible" and "socially-accepted".

9.6 CONCLUSIONS

We have presented two families of methods to forecast human trajectories in crowded scenes. The former is based on Social Forces model and has the capacity to encode the physics behind navigation. The latter is a fully data driven method based on LSTM and has the capacity to encode complex interactions that one might not be aware of. Given a set of experiments on public datasets, the LSTM-based model outperforms other methods. It can jointly reason across multiple individuals to predict human trajectories in a scene. Future work will study the impact of data driven methods in multiclass settings where several objects such as bicycles, skateboards, carts, and pedestrians share the same space. In addition, human–space interaction will also be studied to forecast abrupt nonlinear behaviors due to the static scene.

REFERENCES

[1] Song C, Qu Z, Blumm N, Barabási A-L. Limits of predictability in human mobility. Science 2010:1018–21.
[2] Kitani K, Ziebart B, Bagnell J, Hebert M. Activity forecasting. In: ECCV. 2012.
[3] Helbing D, Molnar P. Social force model for pedestrian dynamics. Phys Rev E 1995.
[4] Moussaïd M, Perozo N, Garnier S, Helbing D, Theraulaz G. The walking behaviour of pedestrian social groups and its impact on crowd dynamics. PLoS ONE 2010;5(4):e10047.
[5] Luber M, Stork J, Tipaldi G, Arras K. People tracking with human motion predictions from social forces. In: ICRA. 2010. p. 464–9.
[6] Mehran R, Oyama A, Shah M. Abnormal crowd behavior detection using social force model. In: IEEE conference on computer vision and pattern recognition. IEEE; 2009. p. 935–42.
[7] Graves A. Generating sequences with recurrent neural networks. Available from arXiv:1308.0850.
[8] Albrecht K. Social intelligence: the new science of success. John Wiley & Sons; 2006.
[9] Graves A, Jaitly N. Towards end-to-end speech recognition with recurrent neural networks. In: Proceedings of the 31st international conference on machine learning (ICML-14). 2014. p. 1764–72.
[10] Pellegrini S, Ess A, Schindler K, Van Gool L. You'll never walk alone: modeling social behavior for multi-target tracking. In: ICCV. 2009.
[11] Lerner A, Chrysanthou Y, Lischinski D. Crowds by example. Comput Graph Forum 2007;26:655–64 [Wiley Online Library].
[12] Yamaguchi K, Berg AC, Ortiz LE, Berg TL. Who are you with and where are you going? In: CVPR. IEEE; 2011.
[13] Pellegrini S, Ess A, Van Gool L. Improving data association by joint modeling of pedestrian trajectories and groupings. In: ECCV. 2010.
[14] Leal-Taixe L, Pons-Moll G, Rosenhahn B. Everybody needs somebody: modeling social and grouping behavior on a linear programming multiple people tracker. In: ICCV workshops. 2011.

[15] Leal-Taixé L, Fenzi M, Kuznetsova A, Rosenhahn B, Savarese S. Learning an image-based motion context for multiple people tracking. In: CVPR. IEEE; 2014. p. 3542–9.

[16] Choi W, Savarese S. A unified framework for multi-target tracking and collective activity recognition. In: Computer vision – ECCV 2012. Springer; 2012. p. 215–30.

[17] Choi W, Savarese S. Understanding collective activities of people from videos. IEEE Trans Pattern Anal Mach Intell 2014;36(6):1242–57.

[18] Treuille A, Cooper S, Popović Z. Continuum crowds. ACM Trans Graph 2006;25:1160–8.

[19] Antonini G, Bierlaire M, Weber M. Discrete choice models of pedestrian walking behavior. Transp Res, Part B, Methodol 2006;40(8):667–87.

[20] Wang JM, Fleet DJ, Hertzmann A. Gaussian process dynamical models for human motion. IEEE Trans Pattern Anal Mach Intell 2008;30(2):283–98.

[21] Tay MKC, Laugier C. Modelling smooth paths using Gaussian processes. In: Field and service robotics. Springer; 2008. p. 381–90.

[22] Yi S, Li H, Wang X. Understanding pedestrian behaviors from stationary crowd groups. In: Proceedings of the IEEE conference on computer vision and pattern recognition. 2015. p. 3488–96.

[23] Park H, Shi J. Social saliency prediction. In: ICCV. 2015.

[24] Alahi A, Ramanathan V, Fei-Fei L. Socially-aware large-scale crowd forecasting. In: CVPR. 2014.

[25] Bonabeau E. Agent-based modeling: methods and techniques for simulating human systems. Proc Natl Acad Sci USA 2002;99(Suppl. 3):7280–7.

[26] Rodriguez M, Sivic J, Laptev I, Audibert J-Y. Data-driven crowd analysis in videos. In: 2011 IEEE international conference on computer vision (ICCV). IEEE; 2011. p. 1235–42.

[27] Hu W, Xie D, Fu Z, Zeng W, Maybank S. Semantic-based surveillance video retrieval. IEEE Trans Image Process 2007;16(4):1168–81.

[28] Kim K, Lee D, Essa I. Gaussian process regression flow for analysis of motion trajectories. In: 2011 IEEE international conference on computer vision (ICCV). IEEE; 2011. p. 1164–71.

[29] Morris BT, Trivedi MM. Trajectory learning for activity understanding: unsupervised, multilevel, and long-term adaptive approach. IEEE Trans Pattern Anal Mach Intell 2011;33(11):2287–301.

[30] Zhou B, Wang X, Tang X. Random field topic model for semantic region analysis in crowded scenes from tracklets. In: 2011 IEEE conference on computer vision and pattern recognition (CVPR). IEEE; 2011. p. 3441–8.

[31] Morris BT, Trivedi MM. A survey of vision-based trajectory learning and analysis for surveillance. IEEE Trans Circuits Syst Video Technol 2008;18(8):1114–27.

[32] Pirsiavash H, Vondrick C, Torralba A. Inferring the why in images. Available from arXiv:1406.5472.

[33] Kooij JFP, Schneider N, Flohr F, Gavrila DM. Context-based pedestrian path prediction. In: Computer vision – ECCV 2014. Springer; 2014. p. 618–33.

[34] Azorin-Lopez J, Saval-Calvo M, Fuster-Guillo A, Oliver-Albert A. A predictive model for recognizing human behaviour based on trajectory representation. In: 2014 international joint conference on neural networks (IJCNN). IEEE; 2014. p. 1494–501.

[35] Elfring J, Van De Molengraft R, Steinbuch M. Learning intentions for improved human motion prediction. Robot Auton Syst 2014;62(4):591–602.

[36] Kong Y, Kit D, Fu Y. A discriminative model with multiple temporal scales for action prediction. In: Computer vision – ECCV 2014. Springer; 2014. p. 596–611.

[37] Kitani KM, Ziebart BD, Bagnell JA, Hebert M. Activity forecasting. In: Computer vision – ECCV 2012. Springer; 2012. p. 201–14.

[38] Walker J, Gupta A, Hebert M. Patch to the future: unsupervised visual prediction. In: CVPR. 2014.

[39] Ziebart BD, Ratliff N, Gallagher G, Mertz C, Peterson K, Bagnell JA, Hebert M, Dey AK, Srinivasa S. Planning-based prediction for pedestrians. In: IEEE/RSJ international conference on intelligent robots and systems. IEEE; 2009. p. 3931–6.

[40] Hawkins KP, Vo N, Bansal S, Bobick AF. Probabilistic human action prediction and wait-sensitive planning for responsive human–robot collaboration. In: 2013 13th IEEE–RAS international conference on humanoid robots (Humanoids). IEEE; 2013. p. 499–506.

[41] Turek MW, Hoogs A, Collins R. Unsupervised learning of functional categories in video scenes. In: ECCV. 2010.

[42] Li K, Fu Y. Prediction of human activity by discovering temporal sequence patterns. IEEE Trans Pattern Anal Mach Intell 2014;36(8):1644–57.

[43] Huang C, Wu B, Nevatia R. Robust object tracking by hierarchical association of detection responses. In: ECCV. 2008.

[44] Gong H, Sim J, Likhachev M, Shi J. Multi-hypothesis motion planning for visual object tracking. In: Proceedings of the 2011 international conference on computer vision. Washington (DC): IEEE Computer Society; 2011. p. 619–26.

[45] Makris D, Ellis T. Learning semantic scene models from observing activity in visual surveillance. IEEE Trans Syst Man Cybern, Part B, Cybern 2005;35(3):397–408.

[46] Kretzschmar H, Kuderer M, Burgard W. Learning to predict trajectories of cooperatively navigating agents. In: 2014 IEEE international conference on robotics and automation (ICRA). IEEE; 2014. p. 4015–20.

[47] Fouhey DF, Zitnick CL. Predicting object dynamics in scenes. In: 2014 IEEE conference on computer vision and pattern recognition (CVPR). IEEE; 2014. p. 2027–34.

[48] Huang D-A, Kitani KM. Action-reaction: forecasting the dynamics of human interaction. In: Computer vision – ECCV 2014. Springer; 2014. p. 489–504.

[49] Ryoo M. Human activity prediction: early recognition of ongoing activities from streaming videos. In: 2011 IEEE international conference on computer vision (ICCV). IEEE; 2011. p. 1036–43.

[50] Cao Y, Barrett D, Barbu A, Narayanaswamy S, Yu H, Michaux A, Lin Y, Dickinson S, Siskind JM, Wang S. Recognize human activities from partially observed videos. In: 2013 IEEE conference on computer vision and pattern recognition (CVPR). IEEE; 2013. p. 2658–65.

[51] Xie D, Todorovic S, Zhu S-C. Inferring "dark matter" and "dark energy" from videos. In: 2013 IEEE international conference on computer vision (ICCV). IEEE; 2013. p. 2224–31.

[52] Vu T-H, Olsson C, Laptev I, Oliva A, Sivic J. Predicting actions from static scenes. In: Computer vision – ECCV 2014. Springer; 2014. p. 421–36.

[53] Minor B, Doppa JR, Cook DJ. Data-driven activity prediction: algorithms, evaluation methodology, and applications. In: Proceedings of the 21st ACM SIGKDD international conference on knowledge discovery and data mining. ACM; 2015. p. 805–14.

[54] Surana A, Srivastava K. Bayesian nonparametric inverse reinforcement learning for switched Markov decision processes. In: 13th international conference on machine learning and applications (ICMLA). IEEE; 2014. p. 47–54.

[55] Ranzato M, et al. Video (language) modeling: a baseline for generative models of natural videos. Available from arXiv:1412.6604.

[56] Srivastava N, Mansimov E, Salakhutdinov R. Unsupervised learning of video representations using LSTMS. Available from arXiv:1502.04681.

[57] Vondrick C, Pirsiavash H, Torralba A. Anticipating the future by watching unlabeled video. Available from arXiv:1504.08023.

[58] Ryoo M, Fuchs TJ, Xia L, Aggarwal J, Matthies L. Early recognition of human activities from first-person videos using onset representations. Available from arXiv:1406.5309.

[59] Kitani K, Okabe T, Sato Y, Sugimoto A. Fast unsupervised ego-action learning for first-person sports videos. In: 2011 IEEE conference on computer vision and pattern recognition (CVPR). 2011. p. 3241–8.

[60] Hochreiter S, Schmidhuber J. Long short-term memory. Neural Comput 1997;9(8):1735–80.

[61] Chung J, Gulcehre C, Cho K, Bengio Y. Empirical evaluation of gated recurrent neural networks on sequence modeling. Available from arXiv:1412.3555.

[62] Chorowski J, Bahdanau D, Cho K, Bengio Y. End-to-end continuous speech recognition using attention-based recurrent NN: first results. Available from arXiv:1412.1602.

[63] Chung J, Kastner K, Dinh L, Goel K, Courville AC, Bengio Y. A recurrent latent variable model for sequential data. Available from arXiv:1506.02216.

[64] Vinyals O, Toshev A, Bengio S, Erhan D. Show and tell: a neural image caption generator. Available from arXiv:1411.4555.

[65] Karpathy A, et al. Deep fragment embeddings for bidirectional image sentence mapping. In: NIPS. 2014.

[66] Yoo D, Park S, Lee J-Y, Paek A, Kweon IS. AttentionNet: aggregating weak directions for accurate object detection. Available from arXiv:1506.07704.

[67] Donahue J, Hendricks LA, Guadarrama S, Rohrbach M, Venugopalan S, Saenko K, Darrell T. Long-term recurrent convolutional networks for visual recognition and description. Available from arXiv:1411.4389.

[68] Xu K, Ba J, Kiros R, Courville A, Salakhutdinov R, Zemel R, Bengio Y. Show, attend and tell: neural image caption generation with visual attention. Available from arXiv:1502.03044.

[69] Bahdanau D, Cho K, Bengio Y. Neural machine translation by jointly learning to align and translate. Available from arXiv:1409.0473.

[70] Cao C, Liu X, Yang Y, Yu Y, Wang J, Wang Z, Huang Y, Wang L, Huang C, Xu W, et al. Look and think twice: capturing top-down visual attention with feedback convolutional neural networks. In: ICCV. 2015.

[71] Gregor K, Danihelka I, Graves A, Wierstra D. Draw: a recurrent neural network for image generation. Available from arXiv:1502.04623.

[72] Xiao T, Xu Y, Yang K, Zhang J, Peng Y, Zhang Z. The application of two-level attention models in deep convolutional neural network for fine-grained image classification. Available from arXiv:1411.6447.

[73] Ng JY-H, Hausknecht M, Vijayanarasimhan S, Vinyals O, Monga R, Toderici G. Beyond short snippets: deep networks for video classification. Available from arXiv:1503.08909.

[74] Fragkiadaki K, Levine S, Felsen P, Malik J. Recurrent network models for human dynamics. In: ICCV. 2015.

[75] Zheng S, Jayasumana S, Romera-Paredes B, Vineet V, Su Z, Du D, Huang C, Torr P. Conditional random fields as recurrent neural networks. Available from arXiv:1502.03240.

[76] Pinheiro PH, Collobert R. Recurrent convolutional neural networks for scene parsing. Available from arXiv:1306.2795.

[77] Visin F, Kastner K, Cho K, Matteucci M, Courville A, Bengio Y. ReNet: a recurrent neural network based alternative to convolutional networks. Available from arXiv:1505.00393.

[78] Dauphin YN, de Vries H, Chung J, Bengio Y. RMSPROP and equilibrated adaptive learning rates for non-convex optimization. Available from arXiv:1502.04390.

[79] Bergstra J, Breuleux O, Bastien F, Lamblin P, Pascanu R, Desjardins G, Turian J, Warde-Farley D, Bengio Y. Theano: a CPU and GPU math compiler in Python.

[80] Trautman P, Ma J, Murray RM, Krause A. Robot navigation in dense human crowds: the case for cooperation. In: 2013 IEEE international conference on robotics and automation (ICRA). IEEE; 2013. p. 2153–60.

[81] Trautman P, Krause A. Unfreezing the robot: navigation in dense, interacting crowds. In: 2010 IEEE/RSJ international conference on intelligent robots and systems (IROS). IEEE; 2010. p. 797–803.

CHAPTER 10

Deep Learning for Scene-Independent Crowd Analysis

Xiaogang Wang*, Chen-Change Loy[†]

*Department of Electronic Engineering, The Chinese University of Hong Kong, Shatin, Hong Kong
[†]Department of Information Engineering, The Chinese University of Hong Kong, Shatin, Hong Kong

Contents

Group and Crowd Behavior for Computer Vision
DOI: 10.1016/B978-0-12-809276-7.00012-6

10.1 INTRODUCTION

Crowd analysis and scene understanding has drawn a lot of attention recently because it has a broad range of applications in video surveillance [1–17]. Besides surveillance, crowd scenes also exist in movies, TV shows, personal video collections, and also videos shared through social media. Since crowd scenes have a large number of people accumulated with frequent and heavy occlusions, many existing technologies of detection, tracking, and activity recognition, which are only applicable to sparse scenes, do not work well in crowded scenes. Therefore a lot of new research works, especially targeting crowd scenes, have been done in the past years. They cover a broad range of topics, including crowd segmentation and detection [2,18–20,11], crowd tracking [21, 12], crowd counting [4,8,15], pedestrian traveling time estimation [16], crowd attribute recognition [13,22], crowd behavior analysis [1,3,5,23–26,10,14,17,17], and abnormality detection in a crowd [27,6,7].

Many existing works [1,4,3,6,11,14] on crowd analysis are scene-specific, i.e., models trained from a particular scene can only be applied to the same scene. When switching to a different scene, data needs to be collected from the new scene to train the model again. It limits the applications of these works. Recently, people worked toward the goal of scene-independent crowd analysis [10,15,13,22], i.e., once a generic crowd model is trained, it can be applied to different scenes without being retrained. This is nontrivial given the inherently complex crowd behaviors observed across different scenes. As there are so many crowd scenes, how to characterize and compare their dynamics is a big challenge.

Many studies [28,29] show that various crowd systems do share a set of universal properties because some general principles underlie different types of crowd behaviors. Researchers proposed and estimated some generic crowd properties, such as density, collectiveness, stability, uniformity, and conflict, from the computer vision point of view [9,30,10,15]. A more comprehensive set of 94 attributes to characterize the locations, subjects, and events/actions of a crowd were proposed in [13]. In recent years, deep learning has achieved great success in many grand challenges of computer vision. It has also been applied to crowd analysis [20,15,13,22]. In this chapter, we will introduce how to use deep convolutional neural networks (CNNs) for scene-independent crowd counting [15], crowd density estimation [15], and crowd attribute recognition [13,22].

The key for the success of deep learning is the availability of large scale training data. Existing crowd datasets are very limited in size, scene-diversity, and annotations, and are not suitable for training generic deep neural networks applicable to different scenes. Very recently, two large-scale crowd datasets, i.e., the Shanghai World Expo'10 crowd dataset [31] and the WWW crowd dataset [22] were proposed. 2630 video sequences from 235 surveillance cameras with disjoint views were collected from Shanghai 2010 World Expo, in the Shanghai World Expo'10 dataset. Crowd segmentation and general crowd properties of crowd density, collectiveness, and cohesiveness on each crowd segment were annotated on this dataset. Besides, the locations of 199,923 pedestrians in the crowd were annotated for the purpose of crowd counting. The WWW crowd dataset provided 10,000 videos with over 8 million frames from 8257 scenes. 94 crowd attributes were annotated on this dataset. Both datasets are suitable for training deep neural networks for crowd analysis. The details of the two datasets will be provided in Section 10.2 of this chapter.

Although convolutional networks have achieved great success on image recognition, it is much more challenging to learn dynamic feature representations from videos for crowd analysis with CNN. The temporal dimension is different from the spatial dimensions. A straightforward way of treating videos as 3D volumes and directly applying CNN on them cannot get very good results. Moreover, the computational complexity of the training process is much higher than that on images. New network architectures and training strategies need to be developed for crowd analysis.

Section 10.3 will introduce a CNN architecture which jointly estimates pedestrian count in a crowd and crowd density. It is trained alternatively with crowd count and density, which helps the training jumping out of local minimum. To handle an unseen target crowd scene, a data-driven method is presented to fine-tune the trained CNN model for the target scene.

Section 10.4 will introduce a slicing CNN which decomposes 3D feature maps into 2D spatio- and 2D temporal-slices representations. It is capable of capturing dynamics of different semantic units such as groups and objects, learns separated appearance and dynamic representations while keeping proper interactions between them, and exploits the selectiveness of spatial filters to discard irrelevant background clutter for crowd understanding.

10.2 LARGE SCALE CROWD DATASETS

This section will introduce the Shanghai World Expo'10 crowd dataset and the WWW crowd dataset, both of which are much larger than previously proposed crowd datasets in terms of the number of video sequences and scene diversity and also provide rich annotations. They are suitable for training deep neural networks and conducting research on scene-independent crowd analysis.

10.2.1 Shanghai World Expo'10 Crowd Dataset

The Shanghai World Expo'10 crowd dataset [31] was contributed as a large-scale benchmark for crowd understanding. All the videos were shot with actual surveillance cameras from Shanghai 2010 World Expo, which was the world's largest fair site ever with an area size of 5.28 km^2. Over 73 million people visited it during six months, and nearly 250 pavilions were built at the expo site. The abundant sources of these surveillance videos enrich the diversity and completeness of the surveillance scenes. Five challenges

Figure 10.1 Examples of four types of scenes from our dataset: road (first row), queue (second row), square (third row), and mixed (last row).

were posted on this dataset: crowd segmentation, estimation of crowd density, pedestrian count, crowd collectiveness, and crowd cohesiveness.

10.2.1.1 Data Collection

A huge amount of crowd videos were collected from Shanghai 2010 World Expo from June to October in 2010. A total of 2630 video sequences from 235 cameras with disjoint views were selected. Each camera has 10–12 videos, one of which was collected at night, and at least two in each month. Each sequence lasts one minute (3000 frames), and the data size is 40 GB. Cameras were mounted on the top of buildings and had far-field views. The resolutions of videos are 720×576, which is higher than or comparable to existing datasets. The data was collected under various weather conditions: sunny, cloudy, and rainy (pedestrians held umbrellas on rainy days). All the scenes generally fall into four categories: road, square, queue at entrances, and mixture of the previous three types of scenes. Generally, crowds in queue or on road tend to have higher collectiveness, while crowds in queue tend to have higher cohesiveness. Examples are shown in Fig. 10.1.

10.2.1.2 Annotation

A professional labeling company was hired, and 20 labelers were trained for the annotation task. Three frames were uniformly sampled from each sequence for annotation. Before labelers annotated a frame, they first browsed its surrounding frames to observe moving objects. The boundaries of crowd regions were drawn with polygons as shown in Fig. 10.2. Each crowd region was labeled with three attributes: density, collectiveness, and cohesiveness. Each crowd attribute was labeled as one of the three levels: low (1), medium (2), and high (3). The attribute of background regions was always labeled as 0. In order to evaluate the performance of crowd counting, some video frames

Figure 10.2 Examples of four types of scenes from the World Expo'10 dataset. Crowd regions were manually labeled with polygons of different density levels (high = red, medium = yellow, low = green). (For interpretation of the references to color in this figure legend, the reader is referred to the web version of this chapter.)

were selected, and the exact locations of all the pedestrians in these video frames were annotated.

The annotation rule for crowd segmentation is as follows. Every person had his or her own territory which was a circle with a radius of 1 m.[1] If the territories of two persons overlapped, the two persons were connected. A crowd region covered a connected component of multiple persons.

Crowd density was annotated with the widely used Jacobs' method [32] proposed in social science, which classifies density into three levels. It counts the average number (n) of persons in every square meter. A scene is sparse if $n \leq 1$, medium if $1 < n \leq 2$, or dense if $n > 2$. Since crowd is not uniformly distributed in a scene, this rule was empirically modified to make it easier for annotation. Within a segmented crowd region, if the territory of a person includes another $0 \leq m \leq 2$ persons on average, this crowd region was annotated as sparse. Similarly, it was labeled as medium if $2 < m \leq 5$, and dense if $m > 5$. This is consistent with the Jacobs' method [32], since the area of a person's

[1] The "1 m" for each person was empirically determined by the labeler as 2/3 of the person's height.

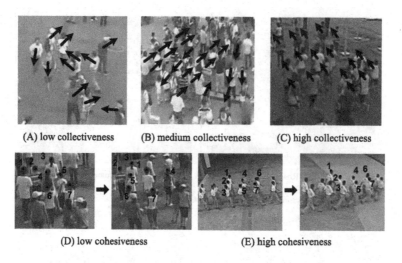

(A) low collectiveness   (B) medium collectiveness   (C) high collectiveness

(D) low cohesiveness   (E) high cohesiveness

Figure 10.3 Illustration of different levels of collectiveness and cohesiveness.

territory is around 3 m². The annotation rule also implicitly considered the crowd size. If a crowd region only had 3 persons, it was always labeled as sparse, even if all three stood tightly within 1 m², because there were no more than two persons in the territory of another person. Examples of density annotations are shown in Fig. 10.2.

Annotation of collectiveness and cohesiveness was more subjective. Fig. 10.3 shows examples of the defined different levels. The collectiveness of Fig. 10.3A is labeled as low, since the pedestrians move in different directions without the same goal. In Fig. 10.3B, a few crowd groups move in opposite directions and its collectiveness is labeled as medium. In Fig. 10.3C, everybody moves in the same direction, and the collectiveness is high.

Cohesiveness measures the stability of local geometrical and topological structures of crowd groups. Fig. 10.3D shows the same crowd at different frames. The topological structure of its members has changed significantly, and therefore the cohesiveness is low. Fig. 10.3E shows an example with high cohesiveness. Note that high collectiveness does not mean high cohesiveness. If a group of people move in the same direction but with very different speed, their local structures cannot maintain stable.

Fig. 10.4 shows the histograms (on the area of crowd regions) of the three attributes for the four type of scenes. According to the statistics, around 75% of regions are background and the remaining 25% are crowds. Most of the crowd regions in this dataset have high density. Generally, road and queue crowd scenes with strict man-made constraints have higher collectiveness and cohesiveness than open scenes such as a square. Especially, in queue scenes, people were kept within some bounds, and most of the crowd regions had high cohesiveness.

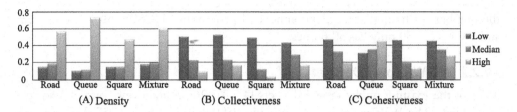

Figure 10.4 The statistics of three attributes in different crowd scenes (blue = low, red = medium, green = high). (For interpretation of the references to color in this figure legend, the reader is referred to the web version of this chapter.)

Table 10.1 Statistics of three datasets: N_f is numbers of frames; N_c is numbers of scenes; R is the resolution; FPS is frames per second; D is the minimum and maximum number of people in the ROI of a frame; and T_p is the total number of labeled pedestrian instances

Dataset	N_f	N_c	R	FPS	D	T_p
UCSD	2000	1	158 × 238	10	11–46	49,885
UCF_CC_50	50	50	—	image	94–4543	63,974
World Expo	4.44 million	106	576 × 720	50	1–253	199,923

Figure 10.5 (A) Example frames of the UCSD dataset. (B) Example frames of the UCF_CC_50. (C) Example frames of the World Expo dataset. The region within the blue polygons are the regions of interest and positions of pedestrian heads labeled by red dots. (For interpretation of the references to color in this figure legend, the reader is referred to the web version of this chapter.)

A subset from the World Expo'10 crowd dataset was selected for the purpose of crowd counting. The subsets includes 1132 annotated video sequences captured by 108 surveillance cameras. Since most of the cameras had disjoint bird views, they covered a large variety of scenes. A total of 199,923 pedestrians were labeled at the centers of their heads in 3980 frames. These frames were uniformly sampled from all the video sequences. The details are listed in Table 10.1 and some instances are shown in Fig. 10.5. Table 10.1 and Fig. 10.5 also make comparison with other crowd counting datasets UCSD [4] and UCF_CC_50 [33]. It is much larger than UCSD and UCF_CC_50 in the numbers of frames, scenes, and annotated pedestrians. UCSD only has one scene and UCF_CC_50 only has 50 frames.

10.2.2 WWW Crowd Dataset

Most of the existing public crowd datasets [4,8,3,5,26] contain only one or two specific scenes. The WWW dataset [13] provides 10,000 videos from 8257 diverse scenes with over 8 million frames, therefore offering a superiorly comprehensive dataset for the research of crowd understanding. The abundant sources of these videos also enrich the diversity and completeness. A comparison of the WWW dataset with other public-available crowd datasets is summarized in Table 10.2.

10.2.2.1 Crowd Video Construction

Collecting Keywords

In order to obtain a large scaled and comprehensive crowd dataset, a set of keywords related to common crowd scenarios (e.g., street, stadium, and rink) and crowd events (e.g., marching, chorus, and graduation) were selected for the sake of searching efficiency and effectiveness.

Collecting Crowd Videos[2]

The gathered keywords were used to search for videos from several public video search engines including Getty Images,[3] Pond5,[4] and YouTube.[5] Besides these three sources, 469 videos were further collected from 23 movies. Videos with blurred motion, synthetic crowd, and extremely short length were removed. In addition, all the duplicated videos were filtered.

10.2.2.2 Crowd Attribute Annotation

Given a video collection of many different crowded scenes, there is an enormous number of possible attributes describing different scenarios, subjects, and events. The capacity to infer attributes allows us to describe a crowded scene by answering "Where is the crowd?", "Who is in the crowd?", and "Why is the crowd here?". Importantly, when faced with a new crowded scene, we can still describe it with these three types of cues (e.g., newlywed couple [Who] are in the wedding [Why] at a beach [Where]).

There are many possible interactions among these attributes. Some attributes are likely to cooccur, whilst some seem exclusive. For example, the scenario "street" attribute is likely to cooccur with subject "pedestrian" when the subject is "walking",

---

[2] This collection covers major existing crowd video datasets such as [10,9,25].

[3] http://www.gettyimages.com/.

[4] http://www.pond5.com/.

[5] http://www.youtube.com/.

Table 10.2 Comparison of some properties of WWW versus other datasets. WWW offers the largest number of videos, scenes, and frames

	CUHK [10]	Collectiveness [9]	Violence [25]	Data-driven [19]	UCF [2]	WWW
# video	474	413	246	212	46	**10,000**
# scene	215	62	246	212	46	**8257**
# frame	60,384	40,796	22,074	121,626	18,196	**> 8 million**
resolution	multiple	670 × 1000	320 × 240	640 × 360	multiple	**640 × 360**

Figure 10.6 A quick glance of the WWW crowd dataset with its attributes. Red represents the locations (Where), green represents the subjects (Who), and blue refers to events/actions (Why). The area of each word is proportional to the frequency of that attribute in WWW dataset. (For interpretation of the references to color in this figure legend, the reader is referred to the web version of this chapter.)

and also likely to cooccur with subject "mob" when the subject is "fighting", but not related to subject "swimmer" because the subject cannot "swim" on "street". In addition, there exist attributes that are grouped hierarchically, e.g., "outdoor/indoor" contains almost all the other attributes of location. Some attributes, like "stadium" and "stage", belong to both "outdoor" and "indoor."

Collecting Crowd Attributes from Web Tags

To build the attribute taxonomy, tags from Pond5 and Getty Images[6] as shown in Fig. 10.7 were collected as a form of wordle.[7] The total number of the retrieved tags

---

[6] Other websites, i.e., YouTube, movies, and existing datasets do not have tag information.
[7] http://www.wordle.net/.

Figure 10.7 Raw tag wordle (partial tag set). Bigger font size of a word suggests higher frequency of a tag appearing in the dataset.

is 7000+. It is laborious and nontrivial to define attributes from these raw tags since the majority of them are not relevant to the problem of interest, even not related to the crowd. In addition, tags with the highest frequency (e.g., people, adult, time, and ethnicity) are also likely to be discarded. Efforts were spent to clean raw tags, and a set of 94 crowd-related attributes were finally constructed and are shown in Fig. 10.6. It includes 3 types of attributes: (1) Where (e.g., street, temple, and classroom), (2) Who (e.g., star, protester, and skater), and (3) Why (e.g., walk, board, and ceremony). A complete list can be found at the project website.[8]

Crowd Attribute Annotation

16 annotators were hired to label attributes in the WWW dataset, and another 3 annotators refined labeling for all videos. All the attributes in the dataset are commonly seen and experienced in daily life, so the annotation did not require special background knowledge. Each annotator was provided with a sample set containing both positive and negative samples of each attribute. They were asked to select possible attributes for each test video in a label tool containing three attribute lists of Where, Who, and Why, respectively. In every round, each annotator was shown a 10-second video clip and was required to label at least one attribute from each attribute list without time constraint.

Fig. 10.8 shows several examples in the WWW crowd dataset. As shown in the first row of Fig. 10.8, not all the video clips can tell a complete story in the form of "somebody" "does something" "somewhere". Therefore, before labeling, "ambiguous" options were added in each attribute list. Totally, the annotators labeled 2855 "ambiguous" among all the marked 980,000 labels, taking 0.1%, 0.2%, and 0.4% in Where, Who, and Why, respectively. Two videos shown in the second row of Fig. 10.8 demon-

---

[8] http://www.ee.cuhk.edu.hk/~jshao/WWWCrowdDataset.html.

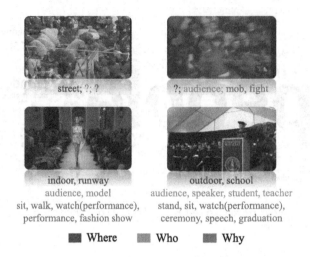

street; ?; ?

?; audience; mob, fight

indoor, runway
audience, model
sit, walk, watch(performance),
performance, fashion show

outdoor, school
audience, speaker, student, teacher
stand, sit, watch(performance),
ceremony, speech, graduation

■ Where   ■ Who   ■ Why

Figure 10.8 Several video examples in the WWW dataset. Both videos in the first row have ambiguous attributes while the other two videos in the second row have multiple attributes in Where, Who, and Why.

strate that a video might have quite a number of attributes, i.e., multiple subjects doing different tasks at different locations in a single video.

10.2.3 User Study on Crowd Attribute

Appearance and motion cues play different roles in crowd scene understanding. In this section, a user study was conducted on the WWW crowd dataset to investigate human performance if only one type of cue was shown. This could serve as a reference to explore the correlation between human perception and computational models.

Eight users were provided with four types of data, including single frame image, background,[9] tracklets, and background with tracklets. The compared ground truth was the set of annotations from whole videos. To avoid bias, every user was provided with all the four types of data and randomly selected 10–15 attributes. Before starting labeling, each annotator was provided with 5–10 positive and negative samples to get familiar with the attributes. Users were informed that their response time would be recorded.

(1) *Response time.* The average response time of all the users was 1.1094 seconds, as shown in Fig. 10.9A. Fig. 10.9B shows that labeling with only tracklets was more laborious, and it was not easy for human to recognize crowd attributes simply from motions without seeing images.

---

[9] The average image of all frames of each video.

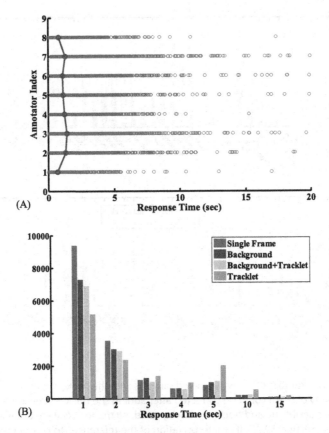

Figure 10.9 Visualization of users' response time. The blue circles in (A) plot the response time of all annotators on labeling tasks, and the red line marks the average response time of each annotator. Panel (B) shows the histograms of response time of different cues. (For interpretation of the references to color in this figure legend, the reader is referred to the web version of this chapter.)

Table 10.3 User accuracy with four types of cue

	Single frame	Background	Background + tracklet	Tracklet	Average
Accuracy	0.82	0.71	0.74	0.41	0.67

(2) *Accuracy*. Table 10.3 shows that with single frames users could achieve much higher accuracy than with only tracklets or background. It means that the appearance of moving people and their poses are useful, but they are blurred on the background image. It was found that the background and tracklet cues were complementary. Fig. 10.10A shows how many samples were wrongly labeled only with the background cue and how many of them were corrected after users had also seen the tracklet cue. Very few failure cases in the first 17 attributes were corrected by the tracklet cues, because these

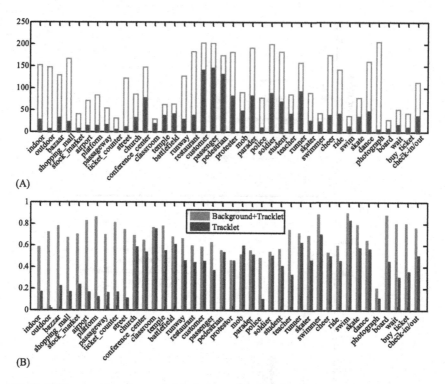

Figure 10.10 (A) The number of wrongly labeled samples with the background cue (indicated by blue bars) and how many of them could be corrected after adding the track let cue. (B) Accuracy comparison between the tracklet cue and tracklet + background. All the results were obtained from the user study described in Section 10.2.3. (For interpretation of the references to color in this figure legend, the reader is referred to the web version of this chapter.)

attributes belong to "where". Tracklets were more effective on the last 23 attributes belonging to "who" and "why". Fig. 10.10B shows that tracklets performed poorly on recognizing attributes belonging to "where".

10.3 CROWD COUNTING AND DENSITY ESTIMATION

Crowd counting is a challenging topic due to severe occlusions, scene perspective distortions and diverse crowd distributions. Since pedestrian detection and tracking has difficulty being used in crowd scenes, most state-of-the-art methods [8,4,34,35] are regression based, and the goal is to learn a mapping between low-level features and crowd counts. However, most recent works are scene-specific, i.e., a crowd counting model learned for a particular scene can only be applied to the same scene. Given an unseen scene or a changed scene layout, the model has to be retrained with new annotations.

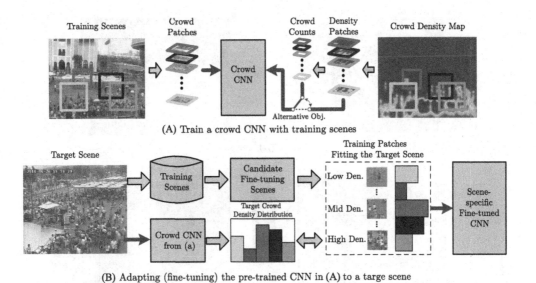

Figure 10.11 Illustration of the cross-scene crowd counting method in [15].

This section will describe a CNN model proposed in [15] for cross–scene crowd counting and density estimation. The network was trained on the Shanghai World Expo'10 crowd dataset. Fig. 10.11 illustrates the overall framework of the proposed CNN. It has the following advantages:

1. The CNN model is trained for crowd scenes by a switchable learning process with two learning objectives, crowd density map and crowd counts. The two different but related objectives can alternatively assist each other to obtain better optima.

2. The target scenes require no extra label in our framework for cross-scene counting. The pretrained CNN model is fine-tuned for each target scene to overcome the domain gap between different scenes. The fine-tuned model is specifically adapted to the new target scene.

3. The framework does not rely on foreground segmentation results because only appearance information is considered in the method. Whether the crowd is moving or not, the crowd texture would be captured by the CNN model, which can obtain a reasonable counting result.

10.3.1 Method

10.3.1.1 Normalized Crowd Density Map for Training

The main objective for the crowd CNN model is a regression problem to recover a mapping $\mathcal{F} : \mathcal{X} \to \mathcal{D}$, where $\mathcal{X}$ is the set of low-level features extracted from training images and $\mathcal{D}$ is the crowd density map of the image. Assuming that the position of each

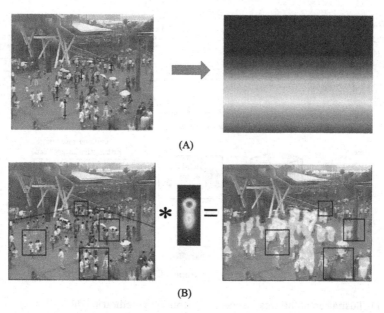

Figure 10.12 (A) Performing the perspective map labeling process. Hot color indicates a high value in the perspective map. (B) The crowd density map and the red box show some training patch randomly cropped from image and density map. The patches cover the same actual area. The ones in the further away regions are smaller and the ones in the closer regions are larger. (For interpretation of the references to color in this figure legend, the reader is referred to the web version of this chapter.)

pedestrian is labeled, the density map is created based on pedestrians' spatial location, human body shape and perspective distortion of images. Patches, randomly selected from the training images, can be treated as training samples, and the density maps of corresponding patches are treated as the ground truth for crowd CNN model. As an auxiliary objective, the total crowd number in a selected training patch can be calculated through the density map integration. Note that the total number will be a decimal, not an integer.

Many works follow [36] and define the density map regression ground-truth as a sum of Gaussian kernels centered on the user annotations. This type of density map is suitable for characterizing the density distribution of circle-like objects such as cells and bacteria. However, this assumption may fail when it comes to the pedestrian crowd, where cameras are generally not in bird-view. An example of pedestrians in an ordinary surveillance camera is shown in Fig. 10.12. It has three visible characteristics: (i) pedestrian images in the surveillance videos have different scales due to perspective distortion; (ii) the shapes of pedestrians are more similar to ellipses than circles; (iii) due to severe occlusions, heads and shoulders are the main guidelines to judge whether there exists a pedestrian at each position. The body parts of pedestrians are not reliable for human annotation. Taking these characteristics into account, the crowd density map is created by the combination of several distributions with perspective normalization.

Perspective normalization is necessary to estimate the pedestrian scales. Inspired by [4], for each scene, several adult pedestrians are randomly selected and labeled from head to toe. Assuming that the mean height of adults is 175 cm, the perspective map M can be approximated through a linear regression as shown in Fig. 10.12A. The pixel value in the perspective map $M(p)$ denotes that the number of pixels in the image represents 1 m at that location in the actual scene. After obtaining the perspective map and the center positions of pedestrian head P_h in the region of interest (ROI), the crowd density map is created as

$$D_i(p) = \sum_{P \in \mathbf{P}_i} \frac{1}{\|\mathbf{Z}\|} (\mathcal{N}_h(p; P_h, \sigma_h) + \mathcal{N}_b(p; P_b, \Sigma)). \tag{10.1}$$

The crowd density distribution kernel contains two terms, a normalized 2D Gaussian kernel $\mathcal{N}_h$ as a head part and a bivariate normal distribution $\mathcal{N}_b$ as a body part. Here P_b is the position of pedestrian body, estimated by the head position and the perspective value. To best represent the pedestrian contour, the variance $\sigma_h = 0.2M(p)$ for the term $\mathcal{N}_h$, and $\sigma_x = 0.2M(p)$, $\sigma_y = 0.5M(p)$ for the term $\mathcal{N}_b$ are chosen. To ensure that the integration of all density values in a density map equals the total crowd number in the original image, the whole distribution is normalized by $\mathbf{Z}$. The crowd density distribution kernel and created density map are visualized in Fig. 10.12(B).

10.3.1.2 Crowd CNN Model

An overview of the crowd CNN model with switchable objectives is shown in Fig. 10.13. The input is the image patches cropped from the ROI in training images. In order to obtain pedestrians at similar scales, the size of each patch at different locations is dependent on the perspective value of its center pixel. Here each patch is constrained to cover a 3 m × 3 m square in the actual scene as shown in Fig. 10.12. Then the patches are warped to 72 pixels × 72 pixels as the input of the crowd CNN model. The crowd CNN model contains 3 convolution layers (conv1–conv3) and three fully connected layers (fc4, fc5, and fc6 or fc7). Conv1 has 32 7 × 7 × 3 filters, conv2 has 32 7 × 7 × 32 filters, and the last convolution layer has 64 5 × 5 × 32 filters. Max pooling layers with 2 × 2 kernel size are used after conv1 and conv2. Rectified linear unit (ReLU), which is not shown in Fig. 10.13, is the activation function applied after every convolution layer and fully connected layer.

An iterative switch process is introduced to alternatively optimize the density map estimation task and count estimation task into the crowd counting task. The main task for the crowd CNN model is to estimate the crowd density map of the input patch. Because two pooling layers exist in the CNN model, the input patch is down-sampled.

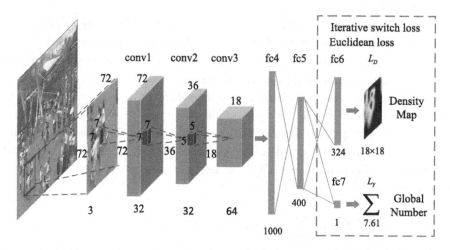

Figure 10.13 The structure of the crowd convolutional neural network. At the loss layer, a density map loss and a global count loss is minimized alternatively.

Therefore, the ground–truth density map is down-sampled to 18×18. Since the density map contains rich and abundant local and detailed information, the CNN model can benefit from learning to predict the density map and obtains a better representation of crowd patches. The total count regression of the input patch is treated as the secondary task, which is calculated by integrating the density map patch. Two tasks alternatively assist each other and obtain a better solution. The two loss functions are defined as:

$$L_D(\Theta) = \frac{1}{N} \sum_i^N \| F_d(X_i; \Theta) - D_i \|^2, \tag{10.2}$$

$$L_Y(\Theta) = \frac{1}{N} \sum_i^N \| F_y(X_i; \Theta) - Y_i \|^2, \tag{10.3}$$

where Θ is the parameter set of CNN model and N is the number of training samples. L_D is the loss between estimated density map $F_d(X_i; \Theta)$ (the output of fc6) and the ground truth density map D_i. Similarly, L_Y is the loss between the estimated crowd number $F_y(X_i; \Theta)$ (the output of fc7) and ground–truth number Y_i. Euclidean loss is adopted in these two objective losses. The loss is minimized using mini-batch gradient descent with standard back-propagation.

The alternative training procedure is summarized in Algorithm 1. L_D is set as the first objective loss to be minimized, since the density map can introduce more spatial information to the CNN model. Density map estimation requires the model to learn a general representation for a crowd. Then after the first objective converges, the model switches to minimize the objective of global count regression. Count regression is more abstract, and its learning converges faster than the task of density map regression. Note that the two objective losses should be normalized to similar scales, otherwise the objective with the larger scale would be dominant in the training process. Empirically, the scale weight of density loss is set to 10, and the scale weight of count loss is 1. The switchable process would not require iterations and the training loss converged after about 6 switch iterations. This switching learning approach can achieve better performance than the widely used multitask learning approach.

Algorithm 1: Training with iterative switch losses.

Input: Training set, namely size-normalized patches with their global number label and density map from the whole training data

Output: Parameters Θ for crowd CNN model

1 set L_D as the first objective;
2 **for** $t = 1$ *to* T **do**
3   BP to learn Θ, until the validation loss drop rate ΔL is less than the threshold ε
4   Switch the objective loss function
5 **end**

10.3.2 Nonparametric Fine-Tuning Method for Target Scene

The crowd CNN model is pretrained based on all training scene data through our proposed switchable learning process. However, each query crowd scene has its unique scene properties, such as different view angles, scales, and different density distributions. These properties significantly change the appearance of a crowd patch and affect the performance of the crowd CNN model. In order to bridge the distribution gap between the training and test scenes, we design a nonparametric fine-tuning scheme to adapt our pretrained CNN model to unseen target scenes. Given a target video from the unseen scenes, samples with similar properties from the training scenes are retrieved and added into training data to fine-tune the crowd CNN model. The retrieval task consists of two steps, candidate fine-tuning scenes retrieval and local patch retrieval.

10.3.2.1 Candidate Fine-Tuning Scene Retrieval

View angle and the scale of a scene are main facts that significantly affect the appearance of a crowd. The perspective map can indicate both the view angle and the scale of the scene as shown in Fig. 10.12A. To overcome the scale gap between different scenes, each input patch is normalized into the same scale, which covers a 3 m × 3 m square in the actual scene according to the perspective map. Therefore, the first step of the nonparametric fine-tuning method focuses on retrieving training scenes that have similar

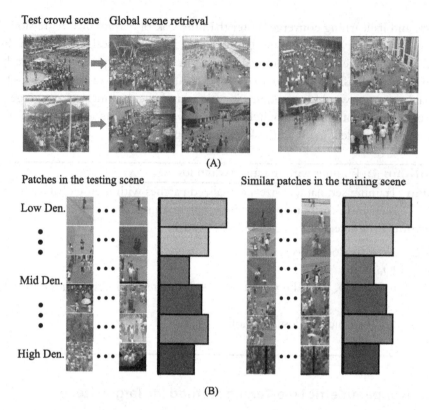

Figure 10.14 Illustration of retrieving local patches similar to those in the test scene to fine-tune the crowd CNN model. (A) Retrieving candidate fine-tuning scenes by matching perspective maps of the training scenes and the test scene. (B) Local patches similar to those in the test scene are retrieved from the candidate fine-tuning scenes.

perspective map with the target scene from all training scenes. Those retrieved scenes are called candidate fine-tuning scenes. A perspective descriptor is designed to represent the view angle of each scene. Since the perspective map is linearly fitted along the y axis, its vertical gradient ΔM_y is used as the perspective descriptor.

Based on the descriptor, for a target unseen scene, the top 20 perspective map similar scenes are retrieved from the whole training dataset as shown in Fig. 10.14A. The retrieved images are treated as the candidate scenes for local patch retrieval.

10.3.2.2 Local Patch Retrieval

The second step is to select similar patches, which have similar density distribution with those in the test scene, from candidate fine-tuning scenes. Besides the view angle and the scale, crowd density distribution also affects the appearance pattern of crowds. Higher density crowd has more severe occlusions, and only heads and shoulders can be observed. On the contrary, in a sparse crowd, pedestrians appear with entire body shapes. Some instances of input patches are shown in Fig. 10.14B. Therefore, the density distribution of the target scene is predicted, and similar patches that match the predicted target density distribution from the candidate fine-tuning scenes are retrieved to obtain ground-truth density maps and counts. For example, for a very dense crowd scene, denser patches should be retrieved to fine-tune the pretrained model to fit the target scene.

With the pretrained CNN model presented in Section 10.3.2, the density and the total count for every patch of the target image can be roughly predicted. It is assumed that patches with similar density map have similar output through the pretrained model. Based on the prediction result, a histogram of the density distribution for the target scene is computed. Each bin is calculated as

$$c_i = \left\lfloor \ln(\hat{y}_i + 1) \times 2 \right\rfloor \tag{10.4}$$

where $\hat{y}_i$ is the integrating count of estimated density map for sample i. Since there rarely exist scenes where more than 20 pedestrians stand in a 3 m × 3 m square, when $\hat{y}_i > 20$, the patch should be clustered into the sixth bin, i.e., $c_i = 6$. Density distribution of the target scene can be obtained from the density map clustering result. Then, patches are randomly selected from the retrieved training scenes, and the number of patches with different densities are controlled to match the density distribution of the target scene. In this way, the proposed fine-tuning method is adopted to retrieve the patches with similar view angles, scales and density distributions. The fine-tuned crowd CNN model achieves better performance for the target scene.

10.3.2.3 Experimental Results

Experiments are evaluated on the crowd counting subset of the Shanghai World Expo'10 dataset. It is split into two parts. 1127 one-minute length video sequences out of 103 scenes are treated as training and validation sets. The test set has 5 one-hour length video sequences from 5 different scenes. There are 120 labeled frames in each test scene which are labeled once for every 30 seconds. The pedestrian number in the test set changes significantly over time ranging from 1 to 220. The existence of large stationary groups makes it hard to detect the foreground area. Thus, most of existing counting methods are not applicable in this dataset, because their methods significantly rely on the segmentation foreground.

The quantitative results of cross-scene crowd counting on the dataset are recorded in Table 10.4. The Mean Absolute Error (MAE) is employed as the evaluation metric. Firstly, LBP features are extracted and ridge regressor (RR) is used to estimate the crowd number, and the results are listed in the top row. The results predicted from the CNN model without fine-tuning as the baseline are shown at the second row. Then the

Table 10.4 Mean absolute errors of the World Expo'10 crowd dataset

Method	Scene 1	Scene 2	Scene 3	Scene 4	Scene 5	Average
LBP+RR	10.4	54.5	27.8	22.6	18.1	26.7
Crowd CNN	10.0	15.4	15.3	25.6	4.1	14.1
Fine-tuned CNN	9.8	13.5	15.0	20.3	4.0	**12.5**
Ke et al. [8]	7.1	49.8	23.5	26.0	15.8	24.4
Luca Fiaschi et al. [37]	6.7	45.1	30.5	33.2	20.6	27.2
Fine-tuned CNN+RR	2.6	32.7	8.8	9.0	3.1	**11.2**

results of the crowd CNN model with data-driven fine-tuning are listed at the third row. These three methods do not use any data from the test scene. For further comparison, two scene-specific methods in [8] and [37] are tested. [8] contains a global number regression method using various hand-crafted features including area, perimeter, edge and LBP, while [37] adopts random regression forest to predict the density map. The compared methods are trained using the first 60 labeled frames for every test scene, and the remaining frames are used as the test set. The proposed cross-scene crowd counting method outperforms the scene-specific methods, even without using any information from test scenes. It is also observed that the data-driven fine-tuning method improves the performance significantly in some test scenes. Similar samples retrieved from training data can help model to better fit the test data. The density estimation results are shown in Fig. 10.15.

It is also observed that some auxiliary labeling in the target scene could boost the performance of our method. As scene-specific information is introduced, most background noise could be eliminated. The predicted density map can be treated as a single feature and ridge regression is used to fit the pedestrian number. Similar to the compared scene-specific method, the first 60 labeled frames are used as training data. The results are further improved for test scene 1 and scene 4 shown in Table 10.4. However, for scene 2, the ridge regression leads to a worse result, because the density distribution in the first 60 training frames has significant differences with the test data.

The iterative switchable learning scheme is also compared with the joint multitask scheme. The joint multitask loss L_J is defined as

$$L_J(\Theta) = L_D(\Theta) + \lambda L_Y(\Theta). \tag{10.5}$$

The average mean absolute errors of the two different losses in the crowd counting dataset are shown in Table 10.5. The iterative switchable training process achieves better performance than the joint multitask loss. Different but related objectives can help each other to obtain better optima through the switchable training objectives. In contrast, the joint multitask scheme requires more computation to obtain an optimal λ than our switchable training process, and the results are also sensitive to the choice of λ.

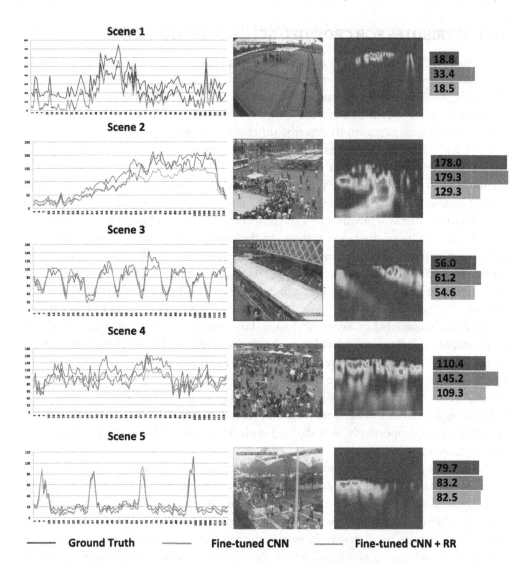

Figure 10.15 The density estimation and counting results in the World Expo'10 crowd counting dataset. (Left) Result curve for each test scene, where the X-axis represents the frame index and the Y-axis means the counting number. (Middle) One sample selected from the corresponding test scene. (Right) The sample's result of the density map estimation and predict count. Best viewed in color.

Table 10.5 Average mean absolute errors (AMSE) via switching training scheme and multitask training scheme

t	1	2	3	4	5	6
AMSE	17.4	15.5	14.9	14.3	14.1	14.3
λ	10	1	0.1	0.05	0.01	0.005
AMSE	50.8	50.8	18.5	15.5	15.3	15.5

10.4 ATTRIBUTES FOR CROWDED SCENE UNDERSTANDING

Understanding crowd behaviors and dynamic properties is a crucial task that has drawn remarkable attention in video surveillance research [5,38–40,10,41,26,9,23,27,21,19,2, 18,42,43]. Most of the above studies [18,5,41,26,38,39,27,23,42,43] on crowd understanding are scene-specific, that is, the crowd model is learned from a specific scene and thus is poor in generalization to describe other novel scenes.

Can we use attributes to characterize generic properties across crowd scenes? Attribute based representations of objects [44–46], faces [47–50], actions [51–53], and scenes [54–57] have been widely employed as an alternative or complement to categorical representations as they characterize the target subject by several attributes rather than discriminative assignment into a single specific category, which is too restrictive to describe the nature of the target subject. In the context of crowd analysis, studies [58, 59] have shown that different crowd systems share similar principles that can be characterized by some common properties or attributes. Indeed, attributes can express more information in a crowd video as they can describe a video by answering "Who is in the crowd?", "Where is the crowd?", and "Why is the crowd here?", but not merely define a categorical scene label or event label to it. For instance, an attribute-based representation might describe a crowd video as the "conductor" and "choir" perform on the "stage" with "audience" "applauding", in contrast to a categorical label like "chorus". In this section, we consider the modeling of these attributes via deep Convolutional Neural Network (CNN). We will make use of the large-scale WWW Crowd Dataset [13, 60] that we introduced in Section 10.2. The dataset contains 10,000 videos from 8257 crowded scenes. Importantly, it is defined with 94 meaningful attributes as high-level crowd scene representations.

Recognizing the rich attributes across different crowded scenes requires learning both the appearance and dynamic representations. The reason is that many attributes, e.g., "applauding", can only be effectively recognized from motion information. Convolutional Neural Networks have shown their remarkable potential in learning appearance representations from images. However, the learning of dynamic representation, and how it can be effectively combined with appearance features for video analysis, remains an open problem.

In existing approaches, a video is treated as a 3D volume and a 2D CNN is simply extended to 3D CNN [61], mixing the appearance and dynamic feature representations in the learned 3D filters. Instead, appearance and dynamic features should be extracted separately, since they are encoded in different ways in videos and convey different information. Ideally, as in most activity analysis studies, objects (i.e., groups or individuals) of interest should be segmented from the background, they should be further allocated into different categories, and tracking should be performed to capture the movements of objects separately. One can then jointly consider the extracted dynamics for global understanding. Unfortunately, this typical pipeline is deemed too challenging for crowd

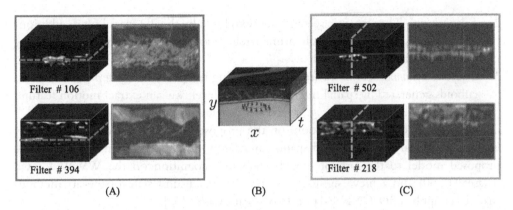

(A)   (B)   (C)

Figure 10.16 From a crowd video volume *ice ballet performance* in (B), several representative semantic feature cuboids and their temporal slices (*xt* and *yt*) are shown in (A) and (C). The temporal slices in the first row of (A) and (C) represent the dynamic patterns of *dancers*. Slices for the background visual patterns are visualized in the second row, where the pattern in (A) corresponds to *background scene* and that in (C) is *audience*.

videos. Alternative solutions include sampling frames along the temporal direction and fusing their 2D CNN feature maps at different levels [62], or feeding motion maps obtained by existing tracking or optical flow methods [63,64]. While computationally more feasible than 3D CNNs, these methods lose critical dynamic information at the input layer.

In this section, we wish to show that appearance and dynamic information can be effectively extracted at a deeper layer of CNN that conveys richer semantical notion (i.e., groups and individuals). In our model design, appearance and dynamics have separate representations yet they interact seamlessly at semantic level. We name our model as *Slicing CNN* (S-CNN). It consists of three CNN branches, each of which adopts different 2D spatio- or temporal-filters. Specifically, the first S-CNN branch applies 2D spatio-slice filters on video volume (*xy*-plane) to extract 3D feature cuboids. The other two CNN branches take the 3D feature cuboids as input and apply 2D temporal-slice filters on the *xt*-plane and *yt*-plane of the 3D feature cuboids, respectively. An illustration of the model is shown in Fig. 10.18.

This design brings a few unique advantages to the task of crowd understanding.

1. *Object-aware* – A 3D feature cuboid generated by a 2D spatial filter records the movement of a particular semantic unit (e.g., groups or individual objects). An example is shown in Fig. 10.16, the feature map from a selected filter of a CNN hidden layer only shows high responses on the ice ballet dancers, while that from another filter shows high responses on the audience. Segregating such semantic classes in a complex scene is conventionally deemed challenging if not impossible for crowd video understanding.

2. *Selectiveness* – The semantic selectiveness exhibited by the 2D spatial filters addition-
    ally guides us to discriminatively prune irrelevant filters such as those corresponding
    to the background clutter.
3. *Temporal dynamics at semantic-level* – By applying temporal-slice filters to 3D feature
    cuboids generated by spatial filters at semantic-level, we can extract motion features
    of different semantic units, e.g., speed and acceleration in x- and y-directions.

We conduct empirical evaluations on the proposed deep structure and thoroughly
examine and analyze the learned spatio- and temporal-representations. We apply the
proposed model to the task of crowd attribute recognition on the WWW Crowd
dataset [13,60] and achieve significant improvements against state-of-the-art methods
that either apply a 3D-CNN [61] or Two-stream CNN [63].

10.4.1 Related Work

Compared to applying CNN to the static image analysis, there are relatively few works
on video analysis [61,65,63,64,13,62,66,67]. A 3D-CNN extends appearance feature
learning in a 2D CNN to its 3D counterpart to simultaneously learn appearance and
motion features on the input 3D video volume [65,61].

It has been reported effective on the task of human action recognition. However, to
capture long-term dependency, larger filter sizes and more layers need to be employed
and the model complexity increases dramatically. To reduce model complexity, Karpathy et al. [62] studied different schemes of sampling frames and fused their features at
multiple stages. These approaches did not separate appearance and dynamic representations. Nevertheless, traditional activity studies always segment objects of interests first
and perform tracking on multiple targets that capture movements of different objects
separately [68–70]. It shows that space and time are not equivalent components and
thus should be learned in different ways. Ignoring such prior knowledge and learning
feature representation blindly would not be effective. Alternatively, two-branch CNN
models [64,63,13,60] have been proposed to extract appearance and dynamic cues separately with independent 2D CNNs and combine them in the top layers. The input
of the motion branch CNN is either 2D motion maps (such as optical flow fields [63]
and dynamic group motion channels [13,60]). Different from 3D convolutions, a two-
branch CNN is at the other extreme, where the extractions of appearance and dynamic
representations have no interactions. These variants are of low cost in memory and calculation, but they inevitably sacrifice the descriptive ability for the inherent temporal
patterns.

Albeit video-oriented CNNs have achieved impressive performances on video related tasks, alternative video representations other than spatial-oriented inputs are still
underexplored. Besides representing a video volume as a stack of spatial xy-slices cut
along the dimension t, previous works have shown that another two representations
of xt-slices in dimension y and yt-slices in dimension x can boost feature learning of

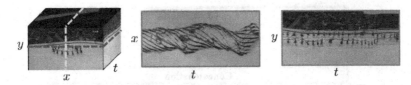

Figure 10.17 The slices over a raw video volume may inevitably mix the dynamics of different objects. For the raw video volume on the left, the *xt*-slice in the middle represents the dynamics of both the *dancers* and *background scene* (i.e., *ice rink*), while the *yt*-slice capture the dynamics of *audience*, *dancers*, as well as *ice rink*.

both appearance and dynamics on a variety of video-tasks [71–76]. Even though they extract the motion feature slices directly from video volumes, they ignore the possibility that multiple objects or instances presented in one slice may occupy distinct motion patterns. Therefore, their dynamic feature representation may mix the motion patterns from different objects and thus fail to describe a particular type of motion patterns. An example is shown in Fig. 10.17. Moreover, the internal properties and connections among different slices were not well learned but just handled independently.

The proposed Slicing CNN model overcomes the limitations listed above. With innovative model design, appearance and dynamic information can be effectively learned from semantic levels, separately and interactively. In addition, the proposed model is capable of extracting appearance and dynamic information from long-range videos (i.e., 100 frames) without sampling or compression.

10.4.2 Slicing Convolutional Neural Network

In this section, we introduce an end-to-end model named as *Slicing CNN* (S-CNN) consisting of three branches [22]. We first learn appearance features by a 2D CNN model on each frame of the input video volume, and obtain a collection of semantic feature cuboids. Each feature cuboid captures a distinct visual pattern, or an object instance/category. Based on the extracted feature cuboids, we introduce three different 2D spatio- and temporal-filters (i.e., *xy*-, *xt*-, and *yt*-) to learn the appearance and dynamic features from different dimensions, each of which is followed by a 1D temporal pooling layer. Recognition of crowd attribute is achieved by applying a classifier on the concatenated feature vector extracted from the feature maps of *xy*-, *xt*-, and *yt*-branch. The complete S-CNN model is shown in Fig. 10.18, and the detailed architecture of the single branch (i.e., S-CNN-*xy*, S-CNN-*xt*, and S-CNN-*yt*) is shown in Fig. 10.21. Their implementation details can be found in Section 10.4.3.

10.4.2.1 Semantic Selectiveness of Feature Maps

Recent studies have shown that the spatial filters in 2D CNNs on image-related tasks posses strong selectiveness on patterns corresponding to object categories and object

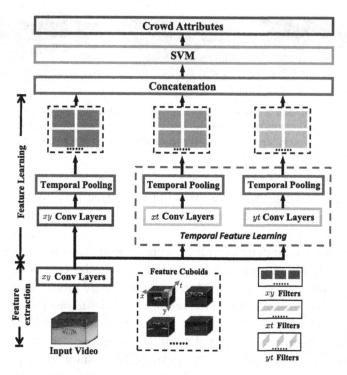

Figure 10.18 The architecture of the three-branch S-CNN model (i.e., S-CNN). The three branches share the same feature extraction procedure in the lower layers while adopt different 2D spatio- and temporal-filters (i.e., *xy-, xt-, yt-*) in feature learning. A classifier (e.g., SVM) is applied to the concatenated features obtained from the three branches for crowd attribute recognition.

identities [77]. Specifically, the feature map obtained by a spatial filter at one intermediate layer of a deep model records the spatial distribution of visual pattern of a specific object. From the example shown in Fig. 10.19, convolutional layers of the VGG model [78] pretrained on ImageNet depict visual patterns in different scales and levels, in which the `conv4_3` layer extracts the semantic patterns in object level. For instance, the filter #26 in this layer precisely captures ice ballet dancers in all frames. Further examining the selectiveness of the feature maps, Fig. 10.20A–C demonstrates that different filters at `conv4_3` layer are possibly linked to different visual patterns. For example, filter #5 indicates the pedestrians on the crosswalk and filter #211 means extremely dense crowd; both of them extract patterns related to crowd. While filter #297 and #212 correspond to background contents like trees and windows of building.

Motivated by the aforementioned observations, we could actually exploit such feature cuboids to separately monitor the movements of different object categories, both spatially and temporally, while reducing the interference caused by the background clutter and irrelevant objects.

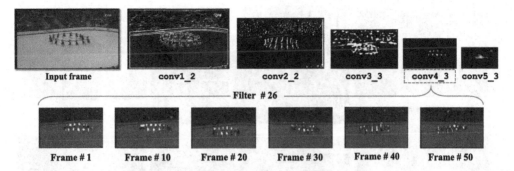

Figure 10.19 Feature responses of selective filters from different convolutional layers of the VGG model, in which `conv4_3` layer owns the best description power for semantic visual patterns in object level. This semantic feature maps precisely capture the *dancers* in ice ballet at all frames presented.

10.4.2.2 Feature Map Pruning

The selectiveness of feature cuboids allows us to design models on a particular set of feature cuboids so as to capture crowd-related dynamic patterns and reject motions from irrelevant background contents. As shown in Fig. 10.20, some feature maps rarely respond to the subjects in crowd but mainly to background regions. How to efficiently learn dynamic feature representations from temporal slices obtained from these feature cuboids? Are all the feature cuboids meaningful to learn dynamic patterns? We answer these questions by pruning spatial filters that generate "irrelevant" feature maps and investigate its impact to the attribute recognition performance.

The "relevance" of a feature map is estimated by investigating their spatial distributions over a fixed validation set of images whose foreground crowds are annotated. The annotation is a binary mask estimated by a crowd segmentation method [20], denoted as $\mathbf{S}_i$ for a query image $i \in \mathcal{I}$, which is then resized to match the resolution of the extracted feature maps. We adopt two scores (i.e., affinity score and conspicuous score) to measure the "relevance".

Affinity Score

The affinity score α_i^n measures the overlap ratio of the crowd foreground instances between the mask $\mathbf{S}_i$ and the nth binarized feature map $\mathbf{F}_i^n \in \mathcal{F}_i$,

$$\alpha_i^n = \|\mathbf{1}_{[\mathbf{F}_i^n > 0]} \bullet \mathbf{S}_i\|_1 / \|\mathbf{S}_i\|_1, \qquad (10.6)$$

where $\mathbf{1}_{[\cdot]}$ is an indicator function that returns 1 when its input argument is true. $\bullet$ denotes the element-wise multiplication.

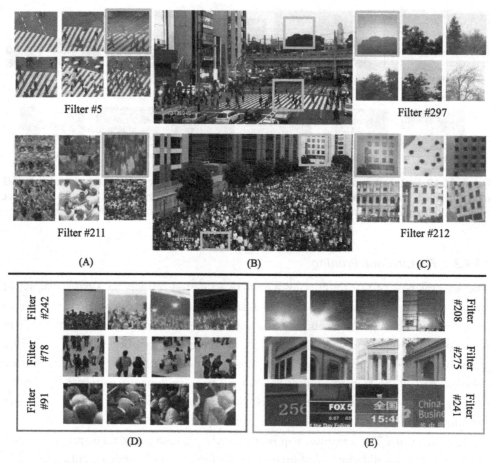

Figure 10.20 Semantic selectiveness of visual patterns by the spatial filters in `conv4_3` layer of the VGG model. The orange patches in (A) and (C) mark the receptive fields of the strongest responses with a certain filter on the given crowd images in (B). The top five receptive fields from images in WWW crowd dataset that have the strongest responses of the corresponding filters are listed aside. (D) and (E) present patches that have strongest responses for the reserved spatial filters and pruned spatial filters.

Conspicuous Score

The conspicuous score κ_i^n calculates the feature's energy inside the crowd foreground annotated in the mask $\mathbf{S}_i$ against its overall energy,

$$\kappa_i^n = \|\mathbf{F}_i^n \bullet \mathbf{S}_i\|_1 / \|\mathbf{F}_i^n\|_1. \tag{10.7}$$

We then construct a histogram $\mathbf{H}$ with respect to the filters in a certain layer. For filter #n, if the feature map $\mathbf{F}_i^n$ satisfies either $\alpha_i^n > \tau_\alpha$ or $\kappa_i^n > \tau_\kappa$, given two thresholds

τ_α and τ_κ, we have the value of its histogram bin as

$$\mathbf{H}(n) = \mathbf{H}(n) + \mathbf{1}_{[\alpha_i^n > \tau_\alpha \cup \kappa_i^n > \tau_\kappa]}, \quad \forall i \in \mathcal{I}. \tag{10.8}$$

By sorting $\mathbf{H}(n)$ in a descending order, we retain the first r spatial filters but prune the left filters. The reserved filters are denoted as $\mathcal{N}_r$.

10.4.2.3 Semantic Temporal Slices

Existing studies typically learn dynamic features from raw video volumes [62] or hand-crafted motion maps [63,13,64]. However, much information is lost at the input layer since it compresses the entire temporal range by subsampling frames or averaging spatial feature maps along the time dimension. Indeed, dynamic feature representations can also be described from 2D temporal slices that cut across 3D volume from another two orthogonal planes, as *xt*- or *yt*-slices shown in Fig. 10.17. They explicitly depict the temporal evolutions of objects, for example, the *dancers* in the *xt*-slice and *audience* in the *yt*-slice.

It is a general case that an *xt*- or *yt*-slice captured from a raw video volume contains motion patterns of multiple objects of different categories, which cannot be well separated since the features that identify these categories always refer to appearance but not motion. For instance, the *yt*-slice in Fig. 10.17 contains motion patterns from *audience*, *dancers*, and ice rink. It is not a trivial task to divide their motion patterns apart without identifying these objects at first.

Motivated by this observation, we propose *Semantic Temporal Slice* (STS) extracted from semantic feature cuboids, which are obtained from the *xy* convolutional layers, as shown in Fig. 10.18. As discussed in the previous subsections, such slices can distinguish and purify the dynamic representation for a certain semantic pattern without the interference from other objects, instances or visual patterns inside one temporal slice. Furthermore, given multiple STSs extracted from different horizontal and vertical probe lines and fed into S-CNN, their information can be combined to learn long-range dynamic features.

10.4.3 S-CNN Deep Architecture

In this section, we provide the architecture details of each branch (i.e., S-CNN-*xy*, S-CNN-*xt*, and S-CNN-*yt*) and their combination (i.e., S-CNN).

10.4.3.1 Single Branch of S-CNN Model

We start with designing a CNN for extracting convolutional feature cuboids from the input video volume. In principle, any kind of CNN architecture can be used for feature extraction. In our implementation, we choose the VGG architecture [78] because of its excellent performance in image-related tasks. As shown in Fig. 10.21, for an input raw

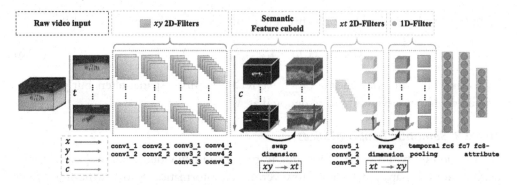

Figure 10.21 Single branch structure (i.e., S-CNN-*xt*). The whole structure is the same as the VGG-16 except the swap layers and the temporal pooling layer. Arrows in different colors denote different dimensions. (i) The first four bunches of convolutional layers in red are 2D convolutions on *xy*-slices, while the last bunch in orange are 2D convolutions on *xt*-slices. Following each bunch of the convolutional layers is a pooling layer. (ii) After the last convolutional layer (i.e., conv5_3), a temporal pooling layer in violet is adopted to fuse cues learned from different *xt*-slices by a 1 × 1 filter. (iii) The first two fully-connected layers both have 4096 neurons while the last one has 94 as the number of crowd attributes. All three branches (i.e., S-CNN-*xt*/-*yt*/-*xy*) use the same structure except with different types of filters. (For interpretation of the references to color in this figure legend, the reader is referred to the web version of this chapter.)

video volume, we first follow the original setting of the lower layers of VGG-16 from conv1_1 to conv4_3[10] to extract spatial feature maps. The size of the feature cuboid $\mathcal{F}_i^s$ of time i is $c \times h_s \times w_s$, where c is the number of feature maps determined by the number of neurons, h_s and w_s denote the size of each feature map in the *xy*-plane. The number of feature cuboid is determined by the input video length τ.

S-CNN-*xy* Branch

The S-CNN-*xy* branch learns spatiotemporal features from the *xy*-plane by *xy*-convolutional filters. Based on the spatial feature cuboids $\{\mathcal{F}_i^s\}_{i=1}^{\tau}$, we continue convolving feature maps with *xy*-filters from conv5_1 to conv5_3, following VGG-16's structure to get the *xy*-temporal feature cuboids with a size of $\tau \times c \times h_t \times w_t$. In other words, there are c *xy* spatiotemporal feature cuboids $\mathcal{F}^{xy}$, each of which is $\tau \times h_t \times w_t$. A 1 × 1 filter is then adopted on each $\mathcal{F}_i^{xy}$ to fuse the temporal information from different frames. The spatiotemporal feature maps $\mathcal{F}^{xy(t)}$ are fed into three fully-connected layers to classify the crowd-related attributes.

---

[10] This structure is used for all experiments except S-CNN-RTS (raw temporal slices from video volume), whose lower layers are not for feature extraction, but also fine-tuned for feature learning.

S-CNN-*xt/-yt* Branch

For the purpose of learning features from $\{\mathcal{F}_i^s\}_{i=1}^T$ by *xt-* or *yt-*branch, we first swap dimensions of the original *xy*-plane to the corresponding *xt-* or *yt-*plane. Take *xt-*branch as an example, as shown in Fig. 10.21, the semantic feature cuboids turn to be $h_s \times c \times \tau \times w_s$ after swapping dimensions. We then substitute the *xy*-convolutional filters used in *xy*-branch with *xt-*filters for conv5_1 to conv5_3 layers. Before temporal pooling at the last stage, again we need to swap dimensions from the *xt-*plane to the *xy*-plane. The following structures are the same as those in the *xy*-branch. The *yt-*branch is similar to the *xt-*branch but contains different convolutional filters.

10.4.3.2 Combined S-CNN Model

After training each branch separately, we fuse the features learned from different spatial and temporal dimensions together by concatenating the spatiotemporal feature maps (i.e., $\mathcal{F}^{xy(t)}$, $\mathcal{F}^{xt(y)}$, and $\mathcal{F}^{(x)ty}$) from three branches with ℓ_1 normalization. Linear SVM is adopted as the classifier for the sake of its efficiency and effectiveness on high-dimensional feature representations. We train an SVM independently for each attribute, thus there are 94 models in total. To train each SVM, we consider videos containing the target attribute as the positive samples and leave all the rest as the negative samples. The complete S-CNN model is visualized in Fig. 10.18.

10.4.4 Experiments

10.4.4.1 Experimental Setting

Dataset

To demonstrate the effectiveness of the proposed S-CNN deep model, we investigate it on the task of crowd attribute recognition with the WWW Crowd Dataset, which is a comprehensive crowd dataset collecting videos from movies, surveillance, and the web. It covers 10,000 videos with 94 crowd attributes including places (Where), subjects (Who), and activities (Why). Following the original setup in [13], we train the models on 720 videos and use a set of 936 videos as validation, while test the results over the rest 1844 videos. These sets have no overlap on scenes to guarantee the attributes are learned scene-independently.

Evaluation Metrics

We adopt both Area Under ROC Curve (AUC) and Average Precision (AP) as the evaluation metrics.[11] AUC is a popular metric for classification and its lower-bound is fixed to 0.5. It fails to carefully measure the performance if the ratio between the positive and negative samples is extremely unbalanced, which is just the case we confront. AP is

---

[11] [13] only uses AUC for evaluation.

effective to evaluate the multiattribute detection performance, which is lower bounded by the ratio of positive samples over all the samples. Its lower bound can be written as $mAP_{lb} = \frac{1}{N_{attr}} \sum_{k=1}^{N_{attr}} |\mathcal{T}_k|/|\mathcal{T}|$, where N_{attr} is the number of attributes, $\mathcal{T}$ is the test set, $\mathcal{T}_k$ is the set of samples with the attribute indexed by k. In our experiments, the theoretical lower bound is 0.067.

Model Pre-Training

As a common practice in most deep learning frameworks for visual tasks, we initialize the proposed S-CNN models with the parameters pretrained on ImageNet. This is necessary since VGG requires diverse data to comprehensively tune its parameters. Although WWW crowd dataset has million of images, the diversity of scenes is low (i.e., around 8000). Specifically, we employ the VGG-16 model with 13 convolutional (`conv`) layers and 3 fully-connected (`fc`) layers. All `conv` layers in S-CNN models are initialized with the pre-trained model while three `fc` layers are randomly initialized by Gaussian distributions. We keep the first two `fc` layers with 4096 neurons followed by Rectified Linear Units (`ReLUs`) and `Dropout` while the last `fc` layer with 94 dimensions (attributes) followed by a cross-entropy loss function.

10.4.4.2 Ablation Study of S-CNN

Level of Semantics and Temporal Range

The unique advantage of S-CNN is that it is capable of learning temporal patterns from semantic layer (higher layer of deep network). In addition, S-CNN can naturally accommodate larger number of input frames due to its effective network design, thus capable of capturing long-range dynamic features.

To understand the benefits of learning long-range dynamic features from semantic level, we compare the recognition performance of the proposed S-CNN models based on semantic temporal slices (STS) extracted from layer `conv4_3` and raw temporal slices (RTS) extracted directly from the video volume. The video length τ has three ranges: 20, 50, and 100 frames, denoted as $S(/R)TS_{[\tau]}$. Due to the hardware limitation of our current implementation, we cannot afford $RTS_{[100]}$ with full spatial information.

Low-level vs. semantic-level temporal slices. In comparison with the results by $RTS_{[\tau]}$, $STS_{[\tau]}$ ($\tau = 20, 50$) is superior especially in mAP scores, as shown in Table 10.6. The results of *xt/yt*-semantic slices in Table 10.7 also reveal that the feature learning stage discovers motion patterns for semantic visual patterns, and they act well as the proxies to convey the motion patterns.

Short-range vs. long-range feature learning. As shown in Table 10.7, $STS_{[100]}$ performs the best and beats the other variants under both evaluation metrics. The result demonstrates that the learned long-range features can actually increase the recognition power to find the crowd attributes that distinctively respond to long-range dynamics but are less likely

Table 10.6 Results of S-CNN-*xys* learned from raw- and semantic-level with different temporal ranges (**bold** indicates the best)

	Mean AUC			Mean AP		
τ	20	50	100	20	50	100
RTS-*xy*	91.06	92.28	—	49.56	52.05	—
STS-*xy*	91.76	92.39	**92.52**	54.97	55.31	**55.67**

Table 10.7 Results of S-CNNs learned from semantic-level with short- and long-range temporal slices. Results in **bold** are the best

		Mean AUC			Mean AP		
Methods	τ	*xy*	*xt*	*yt*	*xy*	*xt*	*yt*
STS	20	91.76	91.11	90.52	54.97	52.38	50.08
STS	100	**92.52**	**93.33**	**92.62**	**55.67**	**59.25**	**57.57**

Table 10.8 Results of $STS_{[100]}$ learned from different number of semantic neurons. Results by single-branch models (*xy*, *xt* and *yt*) and the complete model (*xyt*) are presented

	Mean AUC				Mean AP			
$|\mathcal{N}_r|$	*xy*	*xt*	*yt*	*xyt*	*xy*	*xt*	*yt*	*xyt*
100	91.02	91.70	91.16	92.31	46.83	50.32	48.18	53.14
256	92.61	92.49	92.22	93.49	54.68	54.13	53.26	60.13
256_rnd	91.40	92,32	90.21	92.69	51.87	53.12	46.38	57.03
512	92.52	93.33	92.62	94.04	55.67	59.25	57.57	62.55

to be identified by appearance alone, such as "performance" and "skate." See examples in Fig. 10.22A.

Pruning of Features

Feature pruning is discussed in Section 10.4.2.2. Here we show that by pruning features that are less relevant to the characteristics of crowd, it is promising to observe that the pruned irrelevant features cuboids do not make a significant drop on the performance of crowd attribute recognition. In particular, we prune 412 and 256 feature cuboids respectively out of the total set (i.e., 512) at the layer `conv4_3` with respect to the score defined in Section 10.4.2.2, and retrain the proposed deep models under the same setting as that of $STS_{[100]}$.[12] Their mAUC and mAP are reported in comparison with the results by the default $STS_{[100]}$ in Table 10.8.

Compared with the default model $STS_{[100]}$ with $|\mathcal{N}_r| = 512$, the models with $|\mathcal{N}_r| = 256$ (i) approach to the recognition results by $STS_{[100]}$, (ii) outperform $STS_{[100]}$

---

[12] Without other notations, $STS_{[100]}$ denotes the 100 frames-based S-CNN without feature cuboids pruning.

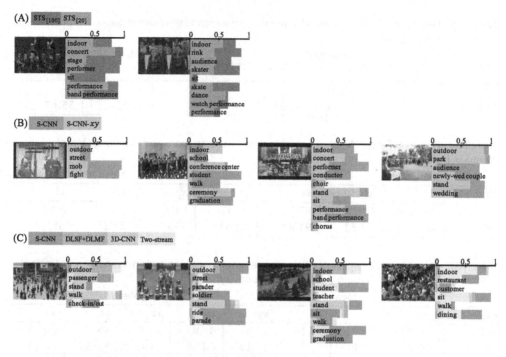

Figure 10.22 Qualitative recognition results on ground truth attributes annotated for the given examples. (A) S-CNN-STS learns on 100 and 20 frames. (B) Results by feeding temporal branch (S-CNN) and without feeding (S-CNN-*xy*). (C) Comparison between S-CNN and state-of-the-art methods. Different colors represent different methods. Bars represent the prediction probabilities. Best viewed in color.

on 13 attributes, 7 of which belong to "why" (e.g.,, "board", "kneel", and "disaster"), and (iii) save about 3% on memory and 34% on time. With 100 feature cuboids remaining, the proposed S-CNN can still perform well, and is superior to the state-of-the-art methods (i.e., DLSF+DLMF [13] and 3D-CNN [61]), even with a single branch. For example, the *xt*-branch has 50.32% mAP which improves the values of 9.1% and 11.2% from DLSF+DLMF and 3D-CNN, respectively, and approaches 51.84% obtained by the Two-stream [63]. To further demonstrate the proposed pruning strategy, we randomly pruned half of the filters ($|\mathcal{N}_r| = 256$_rnd) for the comparison. As observed from Table 10.8, the proposed pruning method performs much better than random pruning, suggesting the effectiveness of the proposed pruning strategy.

The results demonstrate: (i) the relevant spatial features are always companied with top ranks in $\mathbf{H}(n)$, proving the effectiveness of the proposed criteria; and (ii) spatial and dynamic representations can be represented by sparse, yet effective feature cuboids. A small fraction of semantic feature cuboids are enough to fulfill crowd attribute recognition.

Single Branch Model vs. Combined Model

The combination of appearance and dynamic features indeed composes representative descriptions that identify crowd dynamics. Not surprisingly, the combined model integrating the xy-, xt-, and yt-branches outperforms all single-branch models under both evaluation metrics. Under the setting of semantic temporal slices with a temporal range of 100 frames and keeping all feature cuboids, the combined model S-CNN reports remarkable mAUC score of **94.04%** and mAP score of **62.55%**, which improve the optimal results of single-branch models by **3.3%** (reported by the xt-branch) in mAP. The improvement over mAUC is only 0.71%, but it might attribute to the deficiency of evaluation power. As shown in Table 10.8, the S-CNN with $|\mathcal{N}_r| = 100$ and $|\mathcal{N}_r| = 256$ are also superior to the optimal single branch with improvements of 2.82% and 5.45%, respectively.

Qualitative comparisons between the spatial branch S-CNN-xy and the combined model S-CNN are given in Fig. 10.22B, which further demonstrate the significance of the temporal branches as they help improve the performance for most attributes. In particular, for attributes of motion like "mob" and "fight", "sit", "stand", "walk", etc., S-CNN presents a remarkable discriminative power for identification.

10.4.4.3 Comparison with State-of-the-Art Methods

We evaluate the combined Slicing CNN model (S-CNN) with recent state-of-the-art spatio-temporal deep feature learning models:

1. *DLSF+DLMF* [13]. The DLSF+DLMF model is originally proposed for crowd attribute recognition. It is a two-branch model with a late fusion scheme. We employ their published model with the default setting.

2. *Two-stream* [63]. The Two-stream contains two branches as a spatial net and a temporal net. We follow the setting by inputting 10-frame stacking optical flow maps for temporal net as adopted by both [63] and [13]. Besides, the parameters for temporal nets are also initialized with the VGG-16 model, as that in [64] for action recognition.

3. *3D-CNN* [61]. A 3D-CNN uses 3D kernels to learn both appearance and temporal features simultaneously. It thus requires very large memory to capture long-range dynamics. As [61] applied 3D kernels on hand-crafted feature maps, for a fair comparison, we mimic it by extracting features in lower layers of STS$_{[100]}$, and substitute $3 \times 3 \times 3$ 3D kernels for all 2D kernels after `conv4_3` layer and cut off half kernel numbers.[13]

---

[13] It needs 90G to handle 100 frames by the original number of kernels.

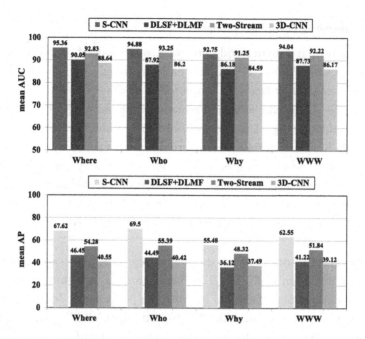

Figure 10.23 Performance comparisons with the referenced methods. The upper one is evaluated by mean AUC and the lower one is by mean AP. The histograms are formed based on the mean scores for attributes of "Where", "Who", and "Why", respectively. "WWW" represents the evaluations on all attributes.

Quantitative Evaluation

As shown in Fig. 10.23, histograms with respect to mAUC and mAP scores are generated to measure the performance on each type of crowd attributes, e.g., "Where", "Who", and "Why", as well as on the complete set "WWW". Clearly, the proposed model outperforms the state-of-the-art methods under both metrics, and shows a large margin (particularly on mAP) over the second best approach in each subcategory. Among the reference methods, the Two-stream presents the best performance in all sub-categories. DLSF+DLMF wins 3D-CNN by the mAUC score on all three attribute types but loses at "Why" by mAP score. The reference methods tend to perform worst on motion-related attributes like "Why" because they can neither capture long-term dynamics as Two-stream or 3D-CNN nor extract dynamic features from specific and hand-craft motion feature maps as DLSF+DLMF. Since the proposed method is able to capture the dynamic feature representations from long-range crowd video and semantically push the features to be crowd-related, its result is thus superior to all other methods. Notice that S-CNN also incorporates the appearance features, which increases the performance of attributes at "Where" and "Who" even further. Even with a prun-

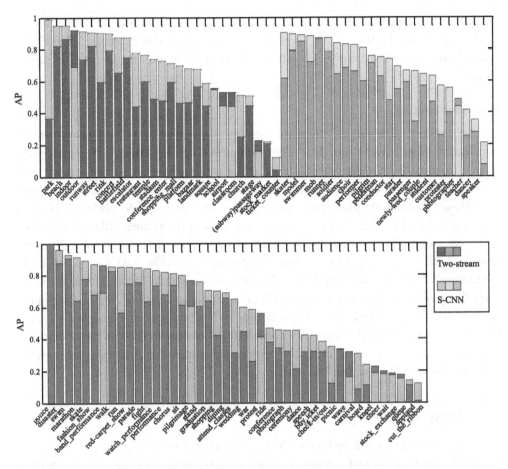

Figure 10.24 Average precision scores for all attributes by S-CNN and Two-stream. The set of bars marked in red, green, and blue refer to "where", "who", and "why", respectively. The bars are sorted according to the larger APs between these two methods. (For interpretation of the references to color in this figure legend, the reader is referred to the web version of this chapter.)

ing of 412 feature cuboids from S-CNN model, it can still reach 53.14% mAP which also outperforms 51.84% of Two-stream [63].

We are also interested in the performance of each attribute. Fig. 10.24 shows the overlapped histograms of average precisions for all attributes by Two-stream and S-CNN. The bars are grouped by their subcategories and sorted in descending order according to the larger AP between these methods at one attribute. It is easy to find that the envelope superimposing this histogram is always supported by the bars of S-CNN with prominent performance gain against Two-stream, while just in 15 attributes the latter wins. Among the failure attributes, most of them contain ambiguities with each other and have low APs for both methods. It means the recognition power is defective to these attributes by the existing deep learning methods. For example, "cheer" and "wave" may be confused with each other, "queue" and "stand" may happen in similar scenes.

Qualitative Evaluation

We also conduct quantitative evaluations for a list of exemplar crowd videos as shown in Fig. 10.22C. The bars are shown as prediction probabilities. Although the probabilities of one attribute do not directly imply its actual recognition results, they uncover the discriminative power of different method as lower probability corresponds to ambiguity or difficulty in correctly predicting one attribute. The proposed S–CNN reliably predicts these attributes with complex or long-range dynamic features, like "graduation" and "ceremony", "parade" and "ride", "check-in/out", etc. Moreover, some attributes that cannot be well defined by motion can also be revealed by S–CNN, for example, "restaurant", "soldier", and "student". The appearance branch of S–CNN indeed captures the inherent appearance patterns belonging to these attributes. Some ambiguous cases do occur, e.g., "outdoor" in top-left and "sit" in bottom-left examples. The top-left instance takes place in a scene of airport/railway station – it is unclear whether the scene is an outdoor or indoor area. The bottom-left instance is a graduation ceremony, in which both "walk" and "sit" coexist.

10.5 CONCLUSION

Scene–independent crowd analysis with deep learning is introduced in this chapter. It covers the topics of crowd counting, crowd density estimation, and crowd attribute recognition. For the success of deep learning, it is critically important to develop large-scale crowd datasets with large scene diversity and rich annotations. The recently developed Shanghai World Expo'10 crowd datasets and WWW crowd datasets are much larger than previous crowd datasets in the numbers of videos, camera views and scenes, and also have richer annotations. To scale the data beyond manually labeled database, in the future one may attempt to leverage realistic data-driven crowd simulations [79] to facilitate the training process.

We show that CNN which alternatively optimizes two objectives of crowd count and crowd density can effectively help training to achieve better local minimum. The proposed data-driven scene matching and fine-tuning can better adapt CNNs to a target scene without extra annotation from the target scene. The proposed slicing CNN can effectively decompose 3D feature maps into 2D spatio- and 2D temporal-slices representations, and therefore can capture dynamics of different semantic units such as groups and objects. It learns separated appearance and dynamic representations while keeping proper interactions between them, and exploits the selectiveness of spatial filters to discard irrelevant background clutter for crowd understanding.

REFERENCES

[1] Wang X, Ma X, Grimson E. Unsupervised activity perception by hierarchical Bayesian models. In: Proc. CVPR. 2007.

[2] Ali X, Shah M. A Lagrangian particle dynamics approach for crowd flow segmentation and stability analysis. In: Proc. CVPR. 2007.

[3] Li J, Gong S, Xiang T. Scene segmentation for behaviour correlation. In: Proc. ECCV. 2008.

[4] Chan AB, Liang ZJ, Vasconcelos N. Privacy preserving crowd monitoring: counting people without people models or tracking. In: Proc. CVPR. 2008.

[5] Wang X, Ma X, Grimson WEL. Unsupervised activity perception in crowded and complicated scenes using hierarchical Bayesian models. IEEE Trans Pattern Anal Mach Intell 2009;31(3):539–55.

[6] Kratz L, Nishino K. Anomaly detection in extremely crowded scenes using spatio-temporal motion pattern models. In: Proc. CVPR. 2009.

[7] Wu S, Moore BE, Shah M. Chaotic invariants of Lagrangian particle trajectories for anomaly detection in crowded scenes. In: Proc. CVPR. 2010.

[8] Chen K, Loy CC, Gong S, Xiang T. Feature mining for localised crowd counting. In: Proc. BMVC. 2012.

[9] Zhou B, Tang X, Wang X. Measuring crowd collectiveness. In: Proc. CVPR. 2013.

[10] Shao J, Loy CC, Wang X. Scene-independent group profiling in crowd. In: Proc. CVPR. 2014.

[11] Yi S, Wang X, Lu C, Jia J. L_0 regularized stationary time estimation for crowd group analysis. In: Proc. CVPR. 2014.

[12] Zhu F, Wang X, Yu N. Crowd tracking with dynamic evolution of group structures. In: Proc. ECCV. 2014.

[13] Shao J, Kang K, Loy CC, Wang X. Deeply learned attributes for crowded scene understanding. In: Proc. CVPR. 2015.

[14] Yi S, Li H, Wang X. Understanding pedestrian behaviors from stationary crowd groups. In: Proc. CVPR. 2015.

[15] Zhang C, Li H, Wang X, Yang X. Cross-scene crowd counting via deep convolutional neural networks. In: Proc. CVPR. 2015.

[16] Yi S, Li H, Wang X. Pedestrian travel time estimation in crowded scenes. In: Proc. ICCV. 2015.

[17] Zhou B, Tang X, Wang X. Learning collective crowd behaviors with dynamic pedestrian-agents. Int J Comput Vis 2015;111(1):50–68.

[18] Chan AB, Vasconcelos N. Modeling, clustering, and segmenting video with mixtures of dynamic textures. IEEE Trans Pattern Anal Mach Intell 2008;30(5):909–26.

[19] Rodriguez M, Sivic J, Laptev I, Audibert JY. Data-driven crowd analysis in videos. In: Proc. ICCV. 2011.

[20] Kang K, Wang X. Fully convolutional neural networks for crowd segmentation. Available from arXiv:1411.4464, 2014.

[21] Ali S, Shah M. Floor fields for tracking in high density crowd scenes. In: Proc. ECCV. 2008.

[22] Shao J, Loy CC, Kang K, Wang X. Slicing convolutional neural network for crowd video understanding. In: Proc. CVPR. 2016.

[23] Hospedales T, Gong S, Xiang T. A Markov clustering topic model for mining behaviour in video. In: Proc. CVPR. 2009.

[24] Zhou B, Wang X, Tang X. Random field topic model for semantic region analysis. In: Proc. CVPR. 2011.

[25] Hassner T, Itcher Y, Kliper-Gross O. Violent flows: real-time detection of violent crowd behavior. In: CVPR workshops. 2012.

[26] Zhou Bolei, Wang Xiaogang, Tang Xiaoou. Understanding collective crowd behaviors: learning a mixture model of dynamic pedestrian-agents. In: Proc. CVPR. 2012.

[27] Mehran R, Oyama A, Shah M. Abnormal crowd behavior detection using social force model. In: Proc. CVPR. 2009.

[28] Toner J, Tu Y, Ramaswamy S. Hydrodynamics and phases of flocks. Ann Phys 2005;318:170–244.

[29] Parrish JK, Edelstein-Keshet L. Complexity, pattern, and evolutionary trade-offs in animal aggregation. Science 1999;284:99–101.

[30] Zhou B, Tang X, Zhang H, Wang X. Measuring crowd collectiveness. IEEE Trans Pattern Anal Mach Intell 2014. http://dx.doi.org/10.1109/TPAMI.2014.2300484.

[31] Zhang C, Kang K, Li H, Xie R, Yang X. Data-driven crowd understanding: the baseline for a large-scale crowd benchmark dataset. IEEE Trans Multimed 2016. http://dx.doi.org/10.1109/TMM.2016.2542585.

[32] Jacobs HA. To count a crowd. Columbia Journal Rev 1967;6:36–40.

[33] Idrees H, Saleemi I, Shah M. Multi-source multi-scale counting in extremely dense crowd images. In: Proc. CVPR. 2013.

[34] Chen K, Gong S, Xiang T, Loy CC. Cumulative attribute space for age and crowd density estimation. In: Proc. CVPR. 2013.

[35] Loy CC, Gong S, Xiang T. From semi-supervised to transfer counting of crowds. In: Proc. ICCV. 2013.

[36] Lempitsky V, Zisserman A. Learning to count objects in images. In: Proceedings of the 23rd international conference on neural information processing systems. 2010. p. 1324–32.

[37] Fiaschi L, Nair R, Koethe U, Hamprecht FA. Learning to count with regression forest and structured labels. In: Proc. ICPR. 2012.

[38] Loy CC, Xiang T, Gong S. Multi-camera activity correlation analysis. In: Proc. CVPR. 2009.

[39] Loy CC, Xiang T, Gong S. Incremental activity modeling in multiple disjoint cameras. IEEE Trans Pattern Anal Mach Intell 2012;34(9):1799–813.

[40] Mahadevan V, Li W, Bhalodia V, Vasconcelos N. Anomaly detection in crowded scenes. In: Proc. CVPR. 2010.

[41] Andrade EL, Blunsden S, Fisher RB. Modelling crowd scenes for event detection. In: Proc. ICPR. 2006.

[42] Loy CC, Xiang T, Gong S. Time-delayed correlation analysis for multi-camera activity understanding. Int J Comput Vis 2010;90(1):106–29.

[43] Loy CC, Hospedales TM, Xiang T, Gong S. Stream-based joint exploration–exploitation active learning. In: Proc. CVPR. 2012.

[44] Farhadi A, Endres I, Hoiem D, Forsyth D. Describing objects by their attributes. In: Proc. CVPR. 2009.

[45] Lampert CH, Nickisch H, Harmeling S. Learning to detect unseen object classes by between-class attribute transfer. In: Proc. CVPR. IEEE; 2009. p. 951–8.

[46] Berg TL, Berg AC, Shih J. Automatic attribute discovery and characterization from noisy web data. In: Computer vision – ECCV 2010. Springer; 2010. p. 663–76.

[47] Kumar N, Berg AC, Belhumeur PN, Nayar SK. Attribute and simile classifiers for face verification. In: Proc. ICCV. IEEE; 2009. p. 365–72.

[48] Luo P, Wang X, Tang X. A deep sum–product architecture for robust facial attributes analysis. In: Proc. ICCV. 2013.

[49] Liu Z, Luo P, Wang X, Tang X. Deep learning face attributes in the wild. In: International conference on computer vision (ICCV). 2015.

[50] Huang C, Loy CC, Tang X. Learning deep representation for imbalanced classification. In: Proc. CVPR. 2016.

[51] Fu Y, Hospedales TM, Xiang T, Gong S. Attribute learning for understanding unstructured social activity. In: Proc. ECCV. Springer; 2012. p. 530–43.

[52] Liu J, Kuipers B, Savarese S. Recognizing human actions by attributes. In: Proc. CVPR. 2011.

[53] Yao B, Jiang X, Khosla A, Lin AL, Guibas L, Fei-Fei L. Human action recognition by learning bases of action attributes and parts. In: Proc. ICCV. IEEE; 2011. p. 1331–8.

[54] Patterson G, Hays J. Sun attribute database: discovering, annotating, and recognizing scene attributes. In: Proc. CVPR. 2012.

[55] Oliva A, Torralba A. Modeling the shape of the scene: a holistic representation of the spatial envelope. Int J Comput Vis 2001;42(3):145–75.

[56] Fei-Fei L, Iyer A, Koch C, Perona P. What do we perceive in a glance of a real-world scene? J Vis 2007;7(1):10.

[57] Parikh D, Grauman K. Interactively building a discriminative vocabulary of nameable attributes. In: Proc. CVPR. 2011.

[58] Castellano C, Fortunato S, Loreto V. Statistical physics of social dynamics. Rev Mod Phys 2009;81(2):591.

[59] Chaté H, Ginelli F, Grégoire G, Raynaud F. Collective motion of self-propelled particles interacting without cohesion. Phys Rev E 2008;77(4):046113.

[60] Shao J, Loy CC, Wang X. Learning scene-independent group descriptors for crowd understanding. IEEE Trans Circuits Syst Video Technol 2016. http://dx.doi.org/10.1109/TCSVT.2016.2539878.

[61] Ji S, Xu W, Yang M, Yu K. 3d convolutional neural networks for human action recognition. IEEE Trans Pattern Anal Mach Intell 2013;35(1):221–31.

[62] Karpathy A, Toderici G, Shetty S, Leung T, Sukthankar R, Fei-Fei L. Large-scale video classification with convolutional neural networks. In: Proc. CVPR. 2014.

[63] Simonyan K, Zisserman Andrew. Two-stream convolutional networks for action recognition in videos. In: Conference on neural information processing systems (NIPS). 2014.

[64] Wang L, Xiong Y, Wang Z, Qiao Y. Towards good practices for very deep two-stream ConvNets. Available from arXiv:1507.02159, 2015.

[65] Baccouche M, Mamalet F, Wolf C, Garcia C, Baskurt A. Sequential deep learning for human action recognition. In: International workshop on human behavior understanding. 2011.

[66] Wang L, Qiao L, Tang X. Mofap: a multi-level representation for action recognition. Int J Comput Vis 2015:1–18.

[67] Xiong Y, Zhu K, Lin D, Tang X. Recognize complex events from static images by fusing deep channels. In: Proc. CVPR. 2015.

[68] Wang H, Klaser A, Schmid C, Liu C. Action recognition by dense trajectories. In: Proc. CVPR. IEEE; 2011. p. 3169–76.

[69] Laptev I. On space-time interest points. Int J Comput Vis 2005;64(2–3):107–23.

[70] Wang H, Schmid C. Action recognition with improved trajectories. In: International conference on computer vision (ICCV). 2013.

[71] Adelson EH, Bergen JR. Spatiotemporal energy models for the perception of motion. J Opt Soc Am A 1985;2(2):284–99.

[72] Ngo C, Pong T, Zhang H. Motion analysis and segmentation through spatio-temporal slices processing. IEEE Trans Image Process 2003;12(3):341–55.

[73] Ricquebourg Y, Bouthemy P. Real-time tracking of moving persons by exploiting spatio-temporal image slices. IEEE Trans Pattern Anal Mach Intell 2000;22(8):797–808.

[74] Niyogi S, Adelson EH, et al. Analyzing gait with spatiotemporal surfaces. In: IEEE workshop on motion of non-rigid and articulated objects. 1994.

[75] Zhao G, Pietikainen M. Dynamic texture recognition using local binary patterns with an application to facial expressions. IEEE Trans Pattern Anal Mach Intell 2007;29(6):915–28.

[76] Ma Z, Chan AB. Crossing the line: crowd counting by integer programming with local features. In: Proc. CVPR. 2013.

[77] Wang L, Ouyang W, Wang X, Lu H. Visual tracking with fully convolutional networks. In: International conference on computer vision (ICCV). 2015.

[78] Simonyan K, Zisserman A. Very deep convolutional networks for large-scale image recognition. Available from arXiv:1409.1556, 2014.

[79] Cheung Ernest, Wong Tsan Kwong, Bera Aniket, Wang Xiaogang, Manocha Dinesh. LCrowdV: generating labeled videos for simulation-based crowd behavior learning. Available from arXiv:1606.08998, 2016.

CHAPTER 11

Physics-Inspired Models for Detecting Abnormal Behaviors in Crowded Scenes

Sadegh Mohammadi*, Hamed K. Galoogahi*, Alessandro Perina†, Vittorio Murino*

*Pattern Analysis and Computer Vision Department, Istituto Italiano di Tecnologia, Genova, Italy
†WDG Core Data Science, Microsoft Corp., Redmond, WA, USA

Contents

11.1 INTRODUCTION

Analyzing the visual content of crowd scenes in videos is increasingly becoming an active research area in computer vision, due to its growing demand in security and surveillance applications. The content of a video captured by surveillance cameras can be potentially monitored by expert personnel for retaining public safety and reducing social crimes in crowded places such as airports, stadiums, and malls. However, this is drastically limited by the scarcity of trained personnel and the natural limitation of human attention capabilities to monitor a huge amount of videos simultaneously filmed by multiple surveillance cameras [1]. This hurdle has motivated vision communities to develop methods for automated analysis of crowd scenes recorded by surveillance cameras.

Group and Crowd Behavior for Computer Vision
DOI: 10.1016/B978-0-12-809276-7.00013-8

Hitherto, numerous computer vision techniques have been successfully developed to detect and understand human activities in video data [2]. These techniques, however, were mainly designed for well-defined environments with only one or a few separable individuals filmed under controlled circumstances, which is not the case in crowd scenarios. This assumption makes the success of such techniques often degraded by the presence of severe occlusions, cluttered background, low quality of surveillance data and, most importantly, by the complex interplays among people involved in a crowd [3]. That has opened up a new broad research line which is generally referred to as *crowd scene analysis* in the computer vision literature [4,5].

Analyzing crowd scenes can be categorized into three topics, including (i) crowd density estimation and people counting, (ii) tracking in crowd, and (iii) modeling crowd behaviors [1]. Estimating the number of people in a crowd is the foremost stage for several real-world applications such as safety control, monitoring public transportation, crowd rendering for animation, and crowd simulation for urban planning. Despite many significant works in this area [6–9], automated crowd density estimation still remains an open problem in computer vision due to extreme occlusions and visual ambiguities of human appearance in crowd images [10]. Tracking individuals (or objects) in crowd scenes is another challenging task [11–13]: other than severe occlusions, cluttered background and pattern deformations, which are common difficulties in visual object tracking, the efficiency of crowd trackers is largely dependent on crowd density and dynamics, people social interactions as well as the crowd's psychological characteristics [14].[1]

The primary goal of modeling crowd behaviors is to identify abnormal events such as riot, panic and violence in crowd scenes [15]. Despite recent success in this research field, detecting crowd abnormalities still remains an open and very challenging problem. The biggest issue of crowd anomaly detection lies in the definition of abnormality as it is strongly context dependent [16–18]. For example, riding a bike in a street is a normal action, whereas it is considered abnormal in another scene with a different context such as a park or sidewalk. Similarly, people gathering for a social event is a normal event, while same gathering at the same place to "protest against a law" is an abnormal event. Another challenge for crowd abnormality detection stems from the lack of adequate training samples to learn a well-generalized crowd behaviors. For instance, only 246 clips are provided by the Violence in Crowds dataset [19] for modeling violence in a crowd, which is not comparable with the UCF-101 dataset [20] with 13K clips of human actions. This drastically degrades the generalization power of current crowd models, since they are not capable of capturing the large intra-class variations of crowd behaviors [1].

---

[1] Readers are referred to [5,10] for a full treatment on the tasks of crowd tracking and density estimation.

In this chapter, we will overview some leading techniques in the computer vision literature designed for detecting abnormal behaviors in crowd, with a focus on existing physics-inspired approaches. This is followed by the introduction of substantial derivative in the fluid mechanics, which is further exploited to model individual's motion dynamics for detecting violence in crowd scenarios.

11.2 CROWD ANOMALY DETECTION: A GENERAL REVIEW

Despite the above limitations, several techniques have been developed to automatically detect abnormal behaviors in crowd scenes, which can coarsely be classified in two classes, namely object-based and holistic techniques [5].

Object-based approaches consider a crowd scene as a set of objects (including individuals) colocated in a common space. These approaches often start by detecting and tracking different objects in the scene of interest, and then modeling crowd behavior with respect to the presence of objects, their motion patterns and their interactions with the physical environment and other objects in the scene [21–23]. The performance of object-based techniques, however, is drastically affected by poor object detection performance and tracking failures, which are inevitable in crowd scenes under severe occlusion and cluttered background. Moreover, this class of approaches is limited to low/medium density of crowd and rarely performs well when low quality footage is provided, which is the common case in real world videos captured by surveillance cameras [24,25]. To cope with above challenges, several works made noticeable efforts to circumvent robustness issues [23,22,26]. For example, Zhao and Nevatia [23] proposed to exploit 3D models of human body along with a Bayesian probabilistic framework to segment, detect and track individuals in the observed scene. To avoid the challenging tasks of object detection and tracking in crowds, [22,26] proposed to track low level features for segmenting moving objects in crowd scenes. To this end, they employed the well-known Kanade–Lucas–Tomasi feature tracker (KLT) [27], and applied a space proximity criterion to cluster similar trajectories [22,26] of the tracked features. The goal of clustering was to obtain a one-to-one association between the points belonging to the same individual (object) and their corresponding trajectories. They demonstrated that compared to objects, tracking low-level features are more robust against background clutter, occlusions, and pattern deformations in crowd scenes.

On the other hand, holistic approaches treat a crowd as a single entity and instead of tracking and detecting objects, they model crowd behaviors by exploiting global features extracted from the entire scene. Existing holistic works basically differ in terms of (i) the type of used features, e.g., spatiotemporal pixel gradients [28,29], optical flow, patch trajectories [30,15,31,32], physics-inspired features [33–37], and more recently heuristic based method [38]; (ii) the type of detected abnormality, e.g., panic [39], violence [19,38,40], or escape [41]; (iii) the type of statistics models adapted for learning and classification to detect abnormalities in crowds; such as Markov Random Fields [42, 43], Bayesian networks [41,15], or probabilistic topic models [44–46].

Typical features used in holistic approaches are spatiotemporal gradients [28,29], thanks to the simplicity and fast computation of these approaches as well as their ability to simultaneously capture appearance and motion information in crowd scenes. However, as noted in the crowd literature [15,47], visual information extracted from crowd's appearance is not reliable due to undesired low image quality, occlusion, and background clutter. This has encouraged some works to discard visual information, and only exploit motion information mainly by means of optical flow [15,19], long trajectories [32], and tracklets [47,48]. More advanced holistic techniques employed physics–inspired features derived from physics concepts such as fluid dynamics, interaction forces, and energy to model behavioral patterns in a crowd as an ensemble of moving particles. A detailed discussion of existing physics–inspired approaches will follow in the next section.

11.3 PHYSICS-INSPIRED CROWD MODELS

In the past, several physics–inspired models have been developed to detect crowd anomalies, which can mainly be divided into Social Force Model, Flow Field Model, and Crowd Energy Model, and other physics–inspired approaches such as substantial derivative, which was borrowed from fluid dynamics to model crowd dynamics.

11.3.1 Social Force Models

The Social Force Model (SFM), originally introduced by Helbing et al. [49], is the historically seminal method for modeling crowd behaviors according to a set of predefined physical rules. More specifically, the SFM aimed to represent the interaction force among individuals in a crowd using the repulsive and attractive forces, which was shown to be a significant feature for analyzing crowd behaviors. Motivated by the success of SFM to reproduce crowd moving patterns, 15 years later Mehran et al. [33] adopted the SFM and particle advection scheme for detecting and localizing abnormal behavior in crowd videos. To this purpose, they considered the entire crowd as a set of moving particles whose interaction force was computed using SFM. Then, they mapped the interaction force into the image plane to obtain the force flow of each particle within frame of videos. This force map was used as the basis for extracting features which, along with the random spatiotemporal path sampling and bag-of-words strategy, was used to assign either normal or abnormal label to each frame.

Inspired by [33], several SFM-based techniques have been proposed to detect anomalies in crowd scenarios [36,50,51]. Raghavendra et al. [36,51] applied the particle swarm optimization (PSO) method for optimizing the interaction force of particles, computed by SFM, such that the population of particles drift toward the main motion in the video sequence. Such displacement is driven by the PSO fitness function, which aims at minimizing the interaction force, so as to model the most diffused and typical crowd behavior [51]. These approaches detect and localize anomalies by checking if some particles do not fit the estimated motion distribution. The method proposed in [50] instead aimed to spatiotemporally locate crowd behavior instability using a new velocity field-based social force model. Unlike the traditional SFM that defines the interaction force as a dependent variable of relative geometric position of the particles, this method can provide a better prediction of interactions using the collision probability in a dynamic crowd. Analyzing spatiotemporal instabilities can help to detect potential abnormality in crowd videos and locate the regions where the abnormalities may occur. Despite the interesting performances of the SFM-based models, recent social psychology studies argued that they are too simplified [52] to capture complex crowd behaviors, other than being heavily affected by a poor generalization power, meaning that a model calibrated on a set of empirical observations may often fail to deal with a different set of observations. Several approaches have been proposed to tackle this problem, behavioral heuristics [52] can be considered as the most notable examples. In particular, unlike SFM-based models which aim at describing complex crowd movements by calibrating a set of forces on empirical observations, this class of approaches defines a set of behavioral heuristic which are formulated using concepts such as velocity and acceleration borrowed from Newtonian mechanics. Motivated by [52], recently behavioral heuristic has been successfully used for detecting abnormality in crowds from video sequences [38], which shows its superiority over SFM-models.

11.3.2 Flow Field Models

The goal of flow field models is to capture crowd regular patterns and find how such patterns change over time. In methods of this type, movements deviating from the regular patterns are considered as anomalies [53,32,54]. The application of modeling flow field was first proposed by Ali et al. [53] for the segmentation of high density crowd flows and detection of flow instabilities. Furthermore, they also extended their framework to anomaly detection. In particular, they constructed a finite time Lyapunov exponent (FTLE) field whose boundaries varied according to the changes of the dynamic flow behavior. New Lagrangian coherent structures (LCS) [55] appeared in the FTLE field, exactly at those locations where the changes happened. The crowd instability was measured by any change in the number of flow segments over time. Detected instabilities can be potentially used to localize anomalies at each frame. This work, however, was limited to structured scenes, in which the boundaries of crowd segments were clearly determined. Differently, Wu et al. [32] proposed a chaotic invariant approach for flow modeling and anomaly detection for both structured and unstructured crowd scenarios. More specifically, particle trajectories computed by particle advection were clustered to obtain representative trajectories for the crowd flow. Next, the chaotic dynamics of all representative trajectories were extracted and quantified using chaotic invariant features. Finally, a probability model was learned from these features to classify a query video of a scene as either normal or abnormal.

11.3.3 Crowd Energy Models

These models focus on simulating how crowd motions transfer from normal to abnormal situation (and vice versa) using different concepts of physics energy such as potential energy [56,57], kinetic energy [58], entropy [59], and pressure [37]. Xiong et al. [57] proposed a novel energy model to detect abnormal pedestrian gathering and running in crowd videos. The model is based on the potential energy [60] and kinetic energy. A term called crowd distribution index (CDI) is defined to represent the dispersion, which can later determine the kinetic energy. Finally, the abnormal activities are detected through threshold analysis. Similarly, Cui et al. [56] proposed to explore the relationships between the current behavior state of a subject and its actions. An interaction energy potential function is introduced to represent the current behavior state of a subject, and velocity is used as its actions, and how the current state changes over time in the form of energy potential. Then, Support Vector Machines are used to find abnormal events. Yang et al. [37] proposed a histogram of oriented pressure (HOP) to detect anomalies in a crowd scene, where the SFM and local binary pattern (LBP) are adopted to calculate the pressure. Cross-histogram is utilized to generate the feature vector instead of parallel merging the magnitude histogram and direction histogram, and a support vector machine and median filter are adopted to detect the anomaly. Ren et al. [59] detected abnormal crowd behavior using behavior entropy. The key idea was to analyze the change of scene behavior entropy (SBE) over time, and localize abnormal behaviors according to pixels' behavior entropy distribution in the image space. Experiments reveal that SBE of the frame will rise when running, dispersion, gathering, or regressive walking occurs.

11.3.4 Substantial Derivative

Substantial derivative is an important concept in fluid mechanics which describes the change of fluid elements by physical properties such as temperature, density, and velocity components of flowing fluid along its trajectory [61]. Unlike aforementioned approaches that only use temporal motion patterns as a main source of information, it has a great capability to encompasses spatial and temporal information of motion changes in a single framework [61]. Particularly, we have proposed a method exploiting the spatiotemporal properties of substantial derivative to detect violent behaviors in various crowd scenarios [40]. In the next section we will present a brief overview on proposed techniques on violence detection in computer vision. Then, we will give a comprehensive overview on substantial derivative equation to model pedestrian's motion dynamics in crowds. Finally, we present an exhaustive experimental section.

11.4 VIOLENCE DETECTION

Violence detection in video sequences is not a novel problem. Despite recent improvements, effective solutions for real-world situations are still unavailable. The first work which appeared on this topic is [62]; it focuses on two person fight episodes and uses motion trajectory information of person's limbs for classification. Besides, only focusing on person–person interactions, this method requires the segmentation of the silhouette, consequently, it is not easily exploitable in crowded scenarios.

More robust methods [63,64,19] only focus on visual cues, and they are all based on the "bag-of-words" paradigm. The differences between [63,64,19] lie in the sampling strategy, the feature descriptor, or the classifier used. For example, [63] used STIP detector and descriptor and linear support vector machines. The approach proposed in [64] employed STIP detection and HOG/MoSIFT descriptor along with the histogram intersection kernel while [18] used random sampling and optical flow magnitude. The success of each of these methods depends on the frame quality and the density of the people involved in violence.

From this brief literature review it is clear that the bag-of-words paradigm preforms well, being especially robust in crowded scenarios. Moreover, the motion descriptors tailored for action recognition often fail in the task of abnormality detection in a crowd due to the unpredictable and sudden changes in crowd motions, which are specific characteristics of violence or riots. Our intuition is that a crowd descriptor should capture the *changes in motion* more than the motion itself which can be captured by higher order derivatives.

This chapter overviews the novel computational framework for detecting abnormal events (i.e., violent and panic behaviors) in video sequences exploiting the substantial derivative [40]. In a nutshell, the substantial derivative equation captures two important properties: (i) *local acceleration* which is a velocity change with respect to time at a given point, occurring when the flow is unsteady, and (ii) *convective acceleration* which is associated with spatial gradients of velocity in the flow field. Convective acceleration occurs when the flow is nonuniform and its velocity changes along is trajectory. Particularly, it is useful to capture useful information in the crowd scenario where the size of parties participating in violence is nonuniform and the structure of motion varies drastically.

Our framework is summarized in Fig. 11.1. First, we extract motion information by means of a dense optic flow (particle advection [53] can also be used). Second, following the substantial derivative equation, we compute *local force* and *convective force* between each consecutive pairs of frames. Then, we follow the standard bag-of-words paradigm for each force separately, sampling P patches and encoding them in K centers. Finally, we concatenate the histograms to form the final descriptor.

In the following, first we cover the main concepts of fluid dynamics and the substantial derivative equation, as well as discuss the parallel between fluids and crowds. Then,

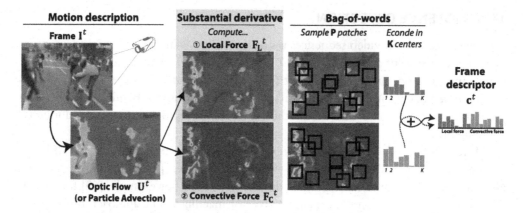

Figure 11.1 Overview of the proposed framework.

we show how the substantial derivative equation can be employed to extract motion primitives, and finally, we present an exhaustive experimental section.

11.4.1 The Substantial Derivative Model

Here, we introduce the main concepts behind the substantial derivative in fluid dynamics, then we discuss about its effectiveness to offer discriminating features to distinguish violent behaviors in crowded scenarios.

11.4.1.1 Substantial Derivative in Fluid Mechanics

Substantial derivative is an important concept in fluid mechanics which describes the change of fluid elements by physical properties such as temperature, density, and velocity components of flowing fluid along its trajectory $(\mathbf{x}, t)$ [61]. In particular, given the velocity components of a certain flowing particle in the x- and y-direction at time t, its velocity flow evolution along its trajectory $\mathbf{U} = U(\mathbf{x}, t)$ can be described as

$$
\begin{aligned}
\frac{D\mathbf{U}}{Dt} &= \frac{\partial \mathbf{U}}{\partial t} + u\frac{\partial \mathbf{U}}{\partial x} + v\frac{\partial \mathbf{U}}{\partial y} \\
&= \frac{\partial \mathbf{U}}{\partial t} + (\mathbf{U} \cdot \nabla)\mathbf{U}
\end{aligned}
\tag{11.1}
$$

where $\frac{D\mathbf{U}}{Dt}$ is the substantial derivative and indicates the *total acceleration* of the flowing particle moving along its trajectory. The term $\frac{\partial \mathbf{U}}{\partial t}$ computes the *local acceleration*. $\frac{\partial \mathbf{U}}{\partial x}$ and $\frac{\partial \mathbf{U}}{\partial y}$ are, respectively, the partial derivatives of velocity field $\mathbf{U}$ to the x- and y-directions. $u = \frac{\partial x}{\partial t}$ and $v = \frac{\partial y}{\partial t}$ are the velocity components of the particle in the x- and y-direction with respect to the time t. $(\mathbf{U} \cdot \nabla)\mathbf{U}$ computes the *convective acceleration* where $\nabla \equiv \frac{\partial}{\partial x} + \frac{\partial}{\partial y}$ is the divergence operator.

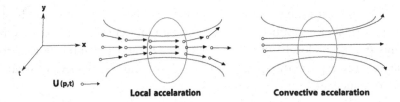

Figure 11.2 An example of local and convective accelerations on a sample frame of video from Violence in Crowds dataset [19]. The local acceleration measures instantaneous rate of change of each fluid particle, while convective acceleration measures the rate of change of the particle moving along its trajectory. Red region indicates the particles that are accelerated as they converge due to the structural change of the environment. (For interpretation of the references to color in this figure legend, the reader is referred to the web version of this chapter.)

The local acceleration captures the change rate of velocity of a certain particle with respect to time and vanishes if its flow is steady. The convective acceleration, on the other hand, captures the change of velocity flow in the spatial space and, therefore, it increases when particles move through the region of spatially varying velocity. In this case, one can say that the local acceleration characterizes the particle velocity field in the temporal domain, while the convective acceleration represents the velocity change due to the spatial variation of the flow particle along its trajectory (see Fig. 11.2 for an example). We incorporate both convective and local accelerations to model the pedestrian motion dynamics in crowd scenes.

11.4.1.2 Modeling Pedestrian Motion Dynamics

In the following, we describe how the substantial derivative can be applied to model pedestrian motion dynamics in a crowd.

Suppose that M pedestrians with masses m_i, $i = 1, \ldots, M$ and corresponding velocities v_i are involved in a crowd. In this case, the total force of each pedestrian i can be represented by $F_i^T = m_i a_i^T$, where $a_i^T = a_i^L + a_i^{Cv}$ is the total acceleration, with a_i^L and a_i^{Cv} being respectively the local and convective accelerations of pedestrian i. m_i is the mass of the pedestrian i. Therefore, the total force of pedestrian/particle i can be expressed as

$$F_i^T = m_i \cdot a_i^T = F_i^L + F_i^{Cv} \tag{11.2}$$

where $F_i^L = m_i \cdot a_i^L$ and $F_i^{Cv} = m_i \cdot a_i^{Cv}$ are indicated as the local and convective forces, respectively caused by the local and convective accelerations.

In abnormal scenarios such as violence, in particular, an individual shows intentional aggressive behaviors against another one with a sudden change of his/her velocity field in the temporal domain (time). This crowd motion pattern can be characterized by the local force F^L. Moreover, the motions of people involved in an abnormal crowd, e.g., violence, are convened by the crowd dynamics and are mainly unpredicted and sudden. These motion changes show the spatial gradients of velocity fields of people within an abnormal crowd situation which can be represented by the convective force F^{Cv}. By integrating the local and convective forces F^L and F^{Cv} into the total force, we can simultaneously capture the spatial structure and temporal changes of motion fields within video sequences.

11.4.1.3 Estimation of Local and Convective Forces from Videos

In this section, we detail the process of estimation of local and convective forces from a video.

As the first step, we computed the optic flow of the video sequence using the algorithm presented in [65] (any other method can be employed). For each frame of the video $\{I^t\}_{t=1}^{N-1}$, the optic flow represents an estimate of the velocity components of each pixel in the x- and y-direction, e.g., $\{v_x^t, v_y^t\}_{t=1}^{N-1}$.

According to Eq. (11.1), the *local acceleration* a^L is the derivative of the velocity with respect to time. By considering a unit time change (per frame), the local acceleration in the x- and y-directions of two consecutive optical flows can be computed by

$$a_x^t = v_x^t - v_x^{t-1} \quad \text{and} \quad a_y^t = v_y^t - v_y^{t-1}. \tag{11.3}$$

Given the two components a^x and a^y, we extract the magnitude of a^L as $a^L = \sqrt{(a_x)^2 + (a_y)^2}$, this is shown in Fig. 11.3.

The *convective acceleration* a^{Cv} captures the spatial evolution of a particle along its trajectory. This requires tracking each particle (individuals in our case) both in the spatial and temporal domain. Tracking individuals, however, is a very challenging task especially in crowded scenarios, and, like in previous work, we resort to particle advection [33]. Following the standard procedure, we placed a homogeneous grid of particles over the video frames and we "advected" them according to the average optic flow over a fixed window of time and as well as space. This is done by a weighted average using a Gaussian kernel. Using the described process, each particle moves with the average velocity of its neighborhood, resembling the collective velocity of a group of people in the crowd.

Given the averaged velocity components that move each particle, e.g., $\{\bar{v}_x^t, \bar{v}_y^t\}_{t=1}^{N-1}$, we compute their spatial derivatives in the x- and y-directions, namely the convective acceleration components

$$\bar{a}_x = \left(\frac{\partial \bar{v}_x}{\partial x} + \frac{\partial \bar{v}_y}{\partial y}\right) \cdot \bar{v}_x \quad \text{and} \quad \bar{a}_y = \left(\frac{\partial \bar{v}_x}{\partial x} + \frac{\partial \bar{v}_y}{\partial y}\right) \cdot \bar{v}_y. \tag{11.4}$$

Finally, the magnitude of the convective acceleration is computed by $a^{Cv} = \sqrt{(\bar{a}_x)^2 + (\bar{a}_y)^2}$. The convective acceleration for a particular frame is shown in Fig. 11.3;

Video frame   **Local force**   **Convective force**

Figure 11.3 Examples of computed local and convective force fields for a video sequence. The image on the left is the video frame, the heatmap in the middle is computed for the local force, and on the right we show convective force underplayed over original frame. Red pixels correspond to the highest force values. (For interpretation of the references to color in this figure legend, the reader is referred to the web version of this chapter.)

for a better visualization we computed it using dense optic flow (e.g., no particle advection) using the same procedure of local acceleration.

Finally, following prior work [33], if we assume that all individuals in a crowd have mass $m_i = 1$, then the local and convective forces are respectively equal to the local and convective accelerations, $F^L = a^L$ and $F^{Cv} = a^{Cv}$.

Given convective and local forces computed for each video, we applied the standard bag-of-words method separately for the local and convective forces. For each video we randomly sampled P patches of size $5 \times 5 \times 5$ and we learned a visual dictionary of size $K = 500$ cluster centers using K-means.[2] In the bag-of-words assumption, each video is encoded by a bag; to compute such bags we assigned each of the P patches to the closest codebook and we pooled together all the patches to generate a histogram over the K visual words. The final descriptor is simply computed by concatenating the histograms of local and convective forces. With a little abuse of notation, in the experiments we will refer to these histogram-descriptors computed from local and convective force as F^L and F^{Cv}, and to the final descriptor as $F^L|F^{Cv}$.

11.5 EXPERIMENTAL RESULTS

We extensively evaluated proposed approach on eight benchmarks, consisting of five standard datasets, Violence in Crowds [19], Violence in Movies [64], Behave [66], Hockey [64], and UCF-101 [20], along with three more benchmarks collected from web sources (i.e., www.YouTube.com) which we named Panic1, Panic2, and Riot in Prison. Fig. 11.4 shows a few samples of the benchmarks. We randomly sampled 3D patches from the computed force maps, and used them for video representation using

---

[2] To employ K-means, we rasterized each patch in a vector of size 125 and we used Euclidean distance.

Figure 11.4 Sample frame of videos taken from Violence in Movies [64], Hockey [64], Violence in Crowds [19], UCF101, Behave [66], Panic1, Panic2, and Riot in Prison. The first row is related to the normal behavior, while the second row is related to the abnormal behavior such as violence and panic.

the bag-of-words paradigm. To generate codewords, we used K-means clustering with 500 cluster centers as visual words. However, using all the random patches to create codewords is computationally expensive, therefore, we selected a subset of the random patches. Then, given the computed cluster centers from the previous step, we assigned each extracted 3D sampled patches to their nearest cluster center using Euclidean distance metric. Subsequently, the resulting histogram of visual word occurrences was fed into an SVM with the histogram intersection kernel for all the standard benchmarks except for Behave dataset [66], and for the other datasets we used LDA model. This is mainly due to the fact that Violence in Crowds [19], Violence in Movies [64], and Hockey [64] contain positive and negative samples, both in training and testing time. Whilst in the other datasets, we assumed that we did not have access to the negative samples, we only relied on positive samples for learning the model, and we tested over the positive and negative samples.

Finally, it is worth mentioning that in order to compensate for the effect of random sampling we repeated each experiment five times and we report the average accuracy over the five runs with its standard deviation.

11.5.1 Datasets

Violence in Crowds [19] is the only dataset specifically assembled for violent scene classification. The sequences were mainly collected from the YouTube, in which videos were recorded in different imaging conditions: camera motion, perspective distortion, various points of view, to name a few. In total, data consists of 246 video sequences, equally divided into violent and nonviolent scenes, 123 each. Originally, the dataset is divided into 5 different folds including half-normal and half-violent scenarios. For the evaluation, we trained a nonlinear SVM on a training set and report the average accuracy over five folds and five repetitions (i.e., to compensate for the effect of random sampling).

Violence in Movies [64] contains 2 classes and 200 videos taken from action recognition datasets and Hollywood action movies. The videos are equally divided into two groups of person-on-person fights (e.g., box competition) and nonfight videos. The dataset is relatively challenging due to the camera motion, clutter background, large variation, etc. For the evaluation, as in [64], we used 5-fold cross-validation.

Hockey [64] dataset constitutes 1000 video sequences collected from the National Hockey League (NHL) from different hockey competitions. There are 2 action categories: violent and nonviolent classes. In total, there are 1000 video sequences divided into 500 normal and 500 fight scenarios. Similar to the original set-up [64], we used 5-fold cross-validation using SVM classifier.

UCF101 is a benchmark which has been widely used for action recognition task. It is well-known to be the largest dataset both in terms of action diversity and imaging conditions. Particularly, it contains 101 action categories, with the presence of viewpoints, camera motion, cluttered background, etc. Similar to [67], we divided video sequences into two groups of fight and nonfight considering "punching" and "Sumo" as violent behaviors and the rest as normal activities. However, we also borrowed some violence sequences from Violence in Crowd [19] and Violence in Movies [64] as acts of violence. In total, we obtained 42278 nonfight clips and 500 fight clips. Finally, in order to keep a balance between training and testing, we randomly sampled 500 videos from normal video sequences. Therefore, the final collection contains 500 normal and 500 abnormal sequences. Compared to Violence in Movies and Hockey datasets, this dataset is more challenging since the definition of abnormality is not only limited to person-on-person fights (e.g., Violence in Movies dataset) or only punching activities (e.g., Hockey dataset). We use 5-fold cross-validation set-up to evaluate the effectiveness of the physics-based approach to detect abnormalities.

Behave [66] dataset was originally assembled for group activity recognition, including walking together, meeting, splitting, ignoring, chasing, running together, and fighting with the number of participants varying from 2 to 7 pedestrians. The dataset contains 20,000 frames. In order to evaluate our method, we consider all the activities excluding fighting and panic as normal activities.

Riot in Prison is a video sequence recorded with a surveillance camera inside a prison. The video starts with a person-on-person fight; then the number of participants increases by involving more prisoners in the fight. Then the situation gets stable after security guard intervention. In total, the dataset contains 3730 frames long, with 1161 being violent frames, and 2569 normal data.

Panic1 is a video sequence recorded with a surveillance camera placed in an outdoor scene, under low light conditions. The videos sequence starts with normal behaviors of pedestrians in the street, but ends with panic behaviors, where individuals run in different directions due to a gun shot. In total, it contains 1742 video frames, including 1582 of normal behavior and 142 frames of panic behavior.

Panic2 is collected from YouTube; it contains 2495 video frames, 2248 of them are normal and 247 show panic.

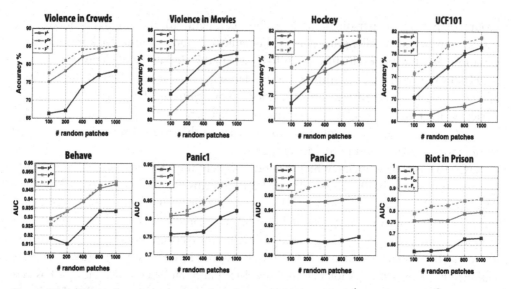

Figure 11.5 Effect of varying number of random patches on local (F^L), convective (F^{Cv}), and total forces (F^T), respectively. (First row) Comparison of average accuracy of proposed forces on Violence in Crowds, and Violence in Movies, Hockey, and UCF101 benchmarks using SVM with 5-fold cross-validation. (Second row) AUCs of proposed forces on Behave, Panic1, Panic2, and Riot In Prison sequences using LDA models. In all the experiments we varied the number of patches, and the results are reported over 5 repetitions.

For the last four datasets, the goal is temporal detection. We divided each sequence into temporally overlapping clips of 15 frames with 5 frame overlap. Then, we described each clip with a bag-of-features, and we tried to detect violence at the clip level. In this case we assumed that we had no access to the abnormal data and we exploited a standard data driven approach: firstly, we learned a Latent Dirichlet Allocation (LDA) to encode the normal data. Therefore, we were able to estimate the likelihood for every clip. Finally, we labeled frames as normal or abnormal based on a fixed threshold on the estimated likelihood. In order to evaluate the effectiveness of the method, we ran each test 5 times and we report the mean AUC along with its 95% confidence interval.

11.5.2 Effect of Sampled Patches

In this experiment, we examined the effect of the number of patches P used for video representation. Fig. 11.5 shows the overall performance when the local, convective, and total forces are varied, which we indicated as F^L, F^{Cv}, and F^T, respectively. For each video, we randomly selected 100, 200, 300, 400, 800, and 1000 patches as features, and performed classification using a standard bag-of-features approach using 500 codewords. Moreover, for the standard datasets such as Violence in Crowds, Violence in Movies, Hockey, and UCF101, which contains negative and positive samples at the training and

Table 11.1 Average AUCs over 5 repetitions and 95% confidence intervals on Behave, Panic1, Panic2, Riot In Prison, and Panic sequences

	Datasets			
Descriptors	**Behave**	**Panic1**	**Panic2**	**Riot in Prison**
Optical flow [65]	0.901 ± 0.03	0.830 ± 0.02	0.89 ± 0.013	0.76 ± 0.05
SFM [33]	0.925 ± 0.00	0.847 ± 0.00	0.89 ± 0.004	0.66 ± 0.02
Jerk [62]	0.913 ± 0.04	0.814 ± 0.01	0.90 ± 0.009	0.65 ± 0.03
Local force	0.933 ± 0.07	0.823 ± 0.01	0.90 ± 0.007	0.68 ± 0.02
Convective force	0.946 ± 0.03	0.885 ± 0.02	0.95 ± 0.002	0.79 ± 0.01
Total force	0.948 ± 0.05	0.891 ± 0.02	0.98 ± 0.005	0.85 ± 0.07

testing steps, we used an SVM classifier with the HIK kernel and we report the results in terms of accuracy. It is also worth mentioning that at the testing time, we used equally many positive and negative samples. While for the other benchmarks, since we assumed that we did not have access to the negative samples, the LDA model was used to learn the normal situation in terms of log-likelihood. We detected the abnormality and report the results in terms of Area Under the Curve (AUC).

Finally, to compensate for the sampling randomness, all the tests were run 5 times. The mean accuracy and AUC with 95% confidence interval are given.

We observe that the best performance is obtained with 1000 random patches per video. In addition, the performance is almost always improving by increasing the number of random patches. This is consistent with the results of random sampling for image classification [68]. Interestingly, one can also observe that in crowded scenarios, especially on Violence in Crowds, Panic1, Panic2, Riot in Prison, and Behave datasets, in which the footages were mainly recorded with the hand-held or surveillance camera, convective force (F^{Cv}) significantly outperformed the local force. Whilst, we observed that the local force (F^L) obtained more appealing results compared with convective force on Violence in Movies, Hockey, and UCF101 datasets, in which the footages are mainly collected from the action movies or under a controlled situation for broadcasting purposes such as boxing and hockey competitions. Therefore, one can observe the importance of the structural information acquired by convective force (F^{Cv}) to disclose underlying dynamic motion patterns that leads to detected abnormalities (panic or violent behavior) in the densely crowded scenes.

However, on all datasets, the best results were consistently achieved by the total force (F^T), which is obtained by considering spatial and temporal information.

Finally, according to Fig. 11.5, all the confidence intervals are very narrow. They demonstrate the effectiveness and consistency of our method in spite of the random sampling for feature extraction.

Table 11.2 Average accuracy over 5 repetitions and 95% confidence intervals on Violence in Crowds, Violence in Movies, Hockey, and UCF-101 sequences

	Datasets			
Descriptors	**Violence in Crowds**	**Violence in Movies**	**Hockey**	**UCF-101**
Optical flow [65]	81.30 ± 0.18	91.31 ± 1.06	84.15 ± 0.18	83.05 ± 0.23
SFM [33]	74.5 ± 0.65	95.51 ± 0.79	81.15 ± 0.32	78.97 ± 0.18
Jerk [62]	74.18 ± 0.85	95.02 ± 0.56	79.62 ± 0.43	78.57 ± 0.83
Local force	78.14 ± 0.92	93.4 ± 0.24	80.35 ± 0.27	79.15 ± 0.55
Convective force	84.3 ± 0.34	92.16 ± 0.13	77.67 ± 0.54	69.87 ± 0.31
Total force	85.43 ± 0.21	96.89 ± 0.21	81.25 ± 0.59	80.90 ± 0.28

11.5.3 Comparison to State-of-the-Art

We compared our descriptors with different physics-based approaches for detecting abnormal behaviors in various scenarios. As baselines, in particular, we considered the bag-of-features representations of the Optical Flow, Jerk [62], and Social Force Model [33]. For all our baselines, we used the very same procedure as in F^L and F^{Cv} setting, with fixed 1000 random 3D patches from each video and 500 codewords.

Tables 11.1 and 11.2 show a comparison of our method with other physics-based approaches. According to the experimental results, F^T consistently obtained good performance in crowded scenes, especially when we do not have access to negative samples at training time (see Table 11.2). Moreover, on Hockey and UCF101, we observed that the local and convective descriptors are not able to match the state-of-the-art. However, experimental results show that the local force outperforms the convective force on Violence in Movies, Hockey, and UCF101, where the footages were mainly collected from action movies. One possible explanation for such good performance of the local force in abnormality detection on action movies confirms the importance of the temporal information to offer discriminating features. However, similar to our previous observation, combination of spatial and temporal information boosts the performance of the classifier to detect abnormalities.

11.6 CONCLUSIONS

In this chapter we gave a comprehensive overview on physics-inspired models to detect abnormal behaviors in crowds. We addressed the ability of the substantial derivative to capture the dynamics of pedestrians based on its spatiotemporal characteristics. The experimental results indicated the importance of spatial information to reveal complex pedestrian dynamics in crowded scenarios. In particular, we demonstrated that the combination of the spatial and temporal motion patterns mostly has a significant effect on the performance of the classifiers. Finally, our descriptor shows its effectiveness not only in various violent situations, but also when considering panic as an abnormal situation.

REFERENCES

[1] Gong S, Loy CC, Xiang T. Security and surveillance. Springer; 2011.

[2] Poppe R. A survey on vision-based human action recognition. Image Vis Comput 2010;28(6):976–90.

[3] Kok VJ, Lim MK, Chan CS. Crowd behavior analysis: a review where physics meets biology. Neurocomputing 2016;177:342–62.

[4] Zhan B, Monekosso DN, Remagnino P, Velastin SA, Xu L-Q. Crowd analysis: a survey. Mach Vis Appl 2008;19(5–6):345–57.

[5] Li T, Chang H, Wang M, Ni B, Hong R, Yan S. Crowded scene analysis: a survey. IEEE Trans Circuits Syst Video Technol 2015;25(3):367–86.

[6] Chan AB, Liang Z-SJ, Vasconcelos N. Privacy preserving crowd monitoring: counting people without people models or tracking. In: IEEE conference on computer vision and pattern recognition. IEEE; 2008. p. 1–7.

[7] Conte D, Foggia P, Percannella G, Tufano F, Vento M. A method for counting people in crowded scenes. In: Seventh IEEE international conference on advanced video and signal based surveillance (AVSS). IEEE; 2010. p. 225–32.

[8] Chen K, Gong S, Xiang T, Loy CC. Cumulative attribute space for age and crowd density estimation. In: 2013 IEEE conference on computer vision and pattern recognition (CVPR). IEEE; 2013. p. 2467–74.

[9] Kilambi P, Ribnick E, Joshi AJ, Masoud O, Papanikolopoulos N. Estimating pedestrian counts in groups. Comput Vis Image Underst 2008;110(1):43–59.

[10] Saleh SAM, Suandi SA, Ibrahim H. Recent survey on crowd density estimation and counting for visual surveillance. Eng Appl Artif Intell 2015;41:103–14.

[11] Rodriguez M, Ali S, Kanade T. Tracking in unstructured crowded scenes. In: 12th IEEE international conference on computer vision. IEEE; 2009. p. 1389–96.

[12] Tang S, Andriluka M, Milan A, Schindler K, Roth S, Schiele B. Learning people detectors for tracking in crowded scenes. In: Proceedings of the IEEE international conference on computer vision. 2013. p. 1049–56.

[13] Kratz L, Nishino K. Tracking with local spatio-temporal motion patterns in extremely crowded scenes. In: 2010 IEEE conference on computer vision and pattern recognition (CVPR). IEEE; 2010. p. 693–700.

[14] Ali S, Shah M. Floor fields for tracking in high density crowd scenes. In: Computer vision – ECCV 2008. Springer; 2008. p. 1–14.

[15] Kratz L, Nishino K. Anomaly detection in extremely crowded scenes using spatio-temporal motion pattern models. In: IEEE conference on computer vision and pattern recognition. IEEE; 2009. p. 1446–53.

[16] Jiang F, Wu Y, Katsaggelos AK. Detecting contextual anomalies of crowd motion in surveillance video. In: 16th IEEE international conference on image processing (ICIP). IEEE; 2009. p. 1117–20.

[17] Leach MJ, Sparks EP, Robertson NM. Contextual anomaly detection in crowded surveillance scenes. Pattern Recognit Lett 2014;44:71–9.

[18] Mohammadi S, Kiani H, Perina A, Murino V. A comparison of crowd commotion measures from generative models. In: Proceedings of the IEEE conference on computer vision and pattern recognition workshops. 2015. p. 49–55.

[19] Hassner T, Itcher Y, Kliper-Gross O. Violent flows: real-time detection of violent crowd behavior. In: 2012 IEEE Computer Society conference on computer vision and pattern recognition workshops (CVPRW). IEEE; 2012. p. 1–6.

[20] Soomro K, Zamir AR, Shah M. UCF101: a dataset of 101 human actions classes from videos in the wild. Available from arXiv:1212.0402.

[21] Rittscher J, Tu PH, Krahnstoever N. Simultaneous estimation of segmentation and shape. In: CVPR, vol. 2. IEEE; 2005. p. 486–93.

[22] Rabaud V, Belongie S. Counting crowded moving objects. In: 2006 IEEE Computer Society conference on computer vision and pattern recognition, vol. 1. IEEE; 2006. p. 705–11.

[23] Zhao T, Nevatia R. Bayesian human segmentation in crowded situations. In: 2003 IEEE Computer Society conference on computer vision and pattern recognition, proceedings, vol. 2. IEEE; 2003. p. II-459.

[24] Marques JS, Jorge PM, Abrantes AJ, Lemos J. Tracking groups of pedestrians in video sequences. In: Conference on computer vision and pattern recognition workshop, vol. 9. IEEE; 2003. p. 101.

[25] Piciarelli C, Micheloni C, Foresti GL. Trajectory-based anomalous event detection. IEEE Trans Circuits Syst Video Technol 2008;18(11):1544–54.

[26] Shi J, Tomasi C. Good features to track. In: 1994 IEEE Computer Society conference on computer vision and pattern recognition, proceedings. IEEE; 1994. p. 593–600.

[27] Tomasi C, Kanade T. Detection and tracking of point features. Pittsburgh: School of Computer Science, Carnegie Mellon University; 1991.

[28] Boiman O, Irani M. Detecting irregularities in images and in video. Int J Comput Vis 2007;74(1):17–31.

[29] Xiang T, Gong S. Video behavior profiling for anomaly detection. IEEE Trans Pattern Anal Mach Intell 2008;30(5):893–908.

[30] Saligrama V, Chen Z. Video anomaly detection based on local statistical aggregates. In: 2012 IEEE conference on computer vision and pattern recognition (CVPR). IEEE; 2012. p. 2112–9.

[31] Krausz B, Bauckhage C. Analyzing pedestrian behavior in crowds for automatic detection of congestions. In: 2011 IEEE international conference on computer vision workshops (ICCV workshops). IEEE; 2011. p. 144–9.

[32] Wu S, Moore BE, Shah M. Chaotic invariants of Lagrangian particle trajectories for anomaly detection in crowded scenes. In: 2010 IEEE conference on computer vision and pattern recognition (CVPR). IEEE; 2010. p. 2054–60.

[33] Mehran R, Oyama A, Shah M. Abnormal crowd behavior detection using social force model. In: 2009 IEEE conference on computer vision and pattern recognition. IEEE; 2009. p. 935–42.

[34] Zhang Y, Qin L, Ji R, Yao H, Huang Q. Social attribute-aware force model: exploiting richness of interaction for abnormal crowd detection. IEEE Trans Circuits Syst Video Technol 2015;25(7):1231–45.

[35] Su H, Yang H, Zheng S, Fan Y, Wei S. The large-scale crowd behavior perception based on spatio-temporal viscous fluid field. IEEE Trans Inf Forensics Secur 2013;8(10):1575–89.

[36] Raghavendra R, Del Bue A, Cristani M, Murino V. Abnormal crowd behavior detection by social force optimization. In: Human behavior understanding. Springer; 2011. p. 134–45.

[37] Yang H, Cao Y, Wu S, Lin W, Zheng S, Yu Z. Abnormal crowd behavior detection based on local pressure model. In: 2012 Asia-Pacific Signal & Information Processing Association annual summit and conference (APSIPA ASC). IEEE; 2012. p. 1–4.

[38] Mohammadi S, Perina A, Kiani H, Murino V. Angry crowds: detecting violent events in videos. In: European conference on computer vision. Springer; 2016.

[39] Haque M, Murshed M. Panic-driven event detection from surveillance video stream without track and motion features. In: 2010 IEEE international conference on multimedia and expo (ICME). IEEE; 2010. p. 173–8.

[40] Mohammadi S, Kiani H, Perina A, Murino V. Violence detection in crowded scenes using substantial derivative. In: 12th IEEE international conference on advanced video and signal based surveillance (AVSS). IEEE; 2015. p. 1–6.

[41] Wu S, Wong H-S, Yu Z. A Bayesian model for crowd escape behavior detection. IEEE Trans Circuits Syst Video Technol 2014;24(1):85–98.

[42] Kim J, Grauman K. Observe locally, infer globally: a space-time MRF for detecting abnormal activities with incremental updates. In: IEEE conference on computer vision and pattern recognition. IEEE; 2009. p. 2921–8.

[43] Zhou S, Shen W, Zeng D, Zhang Z. Unusual event detection in crowded scenes by trajectory analysis. In: 2015 IEEE international conference on acoustics, speech and signal processing (ICASSP). IEEE; 2015. p. 1300–4.

[44] Hospedales T, Gong S, Xiang T. A Markov clustering topic model for mining behaviour in video. In: 12th IEEE international conference on computer vision. IEEE; 2009. p. 1165–72.

[45] Kuettel D, Breitenstein MD, Van Gool L, Ferrari V. What's going on? Discovering spatio-temporal dependencies in dynamic scenes. In: 2010 IEEE conference on computer vision and pattern recognition (CVPR). IEEE; 2010. p. 1951–8.

[46] Wang X, Ma X, Grimson WEL. Unsupervised activity perception in crowded and complicated scenes using hierarchical Bayesian models. IEEE Trans Pattern Anal Mach Intell 2009;31(3):539–55.

[47] Mousavi H, Mohammadi S, Perina A, Chellali R, Murino V. Analyzing tracklets for the detection of abnormal crowd behavior. In: IEEE winter conference on applications of computer vision (WACV). 2015.

[48] Zhou B, Tang X, Wang X. Coherent filtering: detecting coherent motions from crowd clutters. In: Computer vision – ECCV 2012. Springer; 2012. p. 857–71.

[49] Helbing D, Molnar P. Social force model for pedestrian dynamics. Phys Rev E 1995;51(5):4282.

[50] Zhao J, Xu Y, Yang X, Yan Q. Crowd instability analysis using velocity-field based social force model. In: 2011 IEEE visual communications and image processing (VCIP). IEEE; 2011. p. 1–4.

[51] Raghavendra R, Bue AD, Cristani M, Murino V. Optimizing interaction force for global anomaly detection in crowded scenes. In: 2011 IEEE international conference on computer vision workshops (ICCV workshops). IEEE; 2011. p. 136–43.

[52] Moussaïd M, Helbing D, Theraulaz G. How simple rules determine pedestrian behavior and crowd disasters. Proc Natl Acad Sci USA 2011;108(17):6884–8.

[53] Ali S, Shah M. A Lagrangian particle dynamics approach for crowd flow segmentation and stability analysis. In: IEEE conference on computer vision and pattern recognition. IEEE; 2007. p. 1–6.

[54] Loy CC, Xiang T, Gong S. Salient motion detection in crowded scenes. In: 5th international symposium on communications control and signal processing (ISCCSP). IEEE; 2012. p. 1–4.

[55] Shadden SC, Lekien F, Marsden JE. Definition and properties of Lagrangian coherent structures from finite-time Lyapunov exponents in two-dimensional aperiodic flows. Physica D 2005;212(3):271–304.

[56] Cui X, Liu Q, Gao M, Metaxas DN. Abnormal detection using interaction energy potentials. In: 2011 IEEE conference on computer vision and pattern recognition (CVPR). IEEE; 2011. p. 3161–7.

[57] Xiong G, Wu X, Chen Y-l, Ou Y. Abnormal crowd behavior detection based on the energy model. In: 2011 IEEE international conference on information and automation (ICIA). IEEE; 2011. p. 495–500.

[58] Cao T, Wu X, Guo J, Yu S, Xu Y. Abnormal crowd motion analysis. ROBIO 2009;9:1709–14.

[59] Ren W-Y, Li G-h, Chen J, Liang H-z. Abnormal crowd behavior detection using behavior entropy model. In: 2012 international conference on wavelet analysis and pattern recognition (ICWAPR). IEEE; 2012. p. 212–21.

[60] Xiong G, Wu X, Cheng J, Chen Y-L, Ou Y, Liu Y. Crowd density estimation based on image potential energy model. In: 2011 IEEE international conference on robotics and biomimetics (ROBIO). IEEE; 2011. p. 538–43.

[61] Batchelor GK. An introduction to fluid dynamics. Cambridge University Press; 2000.

[62] Datta A, Shah M, da Vitoria Lobo N. Person-on-person violence detection in video. In: 16th international conference on pattern recognition (ICPR). IEEE; 2002. p. 433–8.

[63] de Souza FDM, Chávez GC, do Valle E, Araujo DA, et al. Violence detection in video using spatio-temporal features. In: 23rd SIBGRAPI conference on graphics, patterns and images. IEEE; 2010. p. 224–30.

[64] Nievas EB, Suarez OD, García GB, Sukthankar R. Violence detection in video using computer vision techniques. In: Computer analysis of images and patterns. Springer; 2011. p. 332–9.

[65] Liu C. Beyond pixels: exploring new representations and applications for motion analysis. PhD thesis. 2009.

[66] Blunsden S, Fisher R. The BEHAVE video dataset: ground truthed video for multi-person behavior classification. Ann BMVA 2010;4(4):1–12.

[67] Gracia IS, Suarez OD, Garcia GB, Kim T-K. Fast fight detection. PLoS ONE 2015;10(4):e0120448.

[68] Nowak E, Jurie F, Triggs B. Sampling strategies for bag-of-features image classification. In: Computer vision – ECCV 2006. Springer; 2006. p. 490–503.

CHAPTER 12

Activity Forecasting
An Invitation to Predictive Perception

Kris M. Kitani*, De-An Huang[†], Wei-Chiu Ma[‡]

*The Robotics Institute, Carnegie Mellon University, Pittsburgh, PA, USA
[†]Department of Computer Science, Stanford University, Stanford, CA, USA
[‡]Department of Electrical Engineering and Computer Science, Massachusetts Institute of Technology, Cambridge, MA, USA

Contents

12.1 INTRODUCTION

In the next decade, there will be a paradigm-shift in the focus of computer vision research from image-based recognition, to methods that enable *situational awareness*. Situational awareness is the ability to perceive the potential of the observed world to change. We are particularly interested in developing computational models of situational awareness from observations of human activity in video – which we call *activity forecasting*. Our work in this area of activity forecasting has made significant advances toward building intelligent systems with situational awareness. Prior work has shown various approaches for modeling human activity that explicitly learn how specific environmental variables affect human behavior. Variables such as the passing of time, the physical features of the scene and actions of other people, all affect the projected outcome of how the state of the world will change.

Visual perception can be posed as a two layer process: The first layer being *recognition* and the second layer being *prediction*. Accordingly, a true test of the completeness of machine perception is the ability to both recognize and make predictions about the observed world. The current paradigm in the computer vision community targets the first layer of perception, to recognize the elements of the world. Classical tasks such as

Group and Crowd Behavior for Computer Vision
DOI: 10.1016/B978-0-12-809276-7.00014-X

recognition, reconstruction, and classification aim to map visual observations of environmental elements to geometric or semantic constructs. However, as classical image understanding techniques mature, research efforts will shift toward other challenging problems in perception. The second layer of perception is prediction. In the second layer, perception exhibits the ability to predict (simulate, project, forecast, anticipate, hallucinate) the status of the environment given a change in the variables of the environment. There is a small, yet steadily growing community of computer vision researchers who are shifting their attention to the task of predictive perception. We give a brief review in Section 12.7.

12.2 OVERVIEW

The particular focus of this chapter is to address the task of human activity forecasting using a decision-theoretic approach. We first describe the underlying theory that defines a decision-theoretic model. While a majority of the material presented here can be found in an introductory AI or reinforcement learning text book, we describe the underlying algorithms in terms of human activity forecasting. Our hope is to make the central concepts and algorithms of decision-theoretic models accessible to the vision community. We will cover the Markov decision process, optimal and inverse optimal control, and their maximum entropy variants.

In the latter half, we give three concrete examples of how the decision-theoretic approach can be applied to the task of visual activity forecasting. The task of activity forecasting was first introduced in computer vision research through the task of inferring the future trajectories of people from a single image of a novel scene [1]. To achieve accurate activity forecasting, the approach modeled the effect of the physical environment on the choice of human actions. This was accomplished by the use of state-of-the-art semantic scene understanding combined with ideas from optimal control theory. The unified model also integrated several other key elements of activity analysis, namely, destination forecasting, sequence smoothing, and transfer learning. As a proof-of-concept, the work focused on the domain of trajectory-based activity analysis from visual input. The work proposed was to expand the current scope of vision-based activity analysis by exploring models of human activity that can reason and make predictions about the future. This work is described in Section 12.4.

In multipedestrian scenarios, it is not sufficient to model the effect of the physical scene features on human behavior since the actions of one pedestrian can now influence the environment, namely, the actions of other pedestrian. Expanding on the first task, we focus on modeling and forecasting interactions between multiple pedestrians by using concepts from multiagent game theory [2]. Our hypothesis is that rational pedestrians walking in the presence of other pedestrians will exhibit a heightened level of decision-theoretic planning to avoid collisions. This work examines the effect of social dynamics (e.g., personal spheres of comfort, perceived potential for collision) on pedestrian activities. Our current approach utilizes an instantiation of Brown's classical fictitious play strategy [3] to forecast agent interactions to show that pedestrians exercise moderate levels of strategic planning. Our results show that models that take into account the social-visual dynamics of pedestrian interactions are better at explaining multiagent interactions. This work is described in Section 12.5.

Since our predictive abilities are not limited to trajectory prediction, we also address the task of visual activity forecasting in the context of dual-agent interactions to understand how the actions of one person can be used to predict the actions of another [4,5]. We model dual-agent interactions as an optimal control problem, where the actions of the initiating agent induce a cost topology over the space of reactive poses – a space in which the reactive agent plans an optimal pose trajectory. This optimal control-based representation of human interaction is a fundamentally new way of modeling human interactions for vision-based activity analysis. The technique developed in this work employs a kernel-based reinforcement learning approximation of the soft maximum value function to deal with the high-dimensional nature of human motion and applies a mean-shift procedure over a continuous cost function to infer a smooth reaction sequence. This work is described in Section 12.6.

12.3 ACTIVITY FORECASTING AS OPTIMAL CONTROL

At the heart of our work is a deliberate design choice to model human activities as a sequential process generated by a rational agent. This means that human activity is modeled to be determined through explicit reasoning about the future and the consequences of certain actions on that future – a decision-theoretic model. In the following, we present several classical concepts from reinforcement learning which provides a common thread from which each of the aforementioned tasks are derived. Our aim here is to present the basics of developing a decision-theoretic approach in the context of visual activity forecasting.

12.3.1 Toward Decision-Theoretic Models

The traditional approach to modeling a sequence of human actions assumes a forward model such as a linear Gaussian model, hidden Markov model [6], or more generally a dynamic Bayesian network [7]. The basic assumption is that human activity can be interpreted as a forward propagating sequence of primitives: primitives are positions in the case of pedestrian tracking, or discrete actions in the case of high-level activity recognition. In a forward model, the likelihood of a sequence of states is typically factorized as

$$p(s_{0:T}) = p(s_0) \prod_{t=1}^{T} p(s_t|s_{t-1}), \tag{12.1}$$

a first order Markov chain. An activity is represented as a chain of states s beginning the initial state s_0 and ending at the final state s_T.

The dynamics of the forward model are encoded in terms of the motion model $p(s_{t+1}|s_t)$. In the discrete case, this motion model is typically learned from training data by counting transitions between states. In the case of trajectory tracking, the motion model is often assumed to be a constant velocity model. This assumption is usually true in surveillance videos where the majority of the people are simply in transit according to some underlying floor field induced by the environment.

In the case of activity recognition, the motion model (also called the state transition model) is learned by counting the number of times a certain state s' (a motion primitive) is observed after another state s (a motion primitive). This assumption is often true for sequentially deterministic activities such as assembly tasks or a short motion sequence [6] where the sequence of primitive actions can be enumerated with a grammar. These forward models can also be used to forecast short-term human activity. In fact, this is an integral part of Bayesian filtering for tracking or recognition tasks, where the one-step prediction of the motion model, optionally multiplied with the observation model, is used to obtain the posterior distribution over the current state [8].

A problem arises, however, when the sequential process being modeled is more strategic or less deterministic. Consider the following two cases. In the first case, a person is crossing a street with no traffic. The path of a typical pedestrian would be to cross the street at a constant velocity since there is no need to change direction or speed to avoid traffic. It would be reasonable to model such pedestrian dynamics with a forward model that assumes constant velocity. In the second case, a person is trying to cross a street with heavy traffic from both directions. Each step the pedestrian takes is deliberate, taking into account the predicted position of oncoming cars, while carefully adjusting velocity (possibly stopping at the median strip) to avoid possible collision. The motion of the pedestrian is clearly strategic. The pedestrian takes into account the future consequences of current decisions as the is path planned with care. Forward models would have a hard time modeling such strategic motion because the state transition distribution depends only on the past. Modeling such a strategic sequence of actions calls for a more decision-theoretic model.

One can think of the likelihood of an activity sequence given a decision-theoretic model as conditioning on a new action variable:

$$p(s_{0:T-1}, s_T = g) = p(s_0) \prod_{t=1}^{T} \sum_{a_{t-1}} p(s_t|s_{t-1}, a_{t-1}) p(a_{t-1}|s_{t-1}, s_T = g) p(s_T = g). \tag{12.2}$$

Notice that conditioning on the action variable a has factorized the motion model into a state transition dynamic $p(s_t|s_{t-1}, a_{t-1})$ (it is assumed to be conditionally independent of the goal state g) and a conditional distribution over actions based on the current state and final goal state, $p(a_{t-1}|s_{t-1}, s_T = g)$. This second conditional term is what sets the decision-theoretic model apart from a forward model because it is conditioned on the final goal state $s_T = g$. It is called the *policy* π in reinforcement learning and is typically learned using linear programming or dynamic programming. The policy term is what allows decision-theoretic models take into account the consequences of the future.

12.3.2 Markov Decision Processes and Optimal Control

To model more strategic or deliberate human activities, it is helpful to use methods which can explicitly model the effect of the future on the present. To this end, we use the Markov decision process (MDP) [9] to express the dynamics of a decision-making process. A Markov decision process is defined by a set of states $s \in S$, a set of actions $a \in A$, an initial state distribution $p(s_0)$, a state transition dynamics model $p(s'|s, a)$, a reward function $r(s, a)$ and a discount factor γ.

The state transition dynamics model $p(s'|s, a)$ describes the probability of being in some state s, choosing an action a, and transitioning to state s'. The action a describes bodily motion willed by the person (e.g., extending a leg to walk or retracting the hand to take an object). In most passive vision-based human activity analysis tasks, we can usually assume that a person has full control over his own body. As such, executing a specific action will result in a state transition which is near deterministic. In terms of the MDP, this means that $p(s'|s, a)$ is one for a specific s' as a result of taking action a in state s and zero otherwise. A stochastic state transition dynamics model becomes necessary when the agent being modeled does not have perfect bodily control, such as a robotic system or a person with motor disabilities, where issuing a certain action (e.g., extend right leg) does not guarantee a transition to a desired next state.

The reward function $r(s, a)$ is a concept which is unique to the MDP and more generally to reinforcement learning. In general, the reward is a function of a state–action pair but can also be a function only of the current state $r(s)$ depending on the process being described. The idea is that the person (agent) is receiving a reward (or incurring a cost) for each action and state pair. In the case of modeling human trajectories, one could interpret the reward function to be how much energy is exerted during each state–action transition (i.e., the person is incurring a cost). The concept of the reward is extremely helpful in modeling human activities because it provides an explicit mechanism for encoding human preferences (i.e., a value system) which is distinct from the state transition dynamic. For example, consider the task of modeling the motions of a pedestrian in a park who typically moves according to a constant velocity model but also prefers to walk on the grass when possible. While one could imagine defining a state space over location and environmental properties (grassy) for use with a forward model, the MDP framework allows us to treat the motion model and preference model as separate entities.

The main inference task is to solve for a *policy* using the parameters of the MDP – this is called optimal control or reinforcement learning. Formally, the goal is to maximize the cumulative reward (or an expectation in the stochastic case) of a trajectory s:

$$R(s) = \sum_{t=0}^{\infty} \gamma^t r(s_{t+1}, a_t, s_t) \tag{12.3}$$

where the action a is performed according to the policy $\pi(a|s)$ and the discount factor γ determines the weight of future rewards relative to current rewards. There are a wide variety of reinforcement learning algorithms which can be used to infer the optimal policy such policy iteration, value iteration, Monte Carlo methods, or temporal-difference learning [10]. Here we focus exclusively on value iteration and its maximum entropy variant.

The expected amount of reward over an activity (state–action sequence) starting at s_0 is called the State Value Function,

$$V(s_0) = \mathbb{E}\left[\sum_{t=0}^{\infty} \gamma^t r(s_{t+1}, a_t, s_t)\right]. \tag{12.4}$$

The value function encodes the future (in terms of rewards) when being in a certain state. The value function is often called the "cost to go" from one state to the goal state.

The value function can be estimated using the Value Iteration algorithm, by repeatedly computing the following two lines for every state s:

$$Q(s, a) = r(s, a) + \gamma \sum_{s'} p(s'|s, a) V(s'), \tag{12.5}$$

$$V(s) = \max_a Q(s, a), \tag{12.6}$$

where the state value function V and the state–action value function Q are typically initialized to a large negative value ($-\infty$), the value of the goal state $V(s = g)$ is set to zero, and the reward function r is known (typically normalized values between -1 and 0). The first equation essentially adds the reward to a weighted current estimate of $V(s')$ since the state transition dynamic is deterministic, there is only one state s' that satisfies the choice of a. The state–action value function Q decouples the state value function and a specific action. The second equation takes the action with the maximum Q value and effectively stores the highest possible reward obtained by taking the best action. By iterating these two equations, the value function converges to a set of optimal values for each state.

A deterministic policy $\pi(s)$ can be derived from the value function. The policy describes the optimal action to take when given a state s. The policy is defined as $\pi(s) = \arg\max_a Q(s, a)$. It is important to note here that both values functions are implicitly dependent on the goal state g (i.e., everything must be recomputed if the goal state changes).

To make the idea of the value function more concrete, let us assume that the state is a 2D vector representing a position on the floor plane $s = [x\ \ y]$, otherwise known as grid-world. Our goal is to compute a policy which describes the optimal path for a person to take in this grid world. In this representation, the action a is a transition from one state to another which can be interpreted to be the velocity of the person $a = [\dot{x}\ \ \dot{y}]$. The domain of actions $s \in S$ is typically assumed to be limited to be transitions to the four neighbors of the current state (left, right, up, down). If we assume that the reward function is dependent only on the state, $r(s)$ is simply a numerical value (a reward) received for visiting a state. Given a known goal state s_g, the computed value function is the expected total reward, $V(s) = \mathbb{E}[\gamma r(s_1) + \gamma^2 r(s_2) + \cdots + \gamma^N r(s_N = g)]$. The policy $\pi(s)$ describes the next best action a person should take to obtain the highest reward path to the goal. Applied to the task of trajectory forecasting, we can use the policy to forecast the best sequence of actions that a strategic person would take. Outside of the domain of visual forecasting, this is more widely known as motion planning.

12.3.3 Maximum Entropy Inverse Optimal Control (MaxEnt IOC)

The idea of inverse optimal control is to observe the behavior of an agent in an environment and to mimic that behavior. In the context of human activity forecasting, we observe the behavior of a pedestrian and desire to learn a model so that we can forecast pedestrian behavior. In optimal control, we are given the reward function (and a policy) but in inverse optimal control we are only given demonstrated behavior and must infer the reward function (and a policy). The task of learning a person's reward function from data can be framed as an inverse optimal control problem.

It has been noted [11] that the straightforward use of the MDP for IOC faces problems when there are imperfections in the demonstrated behavior (e.g., suboptimal or noisy estimates of behavior, noisy or unobserved reward factors). To account for such suboptimality, in each of our tasks, we use maximum entropy inverse optimal control (MaxEnt IOC) [12]. Under the maximum entropy model, the reward function can be parametrized as a linear combination of state features,

$$r(s, a; \theta) = \theta^\top f(s, a), \tag{12.7}$$

where θ are the weights of the cost function, $f(s, a)$ is the corresponding vector of features of state s and action a. To be more concrete, in the context of trajectory forecasting, each feature $f_j(s)$ is state dependent and is the response of a certain feature extracted from the image at a state. For example, $f_j(s)$ could denote the output of a grass, pedestrian or car detector centered at state s. The vector $\boldsymbol{\theta}$ is a vector of weights for each feature. By learning the parameters of the reward (cost) function, we are learning how much a feature affects a person's actions. For example, in trajectory forecasting, features such as car, building, or potential collision point will have large weights because they are high cost and should be avoided.

The goal of MaxEnt IOC is then to recover the optimal cost function parameters $\boldsymbol{\theta}$ and consequentially an optimal policy $\pi(a|s)$, by maximizing the entropy of the conditional distribution or equivalently the likelihood maximization of the observations under the maximum entropy distribution:

$$P(\boldsymbol{s}, \boldsymbol{a}; \boldsymbol{\theta}) = \frac{\prod_t e^{r(s_t, a_t)}}{Z(\boldsymbol{\theta})} = \frac{e^{\sum_t \boldsymbol{\theta}^\top f(s_t, a_t)}}{Z(\boldsymbol{\theta})}, \tag{12.8}$$

where $Z(\theta) = \exp \sum_s V(s)$ is the partition function.

To maximize the entropy of Eq. (12.8), we can use exponentiated gradient descent to iteratively minimize the gradient of the log-likelihood $\mathcal{L} \triangleq \log p(\boldsymbol{s}; \boldsymbol{\theta})$. The gradient can be shown to be the difference between the *empirical* mean feature count $\bar{\mathbf{f}} = \frac{1}{M} \sum_m^M f(s_m)$, the average features accumulated over M demonstrated trajectories, and the *expected* mean feature count $\hat{\mathbf{f}}_\theta$, the average features accumulated by trajectories generated by the parameters, $\nabla \mathcal{L}_\theta = \bar{\mathbf{f}} - \hat{\mathbf{f}}_\theta$. We update $\boldsymbol{\theta}$ according to the exponentiated gradient, $\boldsymbol{\theta} \leftarrow \boldsymbol{\theta} e^{\lambda \nabla \mathcal{L}_\theta}$, where λ is the step size and the gradient is computed using a two-step algorithm described next. At test time, the learned weights are held constant and the same two-step algorithm is used to compute the forecast distribution over future actions, the smoothing distribution or the destination posterior.

Backward pass. In the first step (Algorithm 1), we use the current weight parameters $\boldsymbol{\theta}$ and compute the value function (log-partition function) for the distribution over paths between the start and goal, for all possible start locations. The algorithm revolves around the repeated computation of the *state log-partition function* $V(s)$ and the *state–action log-partition function* $Q(s, a)$ defined in Algorithm 1.

The soft value function $V_{\text{soft}}(s)$ can be efficiently updated by a soft analogue of the Bellman equations:

$$Q_{\text{soft}}(s, a) = r(s, a) + \log \sum_{s'} p(s'|s, a) \exp V_{\text{soft}}(s'), \tag{12.9}$$

$$V_{\text{soft}}(s) = \operatorname*{soft\,max}_a Q_{\text{soft}}(s, a). \tag{12.10}$$

Theoretically, this process is repeated until the value function residual converges to a small value ϵ. In practice, the process is typically run for a maximum number of iterations. This process is summarized in Algorithm 1.

Algorithm 1 Backward pass (optimal control).

$V(s) \leftarrow -\infty$
for $n = N, \ldots, 2, 1$ **do**
  $V^{(n)}(s_{goal}) \leftarrow 0$
  $Q^{(n)}(s, a) = r(s, a; \boldsymbol{\theta}) + \log \sum_{s'} p(s'|s, a) \exp V^{(n)}(s')$
  $V^{(n-1)}(s) = \text{soft max}_a \, Q^{(n)}(s, a)$
end for
$\pi_\theta(a|s) = e^{Q(s,a) - V(s)}$

Intuitively, $V_{\text{soft}}(s)$ is now the expected reward over all possible paths that lead to the goal due to the stochastic suboptimality of the policy, instead of the single best path in the case of deterministic dynamics. $Q_{\text{soft}}(s, a)$ can be similarly interpreted as the expected reward over all possible paths that lead to the goal after taking action a from the current state s. The policy derived from the soft value function can be written more generally as $\pi(a|s) = e^{Q_{\text{soft}}(a,s) - V_{\text{soft}}(s)}$. Notice that the policy is stochastic and encodes a distribution over possible actions at each state. This process of deriving the policy based on a known reward function is analogous to optimal control in the language of control theory.

Forward pass. In the second step (Algorithm 2), we propagate an initial distribution $p(s_0)$ according to the learned policy $\pi_\theta(a|s)$, in order to approximate the expected mean feature of the current policy, $\hat{\mathbf{f}}_\theta$. Let $D^{(n)}(s)$ be defined as the *expected state visitation count* which is a quantity that expresses the probability of being in a certain state s at time step n. Initially, when n is small, $D^{(n)}(s)$ is a distribution that sums to one. However, as the probability mass is absorbed by the goal state, the sum of the state visitation counts quickly converges to zero. By computing the total number of times each state was visited $D(s) = \sum_n D^{(n)}(s)$, we are computing the unnormalized marginal state visitation distribution. We can compute the *expected* mean feature count as a weighted sum of feature counts $\hat{\mathbf{f}}_\theta = \sum_s f(s) D(s)$. This process is summarized in Algorithm 2.

Algorithm 2 Forward pass.

$D(s_{initial}) \leftarrow 1$
for $n = 1, 2, \ldots, N$ **do**
  $D^{(n)}(s_{goal}) \leftarrow 0$
  $D^{(n+1)}(s) = \sum_{s',a} P_{s',a}^s \pi_\theta(a|s') D^{(n)}(s')$
end for
$D(s) = \sum_n D^{(n)}(s)$
$\hat{\mathbf{f}}_\theta = \sum_s f(s) D(s)$

Figure 12.1 Given a single pedestrian detection, our proposed approach forecasts plausible paths and destinations from noisy vision-input.

12.4 SINGLE AGENT TRAJECTORY FORECASTING IN STATIC ENVIRONMENT

We first apply the decision-theoretic framework to forecast walking trajectories from a single image as shown in Fig. 12.1. In this case, the state s represents a physical location in world coordinates $s = [x, y]$ and the action a is the velocity $a = [v_x, v_y]$ of the actor. The policy $\pi(a|s)$ maps states to actions, describing which direction to move (action) when an actor is located at some position (state). Given demonstrated walking trajectories, we aim to recover the underlying reward function, and thus obtain the policy that can be used to infer the future action or walking direction of people from images.

As stated in Eq. (12.7), the features define the expressiveness of our reward function, and thus characterize the predictive power of our model. Although reasoning about future actions often requires a large amount of contextual prior knowledge, let us consider the information that can be gleaned from *physical scene features*. For example, in observing pedestrians navigating through an urban environment, we can predict with high confidence that a person will *prefer* to walk on sidewalks more than streets, and will most certainly avoid walking into obstacles like cars and walls. Understanding the concept of human *preference* with respect to physical scene features enables us to perform higher levels of reasoning about future human actions.

We encode information about the environment into the reward function in the following way. Given a scene and the goal of an agent, our reward or cost function is defined as:

$$r(s; \boldsymbol{\theta}) = \boldsymbol{\theta}^\top \boldsymbol{f}(s), \quad \boldsymbol{f}(s) = [f_1(s) \cdots f_K(s)]^\top, \tag{12.11}$$

where each $f_k(s)$ is the response of a physical scene feature, such as the output of a grass, pavement, or car classifier. The physical attributes were extracted using the scene segmentation labeling algorithm of [13]. In total 9 semantics labels were used, including grass, pavement, sidewalk, curb, person, building, fence, gravel, and car. A visualization of the probability maps used as features is shown in Fig. 12.2.

Using this representation of the reward function, we can utilize inverse optimal control to recover the optimal feature weights that can then be applied to any new

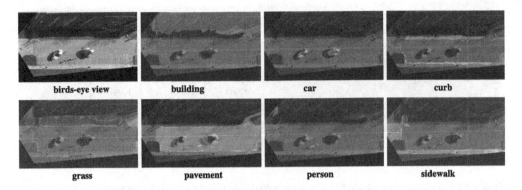

Figure 12.2 Classifier feature response maps. Top left is the original image.

Figure 12.3 Destination forecasting and path smoothing. Our proposed approach infers a pedestrians likely destinations as more noisy observations become available. Concurrently, the *smoothing distribution* (likely paths up to the current time step *t*) and the *forecasting distribution* (likely paths from *t* until the future) are modified as observations are updated.

scene to predict pedestrian trajectories. Fig. 12.3 shows the forecasting results of our method. It is important to note here that a critical assumption has been placed on the nature of the reward function. The features of the scene such as the location of the grass, cars, or pavement are assumed to be static and unchanging. While this assumption of a static scene may hold true in an empty parking lot, in general, we can assume that the environment will contain other dynamic and rational agents (i.e., other people) that have the ability both reason and move.

We evaluate our forecasting performance using the VIRAT ground dataset [14]. We compare how well a learned policy is able to describe a single annotated test trajectory. This can be measured by the negative log-loss (NLL) of a test trajectory *s* given the

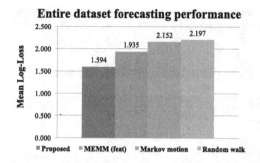

NLL	Proposed	MEMM	MarkovMot
approach	**1.657**	1.962	2.157
depart	**1.618**	1.940	2.103
walk	**1.544**	2.027	2.174
approach	**1.519**	1.780	2.180
depart	**1.519**	1.903	2.115
walk	**1.707**	1.997	2.182

Figure 12.4 Average forecasting NLL for the whole dataset and per activity category.

learned policy $\pi(a|s)$. The negative log-loss,

$$\text{NLL}(s) = \mathbb{E}_{\pi(a|s)}\left[-\log\prod_t \pi(a_t|s_t)\right], \qquad (12.12)$$

is the expectation of the log-likelihood of a trajectory s under a policy $\pi(a|s)$. In our case, this metric measures the probability of drawing the demonstrated trajectory from the learned distribution over all possible trajectories.

We compare against a maximum entropy Markov model (MEMM) that estimates the policy only based on features of the potential next states. The policy is thus computed by:

$$\pi(a|s) \propto \exp\{w_a^\top F(s)\}. \qquad (12.13)$$

This model has no concept of the future beyond a one-step prediction model.

We also compare against a location-based Markov motion model, which learns a policy from observed statistics of states and actions in the training set:

$$\pi(a|s) \propto c(a, s) + \alpha, \qquad (12.14)$$

where $c(a, s)$ is the number of times the action a was observed in state s and α is a pseudo-count used to smooth the distribution via Laplace smoothing.

Fig. 12.4 shows how our proposed model outperforms the baseline models. Our model takes into account the future beyond a one-step prediction, and can better model the human walking trajectory.

12.5 MULTIAGENT TRAJECTORY FORECASTING

In the previous section, we have shown how the decision-theoretic framework can be applied to predict the trajectory of a *single* pedestrian. We encode physical scene fea-

(a) Input Image   (b) Prediction 1   (c) Prediction 2   (d) Our Prediction

Figure 12.5 A scene with multiple agents requires higher levels of reasoning to make better predictions about the future.

tures into the reward function to model human preference when navigating through a *static* environment. These features, however, are not sufficient for reasoning about human behavior when there are *multiple* pedestrians in the environment. As shown in Fig. 12.5, the actions of one pedestrian can influence the actions of other pedestrians, resulting in a complex interdependent decision making process. In order to account for such multiagent scenario and forecast all trajectories simultaneously, we utilize concepts from game theory to computationally mimic the ability to reason about the dynamics of interactive social processes. We take into account the interaction between pedestrians while retaining the efficiency of single agent inference by decomposing the forecasting problem into a sequence of short-term predictions. *Fictitious Play* (FP) [3] is a strategy used in early game theory that models the strategy of each agent in a multiplayer game using an empirical (previously observed) distribution. In FP, each agent alternatively takes an action that best responds to its beliefs about the strategies of other agents, and updates its belief as the game progresses. This has strong parallels to multipedestrian navigation, where each pedestrian first predicts the future movement of other pedestrians then preemptively plans her path according to her beliefs about how other pedestrians will move in the future. Throughout this section, we follow the same notations as defined in Section 12.4.

Formally, at every time step t, each pedestrian i predicts the future movement of others using their respective policy $\pi_j^{(t)}(a|s)$, where $i \neq j$. The predicted action is then encoded into pedestrian i's feature $f_i^{(t)}$, and later used to form the reward function $R_i^{(t)}$ (detailed later). Using the Bellman equations, we can generate pedestrian i's policy $\pi_i^{(t+1)}$ at the next time step using the updated feature $f_i^{(t)}$ and reward $R_i^{(t)}$, and sample an action from it. By repeating the process for every pedestrian at every time step, we can forecast long-term trajectories for multiple pedestrians while considering the dynamics of social interaction. Fig. 12.6 illustrates the procedure where we employ fictitious play to model the interactions between three pedestrians. The three pedestrians, respectively colored in red, green, and yellow, sequentially make predictions and the forecasted distributions of each pedestrian (expressed in the corresponding color) grows over time. Note that features (and thus policies and rewards) are now time-indexed as they are updated

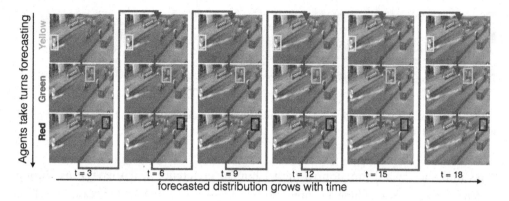

Figure 12.6 Visualization of Fictitious Play with three pedestrians. The orange arrows indicate the order. (For interpretation of the references to color in this figure legend, the reader is referred to the web version of this chapter.)

through the belief generated by fictitious play. For example, a state s may be occupied by others and is dangerous at first, but then becomes safe after all pedestrians passed.

In our multiagent scenario, we parameterize the underlying reward function of each agent as the linear combination of their individual features, as in Eq. (12.11). We consider two types of features: (i) physical scene features which are static, and (ii) social compliance features which can change over time. As discussed earlier, physical features have been proven very effective for single agent forecasting in a static environment. However, in a potentially dynamic multiagent environment, social compliance features are needed to represent social dynamics such as collision avoidance and respecting the personal space of others.

Physical scene features. We utilize a *neighborhood occupancy* feature and a *distance-to-goal* feature. The neighborhood occupancy features help describe pedestrians' avoidance of physical obstacles like a bench or trash can. The distance-to-goal feature helps describe a pedestrians desire to reach his goal quickly.

Social compliance features. As mentioned above, latent rules of social dynamics will influence the path of a pedestrian. If one predicts that other agents will cross the current planned path, social compliance suggests that the pedestrians should take preemptive action to avoid collision or close proximity interactions. We can take into account the short-term motions of other agents as features by forecasting agent trajectories over short-time horizons. To be concrete, given a policy $\pi_n(a|s)$, we can generate a state visitation distribution D_n of pedestrian n for trajectories of length L_n by recursively computing:

$$D_n^{(t)}(s') = \sum_{a,s} P(s'|s, a)\pi_n^{(t-1)}(a|s)D_n^{(t-1)}(s),\qquad(12.15)$$

Figure 12.7 Social compliance features encode the changing dynamics among pedestrians, and are computed by predicting the future movement of others. Red indicates safe and blue indicates dangerous (potential collision). (For interpretation of the references to color in this figure legend, the reader is referred to the web version of this chapter.)

where $D_n^{(0)}(x)$ needs to be initialized to a probability distribution over start locations. Since $D_n^{(t)}(s)$ is defined over the entire state space, it is of the same size as the state space. We can sum visitation counts over time, $\bar{D}_n(s) = \sum_t D_n^{(t)}(s)$, to generate a cumulative distribution over states that may be occupied by pedestrian n. As the cumulative state visitation distribution $\bar{D}_n(s)$ is derived using the past policy $\pi_n^{(t-1)}(a|s)$, it represents the states that are likely to be occupied by pedestrian n based on their past behavior (or empirical distribution in the context of FP). By aggregating the cumulative visitation distribution for all pedestrian except n, we can obtain a predicted occupancy map of all pedestrians in the environment, which we treated as the social compliance feature of pedestrian n: $f_{n,social}^{(t)}(s) = \bar{D}_{\neg n}(s) = \sum_{m \neq n} \bar{D}_m(s)$. Fig. 12.7 illustrates the process.

Note that as social compliance features expressed the changing dynamics among agents, the feature vector $f_n^{(t)}(s) = [f_{physical}(s), f_{n,social}^{(t)}(s)]$ (and as a consequence, the reward function) is time-indexed and depends on the predicted motion of other pedestrians. In this way, we have overcome the limitations of the single agent static scene assumption and have enabled long term trajectory forecasting in complex multiagent scenarios. Selected qualitative results of predicted trajectories are shown in Fig. 12.8. Each pedestrian is marked with a colored bounding box, and a corresponding forecasting distribution is shown in the same color. The more saturated the color, the higher the probability. Many of the predicted trajectories are smoothly curved (and do not exhibit abrupt changes in trajectories), indicating that the proposed approach mimics human behavior of taking preemptive actions to avoid collisions. We note that we consider *all* pedestrians for quantitative experiments but only visualize forecast distributions for a limited number of pedestrian to improve visualization.

Our discussion to this point has focused primarily on forecasting trajectories in low dimensional spaces, that is, 2D trajectories over a quantized floor plane representation. Human behavior is indeed more complex than simple 2D trajectories, and we desire to have an activity forecasting framework that can generalize to more complex representations of human behavior. There is, however, a fundamental computational bottleneck to the approach we have discussed thus far. The dynamic programming algorithm used to

Figure 12.8 Selected qualitative results of our multiagent model, where smoothly curving paths indicate predictive collision avoidance. (For interpretation of the references to color in this figure, the reader is referred to the web version of this chapter.)

compute the value function can only be applied efficiently to low-dimensional spaces, therefore limiting our current activity forecasting approach to low-dimensional problems. In the next section, we describe a procedure for extending the decision-theoretic principles to high-dimensional spaces to enable more expressive forms of activity forecasting.

12.6 DUAL-AGENT INTERACTION FORECASTING

We now extend the decision-theoretic framework to higher dimensional human behavior beyond 2D trajectories and explore the task of activity forecasting in the context of dual-agent interactions to understand how the actions of one person can be used to predict the actions of another. The setting is shown in Fig. 12.9. We fix the right-hand side (RHS) for the initiating agent, whose actions are observed as a video. The left-hand side (LHS) is for the reacting agent, whose action (image) sequence is the target of forecasting. We note here that the state space is quite different from the two previous tasks. In the first two tasks, the state space was defined over a spatial state space. In this task, the state space is a combined spatiotemporal state space. Finding a trajectory through this hybrid state space can be interpreted as performing video extrapolation.

Following our decision-theoretic approach, we model dual-agent interactions as an optimal control problem, where the actions of the initiating agent on the RHS induce a cost topology over the space of images of reactive poses – a space in which the reactive agent on the LHS plans an optimal pose trajectory. In this case, the state space is the space of all possible reactive poses on the LHS, and the action is the transition between images of these poses. It is important to note that the state space is both continuous

Figure 12.9 Examples of ground truth, observation, and our simulation result.

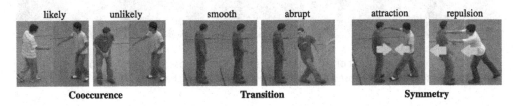

Figure 12.10 We use statistics of human interaction as our features for the cost function.

and high-dimensional. This poses a unique challenge, as it is infeasible to apply the previous framework naively to this large state space. First, we will explain our method for quantizing the state space in order to apply dynamic programming. Next, we will show how to extend this approximation to forecast agent reactions in a continuous space using kernel-based reinforcement learning (KRL). Finally, we present approximate MaxEnt IOC that directly performs function approximation in the high-dimensional continuous space, which improves upon the KRL approach.

The features used in our reward function is crucial to our method in modeling dynamics of human interaction. We use important statistics of human interactions as our features for the cost function (Fig. 12.10). To accumulate the statistics, we perform K-means clustering on all the training frames on the LHS to quantize the continuous state space into K discrete states. For each cluster c ($c = 1$ to K), we use the cluster center X_c as the representation of quantized state c. Given the pose sequence of the initiating agent on the RHS, we use the following features to model dynamics of human interaction:

- *Cooccurrence.* Given a pose on the RHS, we want to know how often a state c occurs on the LHS. This provides a strong clue for simulating human interaction.
- *Transition probability.* We want to know what actions will occur at a state c, which model the probable transitions between consecutive states. For example, at a state c that the agent is moving forward, transition to a jumping back state is less likely.
- *Centroid velocity.* We use centroid velocities to capture the movements of people when they are interacting. The relative velocity of the interacting agents gives us information about the current interaction. For example, in the hugging activity, the

Figure 12.11 Forecasting result of UTI dataset 1. The RHS is the observed initiator, and the LHS is the simulated reaction of the proposed method. The activity is shaking hands.

interacting agents are approaching each other and will have centroid velocities of opposite directions.

Using the above features with MaxEnt IOC on the quantized state space, we are able to obtain the underlying reward function $r(c, a)$ on the quantized state c. This gives us the approximated cost topology induced by the actions of the initiating agent. Based on this $r(c, a)$, we can further derive the value function $V(c)$ for the quantized states, and the associated discretized policy $\pi_\theta(a|c) = e^{V(c_a) - V(c)}$. This, however, can only be used to forecast the reactive sequence on the quantized states, which will be a nonsmooth sequence. Moreover, the state sequence will not be able to generalize to poses outside of the quantized states (Fig. 12.11).

In order to address the quantization error of discretizing the state space, we introduce kernel-based reinforcement learning (KRL) [15] to our problem. Based on KRL, the value function $V_h(x)$ for any pose x in the *continuous* state space is assumed to be a weighted combination of value functions $V(c)$ of the quantized states. This translates our inference from discrete space back to a continuous state space. The value function of a continuous state x is

$$V_h(x) = \frac{\sum_{c=1}^K K_h(x, X_c) V(c)}{\sum_{c=1}^K K_h(x, X_c)}, \tag{12.16}$$

where X_c is representation of quantized state c, and $K_h(\cdot, \cdot)$ is a kernel function with bandwidth h. The kernel function is required for statistical consistency.

Now that we have the value function $V_h^{(t)}(x)$ defined over a continuous state space, the goal of forecasting at test time is to find the pose x^* with the highest value. In other words, the goal is to find the image of a pose that a reactive agent is the most likely to take. In contrast to optimization in the discretized space, it is infeasible to enumerate the values of all the states in continuous space. We leverage the resemblance of our formulation in (Eq. (12.16)) to the well-studied *kernel density estimation* (KDE) in statistics. To optimize the density in KDE, the standard approach is to apply the mean shift procedure, which will converge robustly to the local maximum of the density function. By applying the same procedure, we are able to efficiently find the pose x^* in the continuous pose space that has the highest value during forecasting.

The computational bottleneck for MaxEnt IOC stems from the Value Iteration in Eqs. (12.5) and (12.6), where it is required to iterate through all of the possible states. In large or even continuous state space, this operation makes the algorithm intractable. We have presented a 2-steps approach by (i) computing $V(c)$ at quantized states, (ii) approximating $V(x)$ at all possible states using KRL. This approach, however, only interpolates the value function over a discretized state space. Next, we explain our approach that directly performs IOC on the continuous state space.

The key insight is that the updates in Eqs. (12.5) and (12.6) can be performed by function approximators, rather than iterating through all (quantized) states. More specifically, we use samples in the form of $\{(X_i, A_i, R_i, X_i')\}_{i=1}^{m}$, where X_i is a state in the continuous space, A_i is a possible action from X_i. X_i' is the state we will get by performing A_i at X_i, and R_i is the associated reward. Given these samples, one can estimate $Q(x, a)$ in Eq. (12.6) with $\hat{Q}(x, a)$ by solving a regression problem in which the input variables are $Z_i = (X_i, A_i)$ and the target values are $Y_i = R_i + \hat{V}(X_i')$. The update can thus be written as:

$$\hat{V}(x) = \operatorname*{soft\,max}_{a} \hat{Q}(x, a), \tag{12.17}$$

$$Y_i = R_i + \hat{V}(X_i'), \tag{12.18}$$

$$\hat{Q}(x, a) = \arg\min_{Q} \frac{1}{m} \sum_{i=1}^{m} |Q(X_i, A_i) - Y_i|^2 + \lambda \|Q\|_{\mathcal{H}}^2. \tag{12.19}$$

In particular, a reproducing kernel Hilbert space (RKHS) based formulation is used in Eq. (12.19). This process is also called the Approximate Value Iteration (AVI). The theoretical justification of this approach is provided in [5].

12.7 FINAL REMARKS

We have describe a decision-theoretic approach to activity forecasting, which provides a principled framework for modeling the consequences of taking certain actions and the impact it can have on the future. As a result, our methods are able to generate more informed predictions over future actions. From a broader point of view, we have shown that visual future prediction is a novel application domain for techniques originally developed for motion planning, optimal control, and reinforcement learning. From this perspective, we believe that this opens to new doors to a large body of work that can now be leveraged for various visual prediction tasks.

The approach we have described here is, of course, only one of many possible way to model the predictive perception problem. In addition to decision-theoretic models [16–18], many vision researchers have proposed a variety of alternative approaches for future trajectory prediction such as Dijkstra's shortest path search [19,20], structured prediction [21–23], nonparametric search [24–26], potential fields [27–30], planning with homotopy classes [31], and leveraging neural networks [32]. Some researchers have also start exploring activity forecasting in a more general context, such as predicting the motions of scenes [33,34], or even directly forecasting all possible futures [35–39], which is indicative of the great potential for innovating new approaches for future prediction. In terms of the greater topic of visual awareness, newer tasks such as reasoning in visual question and answering [40–42], learning common sense [43,44], inferring underlying human intent and motivation [41,45,46], and understanding physics [47–53] are examples of visual prediction tasks that go beyond traditional object recognition tasks and will presumably require more complex levels of visual understanding. At present, a remaining challenge for human activity forecasting is to develop the ability to make activity predictions over much longer temporal durations. Whereas the current state-of-the-art approaches can predict actions the order of several of seconds, we hope to see the temporal prediction horizon push into minutes or even hours.

REFERENCES

[1] Kitani KM, Ziebart BD, Bagnell JA, Hebert M. Activity forecasting. In: ECCV. 2012.
[2] Ma WC, Huang DA, Lee N, Kitani KM. A game-theoretic approach to multi-pedestrian activity forecasting. Available from arXiv:1604.01431, 2016.
[3] Brown GW. Iterative solution of games by fictitious play. In: Activity analysis of production and allocation, vol. 13. 1951. p. 374–6.
[4] Huang DA, Kitani KM. Action–reaction: forecasting the dynamics of human interaction. In: ECCV. 2014.
[5] Huang DA, Farahmand AM, Kitani KM, Bagnell JA. Approximate MaxEnt inverse optimal control and its application for mental simulation of human interactions. In: AAAI. 2015.
[6] Yamato J, Ohya J, Ishii K. Recognizing human action in time-sequential images using hidden Markov model. In: CVPR. 1992.
[7] Natarajan P, Nevatia R. Coupled hidden semi Markov models for activity recognition. In: WMVC. 2007.
[8] Russell S, Norvig P. Artificial intelligence: a modern approach. 1995.
[9] Bellman R. A Markovian decision process. J Math Mech 1957.
[10] Sutton RS, Barto AG. Reinforcement learning: an introduction. Cambridge: MIT Press; 1998.
[11] Ziebart BD. Modeling purposeful adaptive behavior with the principle of maximum causal entropy. Ph.D. thesis. 2010.
[12] Ziebart B, Maas A, Bagnell J, Dey A. Maximum entropy inverse reinforcement learning. In: AAAI. 2008.
[13] Munoz D, Bagnell JA, Hebert M. Stacked hierarchical labeling. In: ECCV. 2010.
[14] Oh S, Hoogs A, Perera A, Cuntoor N, Chen CC, Lee JT, Mukherjee S, Aggarwal J, Lee H, Davis L, et al. A large-scale benchmark dataset for event recognition in surveillance video. In: CVPR. 2011.

[15] Ormoneit D, Sen S. Kernel based reinforcement learning. Mach Learn 2002.

[16] Ziebart BD, Ratliff N, Gallagher G, Mertz C, Peterson K, Bagnell JA, Hebert M, Dey AK, Srinivasa S. Planning-based prediction for pedestrians. In: IROS. 2009.

[17] Kuderer M, Kretzschmar H, Sprunk C, Burgard W. Feature-based prediction of trajectories for socially compliant navigation. In: RSS. 2012.

[18] Karasev V, Ayvaci A, Heisele B, Soatto S. Intent-aware long-term prediction of pedestrian motion. In: ICRA. 2016.

[19] Walker J, Gupta A, Hebert M. Patch to the future: unsupervised visual prediction. In: CVPR. 2014.

[20] Xie D, Todorovic S, Zhu SC. Inferring "dark matter" and "dark energy" from videos. In: ICCV. 2013.

[21] Hoai M, De la Torre F. Max-margin early event detectors. In: IJCV. 2014.

[22] Lan T, Chen TC, Savarese S. A hierarchical representation for future action prediction. In: ECCV. 2014.

[23] Koppula HS, Saxena A. Anticipating human activities using object affordances for reactive robotic response. In: PAMI. 2016.

[24] Singh KK, Fatahalian K, Efros AA. Krishnacam: using a longitudinal, single-person, egocentric dataset for scene understanding tasks. In: WACV. 2016.

[25] Park HS, Hwang JJ, Niu Y, Shi J. Egocentric future localization. In: CVPR. 2016.

[26] Ballan L, Castaldo F, Alahi A, Palmieri F, Savarese S. Knowledge transfer for scene-specific motion prediction. In: ECCV. 2016.

[27] Ali S, Shah M. Floor fields for tracking in high density crowd scenes. In: ECCV. 2008.

[28] Tastan B, Sukthankar G. Leveraging human behavior models to predict paths in indoor environments. Pervasive Mob Comput 2011.

[29] Mantini P, Shah SK. Human trajectory forecasting in indoor environments using geometric context. In: ICVGIP. 2014.

[30] Robicquet A, Sadeghian A, Alahi A, Savarese S. Learning social etiquette: human trajectory prediction. In: ECCV. 2016.

[31] Gong H, Sim J, Likhachev M, Shi J. Multi-hypothesis motion planning for visual object tracking. In: ICCV. 2011.

[32] Alahi A, Goel K, Ramanathan V, Robicquet A, Fei-Fei L, Savarese S. Social LSTM: human trajectory prediction in crowded spaces. In: CVPR. 2016.

[33] Pintea SL, van Gemert JC, Smeulders AW. Déja vu: motion prediction in static images. In: ECCV. 2014.

[34] Walker J, Gupta A, Hebert M. Dense optical flow prediction from a static image. In: ICCV. 2015.

[35] Vondrick C, Pirsiavash H, Torralba A. Anticipating visual representations with unlabeled video. In: CVPR. 2016.

[36] Walker J, Doersch C, Gupta A, Hebert M. An uncertain future: forecasting from static images using variational autoencoders. In: ECCV. 2016.

[37] Xue T, Wu J, Bouman KL, Freeman WT. Visual dynamics: probabilistic future frame synthesis via cross convolutional networks. In: NIPS. 2016.

[38] Mathieu M, Couprie C, LeCun Y. Deep multi-scale video prediction beyond mean square error. In: ICLR. 2016.

[39] Vondrick C, Pirsiavash H, Torralba A. Generating videos with scene dynamics. In: NIPS. 2016.

[40] Antol S, Agrawal A, Lu J, Mitchell M, Batra D, Lawrence Zitnick C, Parikh D. VQA: visual question answering. In: ICCV. 2015.

[41] Tapaswi M, Zhu Y, Stiefelhagen R, Torralba A, Urtasun R, Fidler S. MovieQA: understanding stories in movies through question-answering. In: CVPR. 2015.

[42] Zhu Y, Groth O, Bernstein M, Fei-Fei L. Visual7w: grounded question answering in images. Available from arXiv:1605.09526, 2015.

[43] Fouhey D, Zitnick C. Predicting object dynamics in scenes. In: CVPR. 2014.

[44] Vedantam R, Lin X, Batra T, Lawrence Zitnick C, Parikh D. Learning common sense through visual abstraction. In: ICCV. 2015.

[45] Vondrick C, Oktay D, Pirsiavash H, Torralba A. Predicting motivations of actions by leveraging text. In: CVPR. 2016.

[46] Zunino A, Cavazza J, Koul A, Cavallo A, Becchio C, Murino V. Intention from motion. Available from arXiv:1605.09526, 2016.

[47] Hamrick J, Battaglia P, Tenenbaum JB. Internal physics models guide probabilistic judgments about object dynamics. In: Proceedings of the 33rd annual conference of the Cognitive Science Society. Austin (TX): Cognitive Science Society; 2011. p. 1545–50.

[48] Wu J, Yildirim I, Lim JJ, Freeman B, Tenenbaum J. Galileo: perceiving physical object properties by integrating a physics engine with deep learning. In: NIPS. 2015.

[49] Mottaghi R, Bagherinezhad H, Rastegari M, Farhadi A. Newtonian image understanding: unfolding the dynamics of objects in static images. In: CVPR. 2016.

[50] Fragkiadaki K, Agrawal P, Levine S, Malik J. Learning visual predictive models of physics for playing billiards. Available from arXiv:1511.07404, 2015.

[51] Lerer A, Gross S, Fergus R. Learning physical intuition of block towers by example. Available from arXiv:1603.01312, 2016.

[52] Mottaghi R, Rastegari M, Gupta A, Farhadi A. "What happens if..." learning to predict the effect of forces in images. In: ECCV. 2016.

[53] Finn C, Goodfellow I, Levine S. Unsupervised learning for physical interaction through video prediction. Available from arXiv:1605.07157, 2016.

PART 3

Metrics, Benchmarks and Systems

CHAPTER 13

Integrating Computer Vision Algorithms and Ontologies for Spectator Crowd Behavior Analysis

Davide Conigliaro*,†, **Roberta Ferrario**†, **Céline Hudelot**‡, **Daniele Porello**†
*Computer Science Department, University of Verona, Verona, Italy
†Laboratory for Applied Ontology, Institute for Cognitive Sciences and Technologies, National Research Council, Povo, Italy
‡MICS Laboratory, Centrale Supelec, Châtenay-Malabry, France

Contents

13.1 INTRODUCTION

In computer vision, crowd analysis and understanding from a sequence of images have been often defined as the modeling of large masses, within which a single person cannot be finely characterized, due to low resolution, frequent occlusions, or the particular dynamics of the scene. As shown in the various surveys on crowd analysis that may be found in the literature [1–5], this topic has experienced a growing interest in different fields, especially in the last 10 years, mainly motivated by a large and diverse set of real-world applications, such as visual surveillance, crowd management, public space design, or entertainment [1,3].

As illustrated in Fig. 13.1, various approaches have been exploited in crowded scene analysis. Specific features such as optical-flow (e.g., [6–9]), spatiotemporal information (e.g., [10,11]), trajectory (e.g., [12,13]), or texture features (e.g., [14]) can be extracted from video data using dedicated computer vision techniques. Then, crowd models can

Group and Crowd Behavior for Computer Vision
DOI: 10.1016/B978-0-12-809276-7.00016-3

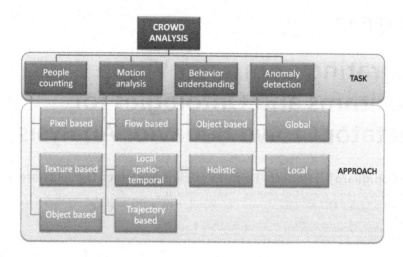

Figure 13.1 Taxonomy for crowd analysis in computer vision.

be built from the extracted features, often based and guided by a priori knowledge on crowd dynamics. At last, these models and some prior knowledge can be used to infer the interpretation of crowded scenes. The objectives of such analysis can be diverse: anomaly or abnormal behavior detection [14–17], people counting [18–20], behavior understanding [21], or motion analysis [10].

All the aforementioned surveys agree on the fact that, due to the complexity both of the visual analysis of crowded scenes and of crowd behavior modeling, specific computer vision technologies and knowledge models on crowds are necessary to tackle the challenging task of crowd analysis and understanding. These computer vision techniques and crowd knowledge models are well presented and reviewed in the aforementioned works. Nevertheless, to the best of our knowledge, crowd knowledge and the resulting models are very often limited to the dynamics of crowds. Such models have been reviewed according to different viewpoints, as, for instance, their inspiration (physical or biological models) in [4], or their level (microscopic or macroscopic) in [3].

Nonetheless, these studies move from the hidden assumption that there is only one kind of crowd, so all crowds move following the same dynamics. However, sociological studies show that at least four categories of large gatherings or crowds can be distinguished: *prosaic* or *casual* crowds, *spectator* or *conventional* crowds, *expressive* crowds, and *demonstration/protest* or *acting* crowds. For a definition of these different types of crowds, the reader can refer to [22] or [23,24]. A complete taxonomy with a specific focus on spectator crowd may be found in Chapter 2. As explained in [25], most of computer vision techniques for crowd analysis are applied to prosaic or demonstration crowds while, on the contrary, only few works consider spectator crowds. Indeed, while the former kinds of crowd usually move freely in the environment, spectator crowds, defined as a collection of people "interested in watching something that they came to see" [26], as, for instance, a sport event, on the contrary, display a very specific and constrained dynamics. In spectator crowds, people stand or sit still on a fixed location for most of the time and execute a limited number of very particular activities, such as applauding, supporting, or discussing with their neighbors. Such activities are highly related to the event or the contest the spectators are watching and its analysis and understanding could be of great importance for both crowd and event monitoring.

So, a first problem with spectator crowds is that, given the contextual and constrained character of their dynamics, their study cannot leverage on the analogy with physical or biological dynamic models. On the other hand, their specific behavior is composed by actions that acquire a social and ritual connotation only in some specific context and would be otherwise meaningless. It seems, thus, that what is needed is a meaning disambiguation instrument, able to describe and represent not only how the crowd looks like and move physically, but also what is the interpretation of a scene in a given social context (for instance, that a person seated in the stands is a spectator or that rhythmically jumping in the stands counts as a sign of exultation). For these reasons, in this paper we propose an ontological model of collective and crowd behavior [27,28], inspired by sociological studies [29,30,22,31] and a general framework to integrate ontological reasoning with computer vision techniques for crowded scenes understanding.

A second problem is the lack of benchmark datasets focused on spectator crowds. A first attempt to fill this gap has been proposed by Conigliaro et al., who created and made publicly available S-HOCK [25], the first dataset on the subject, composed of fully annotated videos of spectator crowds, recorded during an international hockey competition, equipped with computer vision algorithms expressly dedicated to spectator crowd analysis, such as spectators categorization and excitement calculation [32], or basic match event segmentation based on the spectators excitement level [33].

In this paper, building on these previous works, we propose to go deeper into the understanding of crowd behavior by proposing an approach which integrates ontological models of crowd behavior and dedicated computer vision algorithms, with the aim of recognizing some targeted complex events happening in the playground from the observation of the spectator crowd behavior. In order to do that, we first propose an ontology encoding available knowledge on spectator crowd behavior, built as a specialization of the DOLCE foundational ontology, which allows the representation of categories belonging both to the physical and to the social realms. We then propose a simplified and tractable version of such ontology in a new temporal extension of a description logic, which is used for temporally coupling events happening on the playground and spectator crowd behavior. At last, computer vision algorithms provide the input information concerning what is observed on the stands and ontological reasoning delivers the output necessary to perform complex event recognition.

The chapter is structured as follows: in Section 13.2 we illustrate the general vision behind the proposed integration of computer vision and ontologies. In Section 13.3 we present an extension to the DOLCE ontology that includes concepts to represent the domain of spectator crowd. Section 13.4 introduces a temporalized description logic enabling parallel reasoning on actions of the spectator crowd and events happening in the playground, while Section 13.5 concludes the chapter with some final remarks.

13.2 COMPUTER VISION AND ONTOLOGY

As already mentioned in the Introduction, scene understanding is an extremely challenging task for standard approaches to computer vision [34], due to the complexity and the visual variability of scenes that we would like to classify as belonging to the same category.

The general idea behind our proposal is to take inspiration from the cognitive mechanisms that human beings seem to apply, at least intuitively, when they learn to classify objects and scenes in the surrounding environment and try to mimic them in artificial agents, in this case as visual classification systems.

A first elementary mechanism that we, as humans, use is learning from *repeated exposure*, namely being exposed to a number of positive or negative examples of instances of a concept: "this is a kid", "this is a kid running." This is very easy and straightforward for simple or, as we shall call them, *basic* concepts, but not really effective when we are exposed few times to new, complex objects or scenes. Take the example of a volleyball match: for a person who has never attended one of such matches, the scene is, likely, not understandable. In this case what we usually do is tell this person that in a volleyball match there are 12 persons moving on a floor where a rectangle is drawn, six of them located on one side with respect to a net located in the middle of the rectangle and the other six on the other side; that those on one side wear shirts of a different color with respect to those on the other side, and we call all these persons "players", etc.[1]

In order to classify a scene as a volleyball match, the subject obviously needs to know beforehand how a person looks like, how a shirt looks like, how a rectangle looks like, etc. So, in such cases we can say that we learn a complex (or *defined*, as we shall call it) concept, like that of a volleyball match, through a definition which aggregates simpler concepts, that can be in their turn defined through even simpler concepts, down to basic concepts, which are in a sense the atoms of these compositional definitions. Such atoms are those that have been learned through the elementary mechanism of repeated exposure.

---

[1] Here we are obviously talking about how the elements of a scene look like, not about what they or their essential properties are. A discussion about the difference between "genuine" concepts and *look*-concepts applied to the computer vision scenario can be found in [35]. We will not consider this issue in the present work, due to space limitations.

Figure 13.2 Example of a scene with multiple possible interpretations.

We can illustrate this general idea with another example of semantic scene understanding, as depicted in Fig. 13.2. Indeed, various semantic interpretations of this image can be provided, as, for instance:

1. A set of concepts describing the objects of the scene: *polar bear, sea, iceberg*;

2. A sentence describing the content of the scene: *a polar bear on a very small iceberg*;

3. A sentence describing a wider scope interpretation of the scene inferred from its content: *global warming threatening the polar fauna*.

These different interpretations depend on their own objective (the targeted semantic level) and on the available contextual and a priori knowledge on the application domain. In this example, *polar bear*, *sea*, and *iceberg* are **basic concepts** or **atoms** that can be easily recognized with classical computer vision and machine learning approaches. Recent progress on convolutional networks and deep learning also enables to automatically build complex descriptions of the content of such scenes [36–38], but these approaches imply huge learning databases [39,40] and do not enable rich inferences and reasoning processes [41]. As a consequence, we believe that performing high level semantic scene understanding (such as that necessary to obtain the third interpretation of the image in Fig. 13.2) implies building and reasoning on formal models encoding knowledge on the scene (e.g., definition of complex concepts such as *volleyball match* or *global warming*).

In the integrated framework that we are proposing, we would like to develop the intuition that artificial agents can similarly use both mechanisms. Our idea is, therefore, to use low level classification methods, like those of computer vision and pattern recognition to classify basic concepts and knowledge representation and ontological reasoning to classify defined concepts through inference mechanisms that aggregate basic concepts and relations.

Such integrated approach may be used for the recognition of complex objects (or objects for which we do not have good computer vision classifiers), to infer the presence of an object from the identification of its basic properties, but also for the classification of complex scenes, like those concerning spectator crowds, from their constituents.

Some attempts of using semantic and ontology-based approaches for image analysis (see [42,43] and, more recently, [44–47]) and action recognition [48] have already been made. Recent reviews can be found in [49,50]. However, in most of the existing approaches, only the descriptive part of ontologies is used [49]. Ontology-based inferences are rarely deployed to guide scene understanding [51,52,46]. Moreover, the main challenge is the interfacing of the two levels, i.e., filling the gap between the statistical models used in image interpretation and the ontological qualitative representations used to convey high level semantics. This problem is often referred to as the semantic gap [53,54].

Various approaches have been investigated to address this issue. A first approach consists in building a visual concept ontology [55,56], which defines the set of concepts and relations that can be used to visually describe an object or a scene, such as, for instance, color, shape and texture concepts, or spatial relations. The main idea is that these visual concepts can be easily linked to computer vision algorithms or resulting quantitative features and are then used to define application domain concepts. An interesting formalization of this process has been proposed in [35]. Another approach consists in using concrete domains [57] in the description logics framework [58], but results have only be shown on the specific case of spatial relations and reasoning [59].

In the spectator crowd scenario, we can use ontologies for many purposes: first, to classify the same "objects" under different concepts depending on the context, like a person, who can be classified as a spectator if (s)he is sitting on the stands or as a player if (s)he is running on the playground, or an action like jumping, that can be part of an exulting action if performed on the stands or of a shooting on goal action if performed on the playground. Second, they can be used to represent – and thus distinguish – individual and collective actions. Finally, they can be deployed to reason on different actions and events cooccurring on the stands and on the playground.

In this chapter we will propose a framework that, thanks to the integration of computer vision/pattern recognition and ontological approaches, begins to address such issues.

13.3 AN EXTENSION OF THE DOLCE ONTOLOGY FOR SPECTATOR CROWD

DOLCE is a foundational ontology that introduces and describes very general concepts that can be applied across all knowledge domains, in order to foster interoperability among knowledge representation systems. The formal language in which the ontology is defined is first-order logic [60], chosen for its usefulness in enhancing the clarity and the expressive power of the semantic characterizations of concepts. We use DOLCE-

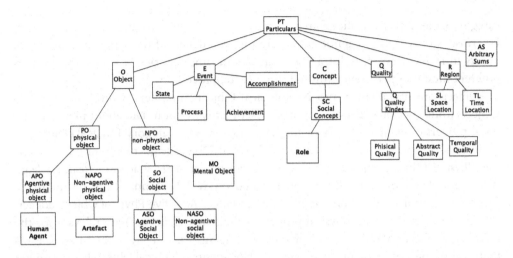

Figure 13.3 Excerpt of DOLCE.

CORE [61], which is a recent and more compact version of DOLCE, addressed to modelers. An excerpt of this ontology is depicted in Fig. 13.3. Our choice of using DOLCE is motivated by the fact that it provides the most general categories that ground in a principled way the knowledge representation of information about crowds as entities; more specifically, the categorization of entities of DOLCE guides the modeling choices for developing an ontology for representing heterogeneous types of crowds.

The uppermost class of DOLCE is that of particulars, which is divided into a number of principled categories. We only discuss few representative cases.

To begin with, the category **Arbitrary Sum** collects (mereological) sums of individuals that do not share a common identity principle (e.g., we may want to talk about the collection of entities constituted by the sum of this book, the number 3 and the Constitution of Italy).

Region includes various kinds of *space*, that is, values of measurable qualities, like a conceptual space or a space of features (e.g., textures, colors). It also includes **Time**, which is conceptualized as temporal location and can be expressed both as instants and as intervals. **Individual Quality** collects qualities that inhere in individuals, for instance, the color of a specific object (e.g., the redness of this rose).

The category of **Concept** collects entities whose belonging to a class may change through time. Here you find *social concepts* that are intended as depending on a community of agents that recognizes them as such (e.g., copyright licenses, contracts, marriages). Important examples of social concepts are those included in the category **Role**: roles are used to give a temporalized and possibly revisable classification of individuals (e.g., the President of the Ministries of Italy). We shall see that roles are crucial in modeling a crowd ontology. The **Event** classifies things that are mainly located in time (e.g., the match Italy–Norway of December 13, 2013), which contrasts with the category **Object**, that classifies things that are mainly located in space (e.g., this chair). For reasons of space, we cannot enter here into the details of the articulated theory of events in DOLCE, we only list a few examples of subcategories of events that express different modes of persistence through time (state, process, etc.).

Objects are first divided into Nonphysical Objects (numbers, words) and Physical Objects (tools, persons), depending on the fact that we want to talk about physical/material properties of such objects or immaterial ones. The category **Physical Object** can be internally separated into Nonagentive Physical Object (e.g., this book) and Agentive Physical Object, objects to which it makes sense to ascribe agentive features (e.g., Adriano Celentano). This distinction is based on whether entities are deemed able to perform intentional actions or not. For instance, **Agentive Physical Object** collects both persons, in case their physical properties (height, weight, age) are relevant for the intended application and artificial or collective agents, in case we want to ascribe intentional properties to them. The category **Nonagentive Physical Object** has a number of subclasses. For instance, **Artifact** refers to physical objects intentionally created or selected on the basis of a function they should accomplish (e.g., this stone used as a paperweight). The category **Image** is significant for bridging ontological information and computer vision techniques, as it is the category that lists the physical objects on which classification, detection, and reasoning is performed (e.g., a picture of a red house which we aim at classifying as such). **Nonphysical Object** subsumes **Mental Object** that classifies, for example, desires, beliefs, and intentions, and **Social Object**, which classifies things like contracts, laws, and agreements. A detailed description of DOLCE may be found in [60,61].

13.3.1 Modeling the Spectator Crowd and the Playground in DOLCE

We can now introduce a number of specific categories for dealing with crowds; they are integrated with DOLCE and they are plugged into the DOLCE-CORE taxonomy as more specific categories. We denote such specific categories by adding the prefix **CO:** (for crowd ontology) to the label of these categories (for example, C would become **CO:C**). Fig. 13.4 shows an excerpt of the proposed ontology. The concept of role from DOLCE is useful to cope with a number of scenarios: we can introduce players, spectators, supporters, teams, supporters groups, e.g., **CO:Spectator**, **CO:Player**, and **CO:Supporter**. We view them as roles that classify persons, but not necessarily; in this way, we allow for reclassification. For instance, a person may be a player only with respect to a given match, while in another one (s)he is a spectator. Moreover, the introduction of the role of spectator (**CO:Spectator**) may be justified by the fact that not necessarily every person in the spectator crowd should be classified as a spectator; for example, the security guards walking on the stands should not be defined as such.

Moreover, the *concept of group* (**CO:Concept-of-Group**) is very important to develop an ontology of crowds [27], as it captures the modes in which collectives of individuals

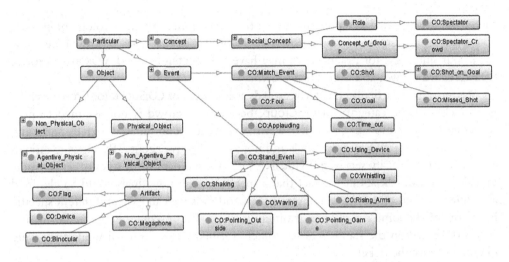

Figure 13.4 Part of the crowd ontology integrated with DOLCE.

are socially recognized (e.g., the supporters of the Trento hockey team). The concept of group classifies aggregates of individuals (or sets) as groups; that is, in ontological terms, it is a role of aggregates of persons, allowing the classification of sets of individuals as groups.

The category **CO:Group** instantiates those entities that we can classify as groups by allowing us to predicate properties of them (e.g., the group of supporters of the Trento hockey team is located on the north side of the bleachers). We view groups in a very general sense, by not assuming any restrictive condition on the cohesion among the members of the group (e.g., the users of Facebook can be considered as a group). By contrast, the category **CO:Organization** refers to groups that can act as a collective agent (and that is why Organization is considered agentive), that is, we can ascribe intentional actions to organizations (e.g., the Trento hockey team as a legally recognized entity). By instantiating the concept of group, we obtain the **CO:Concept-of-Crowd** that classifies sets of individuals as crowds. When instantiating crowds as particulars in our ontology, these are categorized as specific types of groups, that is, we do not ascribe, strictly speaking, intentional actions to crowds. This modeling choice opens the possibility of discussing what types of actions can be ascribed to crowds, whether the ascription is ontologically grounded or merely figuratively appropriate (e.g., movement, applauding, supporting). We leave such discussion for future work and, for the moment, we just assume that a number of actions can be ascribed to crowds, such as applauding, events whose participants are spectator crowds.

Once we have placed crowds within DOLCE, we can develop an analysis of types of crowds by specifying the categories that can be plugged under the node **CO:Crowd** of the DOLCE taxonomy. In this way, we can apply DOLCE conceptually principled analysis of categories for knowledge representation to the development of an ontology of crowds. The category CO:Crowd can be partitioned according to the types of crowd we allow for in our ontology; for instance, we may have CO:Spectator-Crowd, CO:Casual-Crowd, CO:Expressive-Crowd, and CO:Protest-Crowd.

We introduce particular spectator crowds as instances of CO:Spectator-Crowd, which is a subclass of the category of CO:Group, because the crowd is not assumed to possess collective intentionality; in other words, even when coordinated, strictly speaking, crowds still act as a sum of individuals and not as a collective agent.[2] By viewing crowds as concepts that classify sets of individuals – i.e., as instantiations of roles of sets of individuals – we can talk about the fact that a specific set of individuals could be classified as a different kind of crowd at different times, and thus we can expect different specific behaviors of the same individuals at different times. For instance, people who form a spectator crowd on the stands had very likely formed a casual crowd while they were queuing to buy the tickets.

We add a number of types of artifacts, which it is useful to talk about in the case of spectator scenarios: CO:Device, CO:Megaphone, CO:Flag, and CO:Binocular. Those objects usually participate to some of the events happening in the stands that we have singled out.

For the sake of simplicity, in the next paragraph we will not distinguish between events and actions and we will view actions as types of events. Moreover, we can also distinguish joint or collective actions or events, which require groups as participants, from individual events [27].

In order to talk about events that are significant in the case of a spectator crowd scenario, we add the following two subcategories of events, CO:Stand Event and CO:Match Event, to separate events that occur at the level of the stands from events taking place in the match. The category of stand event captures what can happen on the stands and requires spectators as participants, e.g., CO:Shaking, CO:Waving, CO:Applauding, etc. We can model the participants of these events as individuals but also as collectives, e.g., we may say that we can have as participants a single spectator, a spectator crowd, or both.

The category of match event, instead, captures what happens on the play-field, e.g., CO:Foul, CO:Shot, CO:Goal, and CO:Time-out. When analyzing datasets such as S-HOCK, we may also introduce more specific events to be recognized. For instance, CO:Shot can be further divided into CO:Shot-on-goal, a shot that directs the puck toward the net, either going into the net for a goal or being stopped by the goaltender for a save, CO:Missed-shot includes shots that sail wide or high of the net, and shots that hit the goalpost or crossbar; considering these events with the lenses of the social analysis, we built a crowd ontology able to reason about the events of a game–play based on spectators' behavior.

---

[2] We assume that, in order to ascribe intentionality to a collective agent, the latter needs to be endowed with a decisional procedure. See [27] for a detailed description.

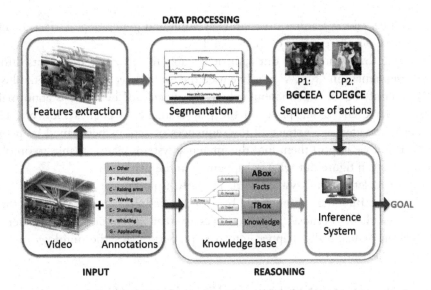

Figure 13.5 Proposed approach for event recognition.

13.3.2 A Tractable Fragment of DOLCE

When designing technologies for image and video understanding, it is in many cases inefficient to use the full power of first-order logic, which is undecidable. Therefore, in the next sections, we shall describe a restricted fragment of first-order logic, based on description logic, that is sufficient to capture a few elements of the concepts required for an ontology of crowds. We shall keep the taxonomy of DOLCE – which is, as we have seen, justified by an independent rationale – by rephrasing it by means of TBox axioms. For this reason, in the following sections, we shall assume that the taxonomy of DOLCE is included in the examples of TBox.

13.4 REASONING ON THE TEMPORAL ALIGNMENT OF STANDS AND PLAYGROUND

In order to recognize hockey related sport events happening in the game-field by only considering the behavior of people in the stands, we propose a system based on an integration of ontology and computer vision, which is structured into two different subsystems, data processing and reasoning. The proposed architecture is illustrated in Fig. 13.5. At the data processing level, the raw-data are processed by computer vision algorithms, generating feature descriptions, i.e., sets of symbolic descriptors which summarize characteristics of data in a quantitative way. On the other hand, the reasoning level is related to the interpretation of visual data. It is built on top of the data processing algorithms, taking features descriptors as input and generating abstract, qualitative descriptions about the content of the visual data.

Following this architecture, our approach is able to recognize the events through a temporal segmentation of the videos (we used the event detection framework illustrated in [33]), identifying salient moments where the spectators' behavior is abnormal, due to a particular event taking place on the play-field, i.e., a goal, a shot, etc. Having segmented the video, we extract all the sequences of actions performed by people in the crowd, and we import them into the inference system, where, thanks to ontological reasoning, we can figure out what is happening in the play-field.

In order to let the full system work properly, a preliminary analysis of the input data is essential. Building the ontology by properly defining classes, as well as the relations between them is crucial. It is also necessary that the outputs of computer vision algorithms are semantically mapped into data properties that the ontology can use, as explained in Section 13.2.

13.4.1 A New Description Logic for Video Interpretation

As already mentioned in Section 13.3.2, we used description logics [58], which are an important paradigm of logic-based knowledge representation and a decidable family of logic-based languages. They also are the most widespread knowledge representation formalism used for encoding ontologies in applications. In our case, as we aim at representing temporal events, we have to consider description logics equipped with time. Since in many application domains time and temporal notions play an important role, several temporal description logics have been introduced in the literature. Surveys are available in [62,63]. The different approaches can be classified according to how time is introduced in the description logic: (i) as a combination of standard description logics with standard temporal logics, such as LTL (linear time temporal logic) or CTL (computational tree logic) [64,65], (ii) as a fixpoint extension of description logics [66], (iii) as an interval-based description logics [67], or (iv) by the use of concrete domains to encode time [68]. Nevertheless, these approaches often result in very expressive description logics and they cannot be used in an operational way for image and video understanding. As a consequence, we propose a new lightweight temporalized description logic, named $\mathcal{ALC}_{Temp}$,[3] built on the basic description logic $\mathcal{ALC}$, dedicated to the representation of video events. We briefly introduce this formalism in the following.

13.4.1.1 $\mathcal{ALC}_{Temp}$: Syntax and Semantics

We consider a signature $\Sigma = (N_C, N_R, N_I, \mathbf{T})$, where N_C is a set of concept names, N_R is a set of role names, N_I is a set of individual names, and $\mathbf{T}$ a totally ordered set of

---

[3] This logic is the result of a collaborative work with Marc Aiguier and Nguyen Nguen of the MICS Laboratory, Centrale Supelec.

times.[4] The sets N_C and N_R are both partitioned into N_C^T and $N_C^{\overline{T}}$ and N_R^T and $N_R^{\overline{T}}$, respectively, where N_C^T and N_R^T are the sets of temporalized concept and role names (i.e., concepts and roles that can change in time), whereas $N_C^{\overline{T}}$ and $N_R^{\overline{T}}$ are the sets of time-independent concept and role names, respectively.[5]

Given a signature Σ, the set of concept expressions $\mathcal{CE}(\Sigma)$ of $\mathcal{ALC}_{Temp}$ is inductively defined as follows:

$$C, D \in \mathcal{CE}(\Sigma),$$

$$C, D ::= \top \mid \bot \mid A \mid \neg C \mid C \sqcap D \mid C \sqcup D \mid \exists r.C \mid \forall r.C \mid \exists TE \mid \exists r.TE \mid \forall r.TE.$$

TE denotes a temporal expression belonging to the set of temporal concept expressions $\mathcal{TCE}(\Sigma)$, inductively defined as follows:

$$C \in \mathcal{CE}(\Sigma), \mathbb{C} \in \mathcal{TCE}(\Sigma),$$

$$\mathbb{C}, \mathbb{D} ::= [C\rangle \mid]C\rangle \mid \mathbb{C} \sqcap^{tr} \mathbb{D},$$

where $tr \in \mathcal{AR} = \{b, m, o, s, d, f\}$.

The concept expressions of $\mathcal{ALC}_{Temp}$ are those of $\mathcal{ALC}$ extended by adding the constructors $\exists.TC$ and $Qr.TC$ with $Q \in \{\exists, \forall\}$ and TC a temporal concept expression. The interpretation domain of a temporal concept is the Cartesian product of a set of individuals and a time structure. The temporal concept $[C\rangle$ intuitively represents the set of individuals that belong to the concept C during an interval of time which begins at the considered time point. Given the considered time point t, $[C\rangle$ is the set of individuals that will belong to the concept C during a time interval placed in the future of t. At last, the six intersection symbols $\sqcap^{\{b,m,o,s,d,f\}}$ correspond to the six basic Allen's temporal relations {*before, meets, overlaps, during, finishes*} illustrated in Table 13.1 [69].

An expression of the form $C \sqsubseteq D$, where C and D are concepts, is called a *concept inclusion*. A finite set of concept inclusions is a *TBox*. Expressions of the form $a : C$ if $C \in N_C^{\overline{T}}$ or $(a, t) : C$, where $a \in N_I$ and C is a concept, are *concept assertions*. For $r \in N_R$ and $a, b \in N_I$, expressions of the form $(a, b) : r$ if $r \in N_R^{\overline{T}}$ or $(a, b, t) : r$ are role assertions. A finite set of concept and role assertions is called an *ABox*. A pair $\mathcal{K} = (\mathcal{T}, \mathcal{A})$ consisting of a TBox $\mathcal{T}$ and an ABox $\mathcal{A}$ is called a *knowledge base*.

In our framework, we are interested in describing the crowd behavior and the different events happening in the game-field and use our framework to encode both background knowledge about the spectator crowd behavior, contained in the TBox,

---

[4] The number of frames in a video is finite and totally ordered by time.

[5] Notice that in this section we will use the terminology of description logics, which is different from that adopted in Section 13.3, which refers to DOLCE. In general terms, DOLCE classes are called concepts here, while DOLCE relations are called roles.

Table 13.1 Allen's 13 basic relations between two intervals

Relation	Abbreviation	Temporal interpretation	Starting and ending point relation
before(X,Y); after(Y,X)	b, a	⊢─X─⊣ ⊢─Y─⊣	$x^+ < y^-$
equal(X,Y)	e	⊢─X─⊣ ⊢─Y─⊣	$(x^- = y^-)$ & $(x^+ = y^+)$
meets(X,Y); met-by(Y,X)	m, mi	⊢─X─⊣ ⊢─Y─⊣	$x^+ = y^-$
overlaps(X,Y); overlapped-by(Y,X)	o, oi	⊢─X─⊣ ⊢─Y─⊣	$(x^- < y^-)$ & $(x^+ > y^-)$ & $(x^+ < y^+)$
starts(X,Y); started-by(Y,X)	s, si	⊢─X─⊣ ⊢──Y──⊣	$(x^- = y^-)$ & $(x^+ < y^+)$
during(X,Y); contains(Y,X)	d, di	⊢─X─⊣ ⊢──Y──⊣	$(x^- > y^-)$ & $(x^+ < y^+)$
finishes(X,Y); finished-by(Y,X)	f, fi	⊢─X─⊣ ⊢──Y──⊣	$(x^- > y^-)$ & $(x^+ = y^+)$

as, for instance, the expected behavior of the crowd during the goal of a given team, and the facts that can be observed on the videos, collected in the ABox. As shown in the *Sequence of actions* in Fig. 13.5, we describe the behavior of each person as a temporal sequence of actions extracted from the annotations. For example, looking at P1:BGCEEA, each letter represents the most frequent action performed by the person P1 during the interval of 1 second (30 frames). This means that in this case the person P1 performed the following actions: *pointing game–applauding–raising arms–shaking flag–shaking flag–other*, for a total duration of 6 seconds. Knowledge on crowd behavior also reveals that, during a goal of a given team, a big part of the spectator crowd supporting this team has the following collective behavior: the crowd first applauds, then raises the arms, and then shakes flags. Thus, on the basis of $\mathcal{ALC}_{Temp}$, we define the following signature Σ:

- $N_C^T = \{CO: GoalBehavior, CO: Applauding, CO: RaisingArms, CO: Shaking, CO: PointingGame, CO: Other\}$,
  $N_C^{\overline{T}} = \{CO: Spectator^6\}$,

---

[6] We consider here that being a spectator is not dependent on the time since the status of a person does not change during all the video frames.

Table 13.2 Example of an ABox and a TBox

ABox	p1: *CO:Spectator*
	p2: *CO:Spectator*
	(p1,*CO:PointingGame*,t_1): *hasActionOf*
	(p1,*CO:Applauding*,t_2): *hasActionOf*
	(p1,*CO:RaisingArms*,t_3): *hasActionOf*
	...
	(p2,*CO:Waving*,t_1): *hasActionOf*
	...
TBox	*CO:Spectator* $\sqsubseteq$ $\top$
	CO:GoalBehavior $\equiv$ *CO:Spectator* $\sqcap\exists$ *hasActionOf*.(([*Applauding*] $\sqcap^m$ (]*Raising*] $\sqcap^o$]*ShakingFlag*])))

- $N_R^T = \{hasActionOf\}$,
  $N_R^{\overline{T}} = \varnothing$,
- $N_I = \{p_1, p_2\}$,
- $\mathbf{T} = \{t_1, \ldots, t_6\}$.

Here, *hasActionOf* is a role that is used to describe who is performing some specific actions. N_C^T is the set of concept names which are influenced by time (properly called "concepts" in DOLCE) and $N_C^{\overline{T}}$ is the set of concept names which are not influenced by time (properly called "classes" in DOLCE). A formal example of ABox and TBox corresponding to Fig. 13.5 constructed with signature Σ is presented in Table 13.2.

The ABox is automatically populated by what is extracted in the video, whereas the TBox encodes the available a priori knowledge. In the example, the GOALBEHAVIOR concept represents the set of individuals who are spectators and who, at the current time t, are applauding for a certain duration, action followed by two overlapping actions consisting of raising arms and shaking a flag.

A temporal interpretation is a pair $\mathcal{I} = (\Delta^{\mathcal{I}}, \cdot^{\mathcal{I}})$, where Δ is a nonempty set and $\cdot^{\mathcal{I}}$ is an interpretation function that maps every $a \in N_I$ to $a^{\mathcal{I}} \in \Delta^{\mathcal{I}}$, and every $C \in N_C$ to a mapping $C^{\mathcal{I}} : \mathbf{T} \to \mathcal{P}(\Delta^{\mathcal{I}})$ such that

$$\forall C \in N_C^{\overline{T}}, \forall t, t' \in \mathbf{T}, C^{\mathcal{I}}(t) = C^{\mathcal{I}}(t');$$

and every $r \in N_R$ to a mapping $r^{\mathcal{I}} : \mathbf{T} \to \mathcal{P}(\Delta^{\mathcal{I}} \times \Delta^{\mathcal{I}})$ such that

$$\forall r \in N_R^{\overline{T}}, \forall t, t' \in \mathbf{T}, r^{\mathcal{I}}(t) = r^{\mathcal{I}}(t').$$

The interpretations of a concept $C \in N_C^T$ and of a role $r \in N_R^T$ change according to each element in $\mathbf{T}$, whereas the interpretations of a concept $C \in N_C^{\overline{T}}$ and of a role $r \in N_R^{\overline{T}}$ do not. In the following, we note $C_t^{\mathcal{I}}$ the interpretation of a concept $C \in \mathcal{CE}(\Sigma)$

Table 13.3 $\mathcal{ALC}_{Temp}$ syntax and interpretation

Constructor	Syntax	Semantics
atomic concept	C	$C_t^{\mathcal{I}} = C^{\mathcal{I}}(t)$
individual	a	$a^{\mathcal{I}} \in \Delta^{\mathcal{I}}$
Top	$\top$	$\Delta^{\mathcal{I}}$
Bottom	$\bot$	$\emptyset$
conjunction	$A \sqcap B$	$A_t^{\mathcal{I}} \cap B_t^{\mathcal{I}}$
disjunction	$A \sqcup B$	$A_t^{\mathcal{I}} \cup B_t^{\mathcal{I}}$
negation	$\neg A$	$\Delta^{\mathcal{I}} \backslash A_t^{\mathcal{I}}$
existential restriction	$\exists r.A$	$\{a \in \Delta^{\mathcal{I}} \mid \exists b \in \Delta^{\mathcal{I}} (a, b) \in r_t^{\mathcal{I}} \wedge b \in A_t^{\mathcal{I}}\}$
universal restriction	$\forall r.A$	$\{a \in \Delta^{\mathcal{I}} \mid \forall b \in \Delta^{\mathcal{I}}, (a, b) \in r_t^{\mathcal{I}} \Rightarrow b \in A_t^{\mathcal{I}}\}$
temporalized concept 1	$[A]$	$\{(a, t') \in (\Delta^{\mathcal{I}} \times \mathbf{T}) \mid \forall t_0, t \leq t_0 \leq t', a \in A_{t_0}^{\mathcal{I}}\}$
temporalized concept 2	$]A]$	$\{(a, t') \in (\Delta^{\mathcal{I}} \times \mathbf{T}) \mid \exists t_1 > t, \forall t_0, t \leq t_0 < t_1, a \notin A_{t_0}^{\mathcal{I}} \wedge \forall t_2,$ $t_1 \leq t_2 \leq t', a \in A_{t_2}^{\mathcal{I}}\}$
temporal existential restriction 1	$\exists r.\mathbb{C}$	$\{a \in \Delta^{\mathcal{I}} \mid \exists (b, t') \in \mathbb{C}_t^{\mathcal{I}} \wedge (a, b) \in r_{t'}^{\mathcal{I}}\}$
temporal universal restriction	$\forall r.\mathbb{C}$	$\{a \in \Delta^{\mathcal{I}} \mid \forall (b, t') \in (\Delta^{\mathcal{I}} \times \mathbf{T}), (a, b) \in r_{t'}^{\mathcal{I}} \Rightarrow (b, t') \in \mathbb{C}_t^{\mathcal{I}}\}$
temporal existential restriction 2	$\exists \mathbb{C}$	$\{a \in \Delta^{\mathcal{I}} \mid (a, t') \in \mathbb{C}_t^{\mathcal{I}}\}$
temporal conjunction	$\mathbb{C} \sqcap^{tr} \mathbb{D}$	$(\mathbb{C} \sqcap^{tr} \mathbb{D})_t^{\mathcal{I}} =$ $\left\{ (a, t') \in (\Delta^{\mathcal{I}} \times \mathbf{T}) \middle\| \begin{array}{l} a \in ((\exists \mathbb{C})_t^{\mathcal{I}} \cap (\exists \mathbb{D})_t^{\mathcal{I}}) \\ \wedge t' \in (Int\{(a, t) : \mathbb{C}\} \cup Int\{(a, t) : \mathbb{D}\}) \\ \wedge TSat(Int\{(a, t) : \mathbb{C}\}\langle tr \rangle Int\{(a, t) : \mathbb{D}\}) \end{array} \right\}$ with $Int\{(a, t) : \mathbb{C}\} = \{t' \in \mathbf{T} \mid (t \leq_{\mathcal{T}} t') \wedge (a, t') \in \mathbb{C}_t^{\mathcal{I}}\}$ and $TSat(Int\{(a, t) : \mathbb{C}\}\langle tr \rangle Int\{(a, t) : \mathbb{D}\})$ corresponding to Allen's relation in Table 13.1

at time t and $\mathbb{C}_t^{\mathcal{I}}$ the interpretation of a concept $C \in \mathcal{TCE}(\Sigma)$ at time t.[7] The semantics of the different concept expressions is illustrated in Table 13.3 and in Fig. 13.6.

At time t, an interpretation $\mathcal{I}$ *satisfies* a concept C if $C_t^{\mathcal{I}} \neq \emptyset$; $\mathcal{I}$ *satisfies* a concept inclusion $C \sqsubseteq D$ if $C_t^{\mathcal{I}} \subseteq D_t^{\mathcal{I}}$; $\mathcal{I}$ *satisfies* a concept assertion of the form $a : C$ or $(a, t) : C$ if $a^{\mathcal{I}} \in C_t^{\mathcal{I}}$ and $\mathcal{I}$ *satisfies* a role assertion of the form $(a, b) : r$ or $(a, b, t) : r$ if $(a^{\mathcal{I}}, b^{\mathcal{I}}) \in r_t^{\mathcal{I}}$. We say that $\mathcal{I}$ is a *model* of a TBox $\mathcal{T}$ or an ABox $\mathcal{A}$ if it satisfies every concept inclusion in $\mathcal{T}$ or every assertions in $\mathcal{A}$. $\mathcal{I}$ is a model of a knowledge base $\mathcal{K} = (\mathcal{T}, \mathcal{A})$ if $\mathcal{I}$ is a model of both $\mathcal{T}$ and $\mathcal{A}$. If there is a model of a KB $\mathcal{K}$ that *satisfies* C, then C is *satisfiable* w.r.t. $\mathcal{K}$.

---

[7] $C_t^{\mathcal{I}}$ is a subset of $\Delta^{\mathcal{I}}$ and $\mathbb{C}_t^{\mathcal{I}}$ is a subset of $\Delta^{\mathcal{I}} \times \mathcal{T}$.

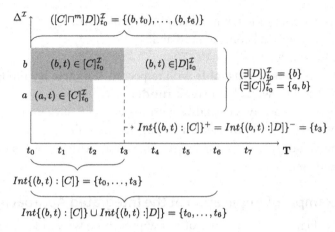

Figure 13.6 Illustration of the semantics of $\mathcal{ALC}_{Temp}$. $Int\{(a,t):\mathbb{C}\}^-$, $Int\{(a,t):\mathbb{C}\}^+$ are the minimal and maximal point of the chain $Int\{(a,t):\mathbb{C}\}$, respectively.

13.4.1.2 Reasoning Services for Video Interpretation

One of the advantages of logic-based formalisms such as description logics that motivates our work is the possibility to infer *implicit knowledge* through reasoning services over the knowledge base. In description logics the most important standard reasoning services are the *concept subsumption* (i.e., deciding whether a concept is more specific than another), the *instance checking* (i.e., verifying whether an individual belongs to a concept), the *concept satisfiability* (i.e., verifying if concepts are satisfiable, meaning that there are individuals belonging to these concepts) and the *consistency checking* (i.e., detecting contradictory information in knowledge bases). Formal definitions of these reasoning services for generic description logics can be found in [58]. In our case, we are interested in answering to the following issues:

- Classification of some objects that can be seen in the scene, as, for instance, detecting if a given person is a spectator or a supporter of a given team or detecting the type of specific action a person is performing.
- Detecting which events are happening in the video, as, for instance, being able to make the distinction between a shot and a goal.
- Detecting if and/or when a targeted event or a targeted sets of events are happening in a video, as, for instance, the detection of the match events in a video sequence in order to make a meaningful summary.
- Detecting and classifying an actor who is performing a specific action, as, for instance, for monitoring supporters who display abnormal behavior.

Some of these questions can be easily modeled as description logics reasoning services. For instance, a classification task can be formalized easily as an instance checking task and detecting if some events occur can be modeled as a concept satisfiability task. In order to take into account the temporality, we extend in $\mathcal{ALC}_{Temp}$ the reasoning services with the notion of temporal concept satisfiability.

Let $\mathcal{K}$ be a knowledge base, $C \in \mathcal{CE}(\Sigma)$:

- A concept C is **weakly satisfiable** with respect to $\mathcal{K}$ if there is a model $\mathcal{I}$ of $\mathcal{K}$ such that $\exists t \in \mathbf{T}$, $C_t^{\mathcal{I}} \neq \emptyset$. We call $\mathcal{I}$ a **weak model** of C.
- A concept C is **strongly satisfiable** with respect to $\mathcal{K}$ if there is a model $\mathcal{I}$ of $\mathcal{K}$ such that $\forall t \in \mathbf{T}$, $C_t^{\mathcal{I}} \neq \emptyset$. We call $\mathcal{I}$ a **strong model** of C.

We will now illustrate how we can use this formalism with an example of application of our integrated approach.

13.4.2 An Example of Application of the Integrated Approach

As illustrated in Fig. 13.5, a first step of our approach consists in building the knowledge base encoding the available knowledge on crowd behavior. As explained in Section 13.4.1, the knowledge base is composed of a TBOX (terminological knowledge) and an ABOX (assertional knowledge). The TBOX stores a set of universally quantified assertions stating general properties of classes and relations. The ABOX comprises assertions on individual objects, factual information. In our framework, the TBOX is a priori built and encodes common sense knowledge on spectator crowd behavior, i.e., the concepts of the crowd ontology defined in Section 13.3.1 and their definition. An excerpt of the TBox is depicted at the bottom of Table 13.2. For the sake of simplicity, in the illustration of the framework we consider the following very simple TBox[8]:

Example 1. $\boxed{TBox\ \mathcal{T}:}$

$Spectator \sqsubseteq \top$
$Applauding \sqsubseteq \top$
$Raising \sqsubseteq \top$
$GoalBehavior \equiv Spectator \sqcap \exists(([Applauding]\sqcap^m]Raising]))$

In our framework, the ABox is automatically built by translated results of the data processing step into the corresponding assertions in $\mathcal{ALC}_{Temp}$. For instance, the following ABox describes what is observed and what is extracted from a video of 4 seconds corresponding to the time $\mathbf{T} = (t_1, t_2, t_3, t_4)$ with our data processing process.

Example 2. $\boxed{ABox\ \mathcal{A}:}$

$p_1 : Spectator,\ p_2 : Spectator,\ p_3 : Spectator$
$(p_1, t_1) : Applauding,\ (p_1, t_2) : Applauding,\ (p_1, t_3) : Applauding,\ (p_1, t_4) : Raising$
$(p_2, t_1) : Applauding,\ (p_2, t_2) : Applauding,\ (p_2, t_3) : Raising,\ (p_2, t_4) : Raising$
$(p_3, t_4) : Raising$

---

[8] Note that the prefix CO has been removed to make the reading easier.

In order to recognize the GOAL event, a viable way is to check the weak satisfiability of the concept GOALBEHAVIOR at anytime of the video. This consists in finding a model $\mathcal{I}$ that satisfies the ABox $\mathcal{A}$, for instance:

- $\Delta^{\mathcal{I}} = \{p_1, p_2, p_3\}$,
- $Spectator_t^{\mathcal{I}} = \{p_1, p_2, p_3\}$ for all $t \in \mathbf{T}$,
- $Applauding_{t_1}^{\mathcal{I}} = \{p_1, p_2\}$, $Applauding_{t_2}^{\mathcal{I}} = \{p_1, p_2\}$, $Applauding_{t_3}^{\mathcal{I}} = \{p_1\}$, $Applauding_{t_4}^{\mathcal{I}} = \emptyset$,
- $Raising_{t_1}^{\mathcal{I}} = \emptyset$, $Raising_{t_2}^{\mathcal{I}} = \emptyset$, $Raising_{t_3}^{\mathcal{I}} = \{p_1\}$, $Raising_{t_4}^{\mathcal{I}} = \{p_1, p_2, p_3\}$,
- $GoalBehavior_{t_1}^{\mathcal{I}} = Spectator_{t_1}^{\mathcal{I}} \cap (\exists([Applauding] \sqcap^m]Raising]))_{t_1}^{\mathcal{I}}$,
- $(\exists([Applauding] \sqcap^m]Raising]))_{t_1}^{\mathcal{I}} = \{a \in \Delta^{\mathcal{I}} \mid (a, t') \in ([Applauding] \sqcap^m]Raising])_{t_1}^{\mathcal{I}}\}$.

Thus,

- $(([Applauding] \sqcap^m]Raising])_{t_1}^{\mathcal{I}})^{\mathcal{I}} = \{(p_1, t_1), (p_1, t_2), (p_1, t_3), (p_1, t_4), (p_2, t_1), (p_2, t_2), (p_2, t_3), (p_2, t_4)\}$

because

- $[Applauding]_{t_1}^{\mathcal{I}} = \{(p_1, t_1), (p_1, t_2), (p_1, t_3), (p_2, t_1), (p_2, t_2)\}$,
- $]Raising]_{t_1}^{\mathcal{I}} = \{(p_1, t_4), (p_2, t_3), (p_2, t_4), (p_3, t_4)\}$.

   Then $GoalBehavior_{t_1}^{\mathcal{I}} = \{p_1, p_2\}$.

It means that the concept GOALBEHAVIOR is weakly satisfiable at t_1, and we can deduce that a goal event happened at t_1.

13.5 CONCLUDING REMARKS

In this chapter we have presented a proposal for a framework integrating computer vision/pattern recognition techniques with ontological reasoning. We believe that such integration can constitute a novel and promising line of research, as it is able to deploy the benefits coming both from data-based and knowledge-based approaches.

First of all, the integrated approach goes in the direction of filling the semantic gap by leveraging on background knowledge stored in the ontology (both general knowledge on the principles governing the domain of interest, contained in the TBox, and inferred knowledge on the situation at hand, obtained by populating the ABox through the execution of reasoning mechanisms) to facilitate classification and recognition of what is seen.

Second, and motivated by what has just been mentioned, the integrated approach is particularly useful when data available for the training are scarce, which is often the case. Moreover, in many situations data are scarce as they refer to anomalous situations that one wants to prevent, as is, for instance, the case for disasters' monitoring. The approach is also more efficient than data-based models when the classification task is directed toward classes that are visually very heterogeneous (as, for instance, chairs).

Finally, the integrated approach is much more generalizable, as what has to be redefined once that the application domain changes is only the TBox. More specifically, if one adopts a principled methodology, anchored on a foundational ontology, as we did with DOLCE, the redefinition of the TBox is driven by modeling guidelines and results in a better interoperability with other domain ontologies.

The most obvious drawback is that the reasoning part can be computationally expensive. The trade-off between expressivity and computability is a long-standing and well-known issue for the ontology community and, of course, the success of the approaches as the one proposed here strongly depends on the technological advances in computational efficiency of information systems.

What can currently be done to address such issue is to work on the application and optimization of the reasoning mechanisms, and the latter is, of course, one of the main future directions of the current work.

ACKNOWLEDGMENTS

D. Conigliaro, R. Ferrario and D. Porello have been supported by the VISCOSO project grant, financed by the Autonomous Province of Trento through the "Team 2011" funding program. C. Hudelot has been supported by the LOGIMA project grant (ANR-2012-CORD-017-03).

REFERENCES

[1] Zhan B, Monekosso DN, Remagnino P, Velastin SA, Xu LQ. Crowd analysis: a survey. Mach Vis Appl 2008;19(5–6):345–57.

[2] Thida M, Yong YL, Climent-Pérez P, Eng HL, Remagnino P. A literature review on video analytics of crowded scenes. In: Intelligent multimedia surveillance. Berlin, Heidelberg: Springer; 2013. p. 17–36.

[3] Li T, Chang H, Wang M, Ni B, Hong R, Yan S. Crowded scene analysis: a survey. IEEE Trans Circuits Syst Video Technol 2015;25(3):367–86.

[4] Kok VJ, Lim MK, Chan CS. Crowd behavior analysis: a review where physics meets biology. Neurocomputing 2015;117:342–62.

[5] Zitouni MS, Bhaskar H, Dias J, Al-Mualla M. Advances and trends in visual crowd analysis: a systematic survey and evaluation of crowd modeling techniques. Neurocomputing 2016;186:139–59.

[6] Ali S, Shah M. A Lagrangian particle dynamics approach for crowd flow segmentation and stability analysis. In: CVPR. 2007.

[7] Mehran R, Oyama A, Shah M. Abnormal crowd behavior detection using social force model. In: IEEE conference on computer vision and pattern recognition. 2009. p. 935–42.

[8] Mehran R, Moore B, Shah M. A streakline representation of flow in crowded scenes. In: ECCV. 2010.

[9] Wu S, Moore BE, Shah M. Chaotic invariants of Lagrangian particle trajectories for anomaly detection in crowded scenes. In: 2010 IEEE conference on computer vision and pattern recognition (CVPR). IEEE; 2010. p. 2054–60.

[10] Kratz L, Nishino K. Tracking pedestrians using local spatio-temporal motion patterns in extremely crowded scenes. IEEE Trans Pattern Anal Mach Intell 2012;34(5):987–1002.

[11] Cong Y, Yuan J, Liu J. Abnormal event detection in crowded scenes using sparse representation. Pattern Recognit 2013;46(7):1851–64.

[12] Zhou B, Wang X, Tang X. Random field topic model for semantic region analysis in crowded scenes from tracklets. In: 2011 IEEE conference on computer vision and pattern recognition (CVPR). IEEE; 2011. p. 3441–8.

[13] Chongjing W, Xu Z, Yi Z, Yuncai L. Analyzing motion patterns in crowded scenes via automatic tracklets clustering. Commun China 2013;10(4):144–54.

[14] Mahadevan V, Li W, Bhalodia V, Vasconcelos N. Anomaly detection in crowded scenes. In: 2010 IEEE conference on computer vision and pattern recognition (CVPR). 2010. p. 1975–81.

[15] Raghavendra R, Bue AD, Cristani M, Murino V. Optimizing interaction force for global anomaly detection in crowded scenes. In: 2011 IEEE international conference on computer vision workshops (ICCV workshops). IEEE; 2011. p. 136–43.

[16] Wu S, Moore BE, Shah M. Chaotic invariants of Lagrangian particle trajectories for anomaly detection in crowded scenes. In: CVPR. 2010. p. 2054–60.

[17] Sodemann AA, Ross MP, Borghetti BJ. A review of anomaly detection in automated surveillance. IEEE Trans Syst Man Cybern, Part C, Appl Rev 2012;42(6):1257–72.

[18] Kong D, Gray D, Tao H. A viewpoint invariant approach for crowd counting. In: ICPR. IEEE Computer Society; 2006. p. 1187–90.

[19] Ryan D, Denman S, Fookes C, Sridharan S. Crowd counting using multiple local features. In: Digital image computing: techniques and applications (DICTA). 2009.

[20] Chan AB, Sheng John Z, Vasconcelos LN. Privacy preserving crowd monitoring: counting people without people models or tracking. In: CVPR, vol. 1. 2008. p. 1–7.

[21] Cristani M, Raghavendra R, Del Bue A, Murino V. Human behavior analysis in video surveillance: a social signal processing perspective. Neurocomputing 2013;100:86–97.

[22] McPhail C. The myth of the madding crowd. 1991.

[23] Blumer H. Collective behavior. In: New outline of the principles of sociology. 1951. p. 166–222.

[24] Goode E. Collective behavior. 1992.

[25] Conigliaro D, Rota P, Setti F, Bassetti C, Conci N, Sebe N, Cristani M. The S-Hock dataset: analyzing crowds at the stadium. In: IEEE conference on computer vision and pattern recognition (CVPR). 2015. p. 2039–47.

[26] Berlonghi A. Understanding and planning for different spectator crowds. Saf Sci 1995;18:239–47.

[27] Porello D, Bottazzi E, Ferrario R. The ontology of group agency. In: FOIS. Front Artif Intell Appl, vol. 267. IOS Press; 2014. p. 183–96.

[28] Rodríguez ND, Cuéllar MP, Lilius J, Calvo-Flores MD. A survey on ontologies for human behavior recognition. ACM Comput Surv 2014;46(4):43:1–33.

[29] Goffman E. Behaviour in public places. New York: Free Press of Glencloe; 1963 [Notes on the social organization of gatherings].

[30] Goffman E. Encounters: two studies in the sociology of interaction. 1961.

[31] Bassetti C. A novel interdisciplinary approach to socio-technical complexity sociologically-driven, computable methods for sport spectator crowds' semi-supervised analysis. In: Carving society: new frontiers in the study of social phenomena. 2015.

[32] Conigliaro D, Setti F, Bassetti C, Ferrario R, Cristani M. Attento: attention observed for automated spectator crowd analysis. In: HBU. 2013.

[33] Conigliaro D, Setti F, Bassetti C, Ferrario R, Cristani M. Viewing the viewers: a novel challenge for automated crowd analysis. In: ICIAP. 2013.

[34] Cristani M, Ferrario R. Statistical pattern recognition meets formal ontology at the University of Verona: towards a semantic visual understanding. Tech. rep., 2015.

[35] Porello D, Cristani M, Ferrario R. Integrating ontologies and computer vision for classification of objects in images. In: Besold TR, Kühnberger KU, editors. Proceedings of the workshop on neural-cognitive integration. KI 2015 38th German conference on artificial intelligence, vol. 1. 2015. p. 1–15.

[36] Karpathy A, Li F. Deep visual-semantic alignments for generating image descriptions. In: IEEE conference on computer vision and pattern recognition. 2015. p. 3128–37.

[37] Vinyals O, Toshev A, Bengio S, Erhan D. Show and tell: a neural image caption generator. In: IEEE conference on computer vision and pattern recognition. 2015. p. 3156–64.

[38] Chen X, Zitnick CL. Mind's eye: a recurrent visual representation for image caption generation. In: IEEE conference on computer vision and pattern recognition. 2015. p. 2422–31.

[39] Chen X, Fang H, Lin T, Vedantam R, Gupta S, Dollár P, Zitnick CL. Microsoft COCO captions: data collection and evaluation server. Available from arXiv:1504.00325, 2015.

[40] Russakovsky O, Deng J, Su H, Krause J, Satheesh S, Ma S, Huang Z, Karpathy A, Khosla A, Bernstein MS, Berg AC, Li F. ImageNet large scale visual recognition challenge. Int J Comput Vis 2015;115(3):211–52.

[41] Bottou L. From machine learning to machine reasoning – an essay. Mach Learn 2014;94(2):133–49.

[42] Town C. Ontological inference for image and video analysis. Mach Vis Appl 2006;17(2):94–115.

[43] Straccia U, Visco G. DLmedia: an ontology mediated multimedia information retrieval system. In: Proceedings of the 2007 international workshop on description logics (DL2007). 2007.

[44] Hudelot C, Atif J, Bloch I. Fuzzy spatial relation ontology for image interpretation. Fuzzy Sets Syst 2008;159(15):1929–51.

[45] Donadello I, Serafini L. Mixing low-level and semantic features for image interpretation – a framework and a simple case study. In: Computer vision – ECCV 2014 workshops, proceedings, part II. 2014. p. 283–98.

[46] Bannour H, Hudelot C. Building and using fuzzy multimedia ontologies for semantic image annotation. Multimed Tools Appl 2014;72(3):2107–41.

[47] Dasiopoulou S, Kompatsiaris I, Strintzis MG. Investigating fuzzy DLS-based reasoning in semantic image analysis. Multimed Tools Appl 2010;49(1):167–94.

[48] Chen L, Nugent C. Ontology-based activity recognition in intelligent pervasive environments. Int J Semantic Web Inf Syst 2009;5(4):410–30.

[49] Bannour H, Hudelot C. Towards ontologies for image interpretation and annotation. In: 9th international workshop on content-based multimedia indexing. 2011. p. 211–6.

[50] Tousch A, Herbin S, Audibert J. Semantic hierarchies for image annotation: a survey. Pattern Recognit 2012;45(1):333–45.

[51] Neumann B, Möller R. On scene interpretation with description logics. Image Vis Comput 2008;26(1):82–101.

[52] Peraldí ISE, Kaya A, Möller R. Logical formalization of multimedia interpretation. In: Knowledge-driven multimedia information extraction and ontology evolution – bridging the semantic gap. 2011. p. 110–33.

[53] Smeulders AWM, Worring M, Santini S, Gupta A, Jain R. Content-based image retrieval at the end of the early years. IEEE Trans Pattern Anal Mach Intell 2000;22(12):1349–80.

[54] Zhao R, Grosky WI. Negotiating the semantic gap: from feature maps to semantic landscapes. Pattern Recognit 2002;35(3):593–600.

[55] Maillot N, Thonnat M, Hudelot C. Ontology based object learning and recognition: application to image retrieval. In: 16th IEEE international conference on tools with artificial intelligence. 2004. p. 620–5.

[56] Hudelot C, Maillot N, Thonnat M. Symbol grounding for semantic image interpretation: from image data to semantics. In: Proceedings of the workshop on semantic knowledge in computer vision. 2005.

[57] Lutz C, Areces C, Horrocks I, Sattler U. Keys, nominals, and concrete domains. J Artif Intell Res 2005;23:667–726.

[58] Baader F, Calvanese D, McGuinness DL, Nardi D, Patel-Schneider PF, editors. The description logic handbook: theory, implementation, and applications. Cambridge University Press; 2003.

[59] Hudelot C, Atif J, Bloch I. ALC(Ʊ): a new description logic for spatial reasoning in images. In: Computer vision – ECCV 2014 workshops. 2014. p. 370–84.

[60] Masolo C, Borgo S, Gangemi A, Guarino N, Oltramari A. WonderWeb deliverable D18. Tech. rep. CNR; 2003.

[61] Borgo S, Masolo C. Foundational choices in DOLCE. In: Handbook on ontologies. Springer; 2009. p. 361–81.

[62] Artale A, Franconi E. A survey of temporal extensions of description logics. Ann Math Artif Intell 2000;30(1–4):171–210.

[63] Lutz C, Wolter F, Zakharyaschev M. Temporal description logics: a survey. In: 15th international symposium on temporal representation and reasoning. 2008. p. 3–14.

[64] Baader F, Ghilardi S, Lutz C. LTL over description logic axioms. ACM Trans Comput Log 2012;13(3):21.

[65] Gutiérrez-Basulto V, Jung JC, Lutz C. Complexity of branching temporal description logics. In: ECAI 2012 – 20th European conference on artificial intelligence. Including prestigious applications of artificial intelligence (PAIS-2012) system demonstrations track. 2012. p. 390–5.

[66] Franconi E, Toman D. Fixpoint extensions of temporal description logics. In: Proceedings of the 2003 international workshop on description logics. 2003.

[67] Artale A, Kontchakov R, Ryzhikov V, Zakharyaschev M. Interval temporal description logics. In: Proceedings of the 28th international workshop on description logics. 2015.

[68] Lutz C. Combining interval-based temporal reasoning with general TBoxes. Artif Intell 2004;152(2):235–74.

[69] Allen JF. Maintaining knowledge about temporal intervals. Commun ACM 1983;26:832–43.

CHAPTER 14

SALSA: A Multimodal Dataset for the Automated Analysis of Free-Standing Social Interactions

Xavier Alameda-Pineda*, Ramanathan Subramanian[†], Elisa Ricci[‡,§],
Oswald Lanz[‡], Nicu Sebe*

*Department of Information Engineering and Computer Science, University of Trento, Trento, Italy
[†]Centre for Visual Information Technology, International Institute of Information Technology, Hyderabad, India
[‡]Center for Information and Communication Technology, Fondazione Bruno Kessler, Trento, Italy
[§]Department of Engineering, University of Perugia, Perugia, Italy

Contents

Group and Crowd Behavior for Computer Vision
DOI: 10.1016/B978-0-12-809276-7.00017-5

(A) AMI corpus [40]   (B) SALSA

Figure 14.1 Exemplar images from a round-table meeting (left) and from the SALSA scenario (right). Analyzing FCGs as in SALSA is challenging due to varying illumination, low resolution of faces, extreme occlusions, and crowdedness (best-viewed in color and under zoom).

14.1 INTRODUCTION

Recent research progress in the areas of computer vision, speech processing, wearable computing, and multimodal analysis has enabled the investigation of complex phenomena such as social interactions. Social interactions are commonplace in our daily lives, and so they have been studied extensively by social psychologists for many years [24]. Indeed, the ability to recognize social interactions and infer social cues is critical for a number of applications such as surveillance, social robotics, and behavior analysis [4,23]. Most automated social interaction analysis (ASIA) methods have focused on round-table meetings [41,35] (Fig. 14.1A), where spatial arrangement and sufficient spatial separation between persons enable reliable extraction of behavioral cues for each scene target (or person).

Recently, researchers have shifted their interest into examining unstructured meeting scenes and in particular *free-standing conversational groups* (FCGs), i.e., small groups of two or more co-located persons engaged in ad hoc interactions [24]. FCGs emerge naturally in diverse social occasions, and interacting persons are characterized by mutual scene locations and orientations, resulting in geometric patterns known as F-formations [29].[1] Studying unstructured social scenes (e.g., a cocktail party) is extremely challenging since it involves inferring (i) positions and head/body orientations of targets in the scene, (ii) semantic and prosodic auditory content corresponding to each target, and (iii) F-formations observed at a particular time. Some of the aforementioned problems have been effectively addressed only in controlled environments such as round-table meetings (Fig. 14.1A), where behavioral cues can be reliably inferred through the use of close-range cameras and close-talk microphones. However, all of them remain unsolved in unstructured social settings (Fig. 14.1B), where only distant surveillance cameras can be

---

[1] An F-formation is a set of possible configurations in space that people may assume while participating in a social interaction.

used to capture FCGs [29], and microphones may be insufficient to clearly recognize the speaker(s) at a given time instant due to scene crowding and reverberations [30].

To address the challenges involved in analyzing FCGs we present *SALSA*, a novel dataset[2] facilitating multimodal and Synergetic sociAL Scene Analysis. In contrast to social datasets studying round-table meetings [35,41], or examining FCGs on a small scale [52,63], SALSA contains uninterrupted recordings of an indoor social event involving 18 subjects over 60 min, thereby serving as a rich and extensive repository for the behavioral analysis and social signal processing communities. Four wall-mounted cameras were used for monitoring the scene. In addition, sociometric badges [16] were also worn by participants to record various aspects of their behavior. These badges include a microphone, an accelerometer, Bluetooth and infrared (IR) sensors. The microphone records auditory content, while the accelerometer captures person motion. Bluetooth and IR transmitters and receivers provide information regarding interacting persons, and are useful for inferring body pose under occlusions. Cumulatively, these sensors can synergetically enhance estimation of target locations and their head and body orientations, F-formations, and provide a rich description of FCGs' behavior. In addition to the raw multimodal data, SALSA also contains position, pose, and F-formation annotations, as well as information regarding participants' personality traits.

The social event captured in SALSA comprised two parts, i.e., a *poster presentation* session and a *cocktail party* scene. During the recordings there were no constraints imposed in terms of behavior of participants (i.e., no scripting) or of the spatial layout of the scene. Furthermore, the indoor environment where the event was recorded was prone to varying illumination and reverberation conditions. The geometry of F-formations was also influenced by the physical space where the social interaction was taking place, and the poster session was intended to simulate a semistructured social setting. While it is reasonable to expect observers to stand in a semicircular fashion around the poster presenter, none of the participants were instructed on where to stand or what to attend. Finally, as seen in Fig. 14.1B, the crowded nature of the scene and resulting F-formations gave rise to extreme occlusions, which along with the low-resolution of faces captured by surveillance cameras made visual analysis extremely challenging. Overall, SALSA represents the most challenging dataset for studying FCGs to our knowledge.

We believe that SALSA will enable researchers to go beyond the traditional audiovisual analysis of FCGs, providing a unique opportunity for multimodal analysis of social interactions. As demonstrated in our experimental evaluation (Section 14.5), fusing multiple modalities can alleviate difficulties posed by occlusions and crowdedness to more precisely estimate behavioral cues. We strongly believe that the challenging nature of SALSA will spur intensive ASIA research.

---

[2] The SALSA dataset is available at http://tev.fbk.eu/salsa.

14.2 LITERATURE REVIEW

This section reviews the state-of-the-art in social behavior analysis with specific emphasis on methods analyzing FCGs. We will first discuss unimodal approaches (i.e., vision-, audio-, and wearable sensor-based) and then describe multimodal approaches.

14.2.1 Unimodal Approaches

14.2.1.1 Vision-Based Approaches

Challenges pertaining to surveillance scenes involving FCGs addressed by vision-based methods include detection and tracking of targets in the scene, estimation of social attention direction, and detection of F-formations.

Given the cluttered nature of social scenes involving FCGs, detecting and tracking the locations of multiple targets is in itself a highly challenging task. As extreme occlusions are common in social scenes involving FCGs (see Fig. 14.1), employing information from multicamera views can enable robust tracking [32]. Tracking-by-detection [12] combined with data association using global appearance and motion optimization [11,42] has benefited multitarget tracking. Some works have further focused on identity-preserving tracking over long sequences [9]. Such methods assume a large number of high-confidence detections which might not always be available with FCGs due to persistent long-term occlusions, although recent works have focused on the detection problem under partial occlusion [39,59]. Multitarget tracking is further shown to improve by incorporating aspects of human social behavior [46,22,33]. However, F-formations are special groups, characterized by largely static arrangements of interacting persons constrained by their locations and head/body orientations.

Social attention determination in round-table meetings, where high-resolution face images are available, has been studied extensively. Vision-based approaches typically employ head pose as a proxy to determine social attention direction [54]. In comparison, head pose estimation (HPE) from blurry surveillance videos is much more difficult. Visual attention direction of moving targets is estimated by Smith et al. [53] using position and head pose cues, but social scenes or occlusions are not addressed here. Some works have exploited the relationship between walking direction and head/body pose to achieve HPE from surveillance videos under limiting conditions where no labeled training examples are available, or under occlusions [10,14,48]. Focus-of-attention estimation in social scenes is explicitly addressed by Bazzani et al. [8], who model a person's visual field in a 3D scene using a subjective view frustum.

Recently, the computer vision community has shown some interest in the detection and analysis of dyadic interactions [38,45] and more general groups [15,20,62]. Also, the interest in detecting social interactions and F-formations from video has intensively grown. Cristani et al. [18,50,51] employ positional and head orientation estimates to detect F-formations based on a Hough voting strategy. Bazzani et al. [8] gather information concerning F-formations in a social scene using the Interrelation pattern matrix. F-formations are modeled as maximal cliques in edge-weighted graphs in [27], and each target is presumed to be oriented toward the closest neighbor but head/body orientation is not explicitly computed. Gan et al. [21] detect F-formations upon inferring location and orientation of subjects using depth sensors and cameras. Setti et al. [52] propose a graph-cuts based framework for F-formation detection based on the position and head orientation estimates of targets. Based on experiments performed on four datasets (*IDIAP Poster* [27], *Cocktail Party* [63], *Coffee Break* [18], and *GDet* [8]), their method outperforms six state-of-the-art methods in the literature. Ricci et al. [49] jointly estimate head, body orientations and F-formations in a unified framework, by casting the same as a convex regression problem.

14.2.1.2 Audio-Based Approaches

Studying interactional behavior in unstructured social settings solely using audio or speech-based cues is extremely challenging, as FCGs are not only characterized by speaking activity, but also by nonverbal cues such as head and body pose, gestural and proxemic cues. Furthermore, classical problems in audio analysis become extremely challenging and remain unexplored in crowded indoor environments involving a large number of targets. Indeed, current methodologies for speaker diarization [6], sound source separation [7], or localization [1] address scenarios with few persons. Nevertheless, a few studies on audio-based detection of FCGs have been published. Wyatt et al. [60] tackle the problem of detecting FCGs upon recognizing the speaker using temporal and spectral audio features. Targets are then clustered to determine co-located groups on the basis of speaker labels. More recently, FCG detection and network inference is achieved in [36] by performing speaker recognition, and F-formations are detected based on proximity information obtained using Bluetooth sensors.

14.2.1.3 Wearable-Sensor Based Approaches

Wearable sensors can provide complementary behavioral cues in situations where visual and speech data are unreliable due to occlusions and crowdedness. Hung et al. [26] detect FCGs in social gatherings by measuring motion via an accelerometer. With the increased usage of smartphones, mobile sensor data have also become a viable choice for analysis of social interactions or more complex social systems [19]. Via mobile phones, proximity can be inferred from WiFi and Bluetooth [37]. However, the spatial resolution of these sensors is limited to only a few meters, and the colocation of mobile devices does not necessarily indicate a social interaction between the corresponding individuals. Therefore, Cattuto et al. [13] propose a framework that balances scalability and resolution through a sensing tier consisting of cheap and unobtrusive active RFID devices, which are capable of sensing face-to-face interactions as well as spatial proximity over different scale lengths down to one meter or less. Nevertheless, we note that many of these works address relatively less crowded scenarios, not comparable in complexity to SALSA.

14.2.2 Multimodal Approaches

Multimodal approaches to social interaction have essentially examined small-group interactions such as round-table meetings, and mainly involve audio-visual analysis as detailed below. Examples of databases containing audio-visual recordings and associated annotations are the Canal9 [57], AMI [41], Mission Survival [35], Free Talk [31], and the Idiap WOLF [25] corpora. Robotics is one among other applications for which datasets have been compiled (see [3]). All these data collection efforts have inspired interdisciplinary research in the field of human behavior understanding and led to the emergence of the social signal processing community [58]. Several of these databases have been utilized for isolating traits relating to an individual or a group. Choudhury and Pentland [16] initiated behavior analysis using wearable sensors by developing the *Sociometer*. Recently, Olguín et al. [44] proposed the *Sociometric badge*, which stores (i) motion using an accelerometer, (ii) speech features (rather than raw audio), (iii) position information, (iv) proximity to other individuals using a Bluetooth sensor, and (v) face-to-face interactions via an infrared sensor. Sociometric badges have been used to capture face-to-face communication patterns, examine relationships among individuals, and model collective and organizational behavior [34], detect personality traits and states, and predict outcomes such as productivity and job satisfaction [43]. Another notable ASIA work employing multisensory information is that of Matic et al. [40], who estimate body orientation and interpersonal distance via mobile data and speech activity to detect social interactions.

14.3 SPOTTING THE RESEARCH GAP

This section details some of the limitations of current ASIA works, in terms of datasets used for evaluation and methodologies.

14.3.1 Datasets

Previous ASIA datasets focusing on FCGs have mainly been used to address two research problems: (1) Detecting F-formations and (2) Studying individual and group behavior from multiple sensing modalities. The vast majority of works addressing F-formation detection are vision-based. Table 14.1 presents key figures of the datasets used for F-formation detection: the number (denoted using #) of annotated frames and scene targets; and also enumerates the sensing modalities. Among them, SALSA is unique due to its (i) multimodal nature, (ii) extensive annotations available over a long duration, and (iii) challenging nature of the captured scene.

Table 14.1 Existing datasets related to SALSA facilitating F-formation detection

Name	# targets	# annot. frames	Sensing modalities
Talking Heads	< 5	< 10	Video, Audio
IDIAP Poster [27]	50	~ 100	Video
Coffee Break [18]	10	~ 150	Video
Cocktail Party [63]	8	~ 300	Video
GDet [8]	< 5	~ 400	Video
SALSA	18	~ 1000	Video, Audio, IR, Bluetooth, Accel.

Table 14.2 Existing ASIA datasets related to SALSA: the first five consist of round-table meetings and span hours, while the last four study social networks/behavior and span days/months

Name	Vision	Audio	IR/RFID	Bluetooth	Accel.
AMI [41]	√	√			
Canal9 [57]	√	√			
Mission Survival [35]	√	√			
Free Talk [31]	√	√			
Idiap WOLF [25]	√	√			
SALSA	√	√	√	√	√
Sociometric Corpus [44]	√	√		√	√
Trento Sociom. Corpus [34]		√	√	√	√
Boston Sociom. Corpus [43]		√	√	√	√
Matic et al. [40]		√		√	

Table 14.2 depicts the datasets used for individual and group behavioral analysis. While the first group (top) consists of audio–visual recordings spanning hours and acquired under controlled settings, the second group (bottom) comprises datasets acquired over days/months for studying social networks and group relationships. SALSA again stands out as it records information from both static cameras and wearable sensors, leading to a previously nonexistent and highly informative combination of sensing modalities.

14.3.2 ASIA Methodologies

14.3.2.1 Human Tracking and Pose Estimation

Human tracking and pose estimation in social scenes is challenging for several reasons. Firstly, a person's visual appearance can change considerably across the scene due to varying camera perspective and illumination, as well as with pose and posture. Secondly, large and persistent occlusions are frequent in such scenes, which corrupt observations and estimates. Thirdly, integrating auditory information to aid localization and orientation estimation is also difficult due to its intermittent nature and the adverse impact of reverberations and interfering sources. Beyond inherent complexity, the state-of-the-art is further challenged when raw audio-visual data cannot be recorded for processing or scene-specific optimization, e.g., due to privacy concerns. Accurate FCG behavior analysis requires the correct assignment of observations to sources (targets) over the long run. Identity switches during tracking can corrupt the extraction of aggregated features computed over time to infer functional roles or interaction networks, and long-term identity-preserving multitarget tracking has not been achieved, yet. Furthermore, existing appearance-based pose estimation methods are not adapted to highly cluttered scenes with large and persistent occlusions. Finally, multitarget approaches are not able to robustly and efficiently scale to large groups due to the computational complexity involved.

14.3.2.2 Speech Processing

While numerous research studies have attempted speech, speaker and prosodics recognition under controlled conditions, several issues arise when auditory information is captured via mobile microphones in crowded indoor scenes. Firstly, regular indoor environments are prone to reverberations, which adversely affect many acoustic processing techniques. Secondly, intermittence of the speech signal necessitates speaker diarization prior to processing. Thirdly, the speech signal is also spatially sparse, and source separation techniques are usually required to segment speaker activity. Currently, there are no algorithms addressing source separation or diarization in uncontrolled conditions involving the presence of multiple concurrent sound sources. While multimodal approaches have addressed these problems via audio-visual processing, they still cannot work with large groups of persons and crowded unconstrained indoor environments.

14.3.2.3 F-Formation Detection

Detecting F-formations in unconstrained environments is a complex task. As these are characterized by mutually located and oriented persons, robust tracking and pose estimation algorithms are necessary. However, both multitarget tracking and head/body pose estimation in crowded scenes are difficult as discussed earlier. Even under ideal conditions, F-formation shapes are influenced by (i) the environment's layout, i.e., room shape, furniture, and other physical obstacles, (ii) scene crowdedness, and (iii) attention hotspots such a poster, painting, etc. While existing methodologies typically assume that F-formation members are placed on an ellipse, robust F-formation detection requires accounting for the above factors as well. Also, most algorithms are visually driven, and few multimodal approaches exist to this end.

14.3.3 Why SALSA?

We believe that SALSA can contribute to the advancement of the state-of-the-art in behavior analysis for many reasons.

- Even if multimodal analysis has been found to outperform unimodal approaches and provide a richer representation of social interplays, some key tasks are not yet addressed in a multimodal fashion, e.g., pose estimation and F-formation detection. We believe that SALSA is an optimal testbed for developing multimodal ASIA methods.
- Social interactions have routinely been studied under controlled settings, and there is a paucity of methods able to cope with unconstrained environments involving large groups, crowded spaces and highly dynamic interactions. Interestingly, SALSA specifically focuses on these aspects to facilitate the design of robust algorithms.
- Most existing approaches have independently studied the different behavioral tasks. For instance, knowing the head and body orientations of individuals can help in the estimation of F-formations and vice versa. Similarly, F-formation detection clearly benefits from accurate tracking, which in turn is benefited by the robust detection of F-formations. We believe that SALSA will foster joint or a holistic examination of these challenges.

14.4 THE SALSA DATASET

SALSA represents an excellent test-bed for multimodal human behavior understanding due to the following reasons. Firstly, all behavioral data were collected in a regular indoor space with the participants only requiring to wear portable and compact sociometric badges which ensured naturalistic social behavior. Secondly, due to the unconstrained nature of the scene, the recordings contain numerous artifacts such as varying illumination, visual occlusions, reverberations or interfering sound sources. Thirdly, the recorded event involved 18 persons: such large social groups have rarely been studied in the behavior analysis literature. These participants did not receive any special instructions or scripts prior to the recording, and the resulting social interactions were therefore free-willed and hedonistic in nature. Finally, the social interplay was recorded via four wall-mounted surveillance cameras and the *Sociometric badges*[3] worn by the targets. These badges recorded different aspects of the targets' social behavior such as audio or motion as detailed later. This combination of static cameras and wearable sensors is scarce in the literature, and provides a wealth of behavioral information as shown in Section 14.5. These four salient characteristics place SALSA in a unique position among the various datasets available for studying social behavior.

14.4.1 Scenario and Roles

SALSA was recorded in a regular indoor space and the captured social event involved 18 participants and consisted of two parts of roughly equal duration. The first part consisted

---

[3] http://www.sociometricsolutions.com/.

Table 14.3 Description of the sensors in SALSA. STFT denotes short-time Fourier transform

Sensor	Output	Frequency (Hz)
Vision	4 synchronized images	15
Audio	Amplitude stats & STFT	2 & 30
Infrared	Detected badge's ID	1
Bluetooth	Detected badge's ID	1/60
Accelerometer	Body motion	20

of a *poster presentation* session, where four research studies were presented by graduate students. A fifth person chaired the poster session. In the second half, all participants were allowed to freely interact over food and beverages during a *cocktail party*. It needs to be noted here that while some participants had specific roles to play during the poster presentation session, none were given any instructions on how to act in the form of a script. Consequently, the interaction dynamics correspond to those of a natural social interplay. Obviously, participants with different roles (chair, poster presenter, attendee) are expected to have different interaction dynamics, and these roles were designed to help behavioral researchers working on meeting role recognition.

14.4.2 Sensors

The SALSA data were captured by a camera network and wearable badges worn by targets. The camera network comprised four synchronized static RGB cameras (1024×768 resolution) operating at 15 frames per second (fps). Each participant wore a sociometric badge during the recordings which is a $9 \times 6 \times 0.5$ cm^3 box equipped with four sensors, namely, a microphone, an infrared (IR) beam and detector, a Bluetooth detector, and an accelerometer. The badges were battery-powered and recorded data on a USB card without the need for any wired connection, thus enabling natural social interplay. Table 14.3 presents an overview of the five sensors used.

14.4.3 Ground Truth Data
14.4.3.1 Annotations

In order to fulfill the requirements expected of a systematic evaluation framework, SALSA provides ground-truth annotations, which were performed either manually or semiautomatically over the entire event duration. The annotations were produced in two steps. In the first step, using a dedicated multiview scene annotation tool, the *position*, *head*, and *body* orientation of each target was annotated every 45 frames (3 s). To speed up the annotation process, the total number of targets was divided among three annotators. A target's position, head and body orientation were annotated by the first annotator and then double-checked by the second. Discrepancies between their judgments were resolved by the third annotator. All annotators were clearly instructed on

(A) Pose   (B) F-formation

Figure 14.2 Annotations. Panel (A) shows the annotations of the head and body poses of all people in the scene. Panel (B) shows the annotation of the F-formations.

how to perform the annotations. To facilitate the annotation task, markings from the previous annotated frame were displayed so that only small modifications were needed.

In the second step, annotated positions and head/body orientations were used for deducing F-formations. Annotations were again performed every 45 frames, and we employed the following criteria for detecting F-formations: an F-formation is characterized by the mutual locations and head, body orientations of interacting targets, and is defined by the convex *O-space* they encompass such that each target has unhindered access to its center. A valid F-formation was assumed if the constituent targets were in one of the established patterns, or had direct and unconstrained access to the O-space center in case of large groups (refer to [18] for details). Fig. 14.2A shows the annotation of the head and body poses of all the people in the scene, and highlights the person being currently annotated (visualization of the annotation tool). Fig. 14.2B illustrates five annotated F-formations around four posters (target feet positions are marked with crosses). Considering the two groups in the foreground, the F-formation in front of the poster on the right does not include the FCG with two targets on the left, since neither of them have access to the center of the larger group.

14.4.3.2 Personality Data

SALSA also contains big-five personality trait scores of participants to facilitate behavioral studies. Prior to data collection, all participants filled the *Big Five* personality questionnaire [28]. The Big Five questionnaire owes its name to the five traits it assumes as being representative of personality: *Extraversion* – being sociable, assertive, playful vs. aloud, reserved, shy; *Agreeableness* – being friendly and cooperative vs. antagonistic and fault-finding; *Conscientiousness* – being self-disciplined, organized vs. inefficient, careless; *Emotional Stability* – being calm and equanimous vs. insecure and anxious; and *Creativity* – being intellectual, insightful vs. shallow, unimaginative. In the questionnaire, each trait is investigated via ten items assessed on a 1–7 Likert scale. The final trait scores were computed according to the procedure detailed in [47].

Table 14.4 Mean tracking statistics and per-target occlusion rates for the four views

	Poster	Party	All
Precision (cm)	15.2 ± 0.1	20.1 ± 0.1	17.3 ± 0.1
Failure rate (%)	2.6 ± 0.1	9.6 ± 0.3	5.7 ± 0.2
Frames-to-failure	1644 ± 63	439 ± 12	759 ± 21
Occlusion (%)	*28, 35, 22, 26*	*25, 28, 49, 27*	*27, 32, 34, 27*

14.5 EXPERIMENTS ON SALSA

This section is devoted to the performance evaluation of state-of-the-art methodologies for different behavioral tasks on SALSA.

14.5.1 Visual Tracking of Multiple Targets

Despite many advances in computer vision research, tracking individuals is still an unsolved problem. Specifically, in the case of SALSA, person tracking is challenging due to the presence of extreme and persistent occlusions. Some targets are difficult to distinguish from others using appearance features, and identity-preserving tracking required for multimodal behavior interpretation is further hindered by nonuniform scene illumination even when multiple views are available. State-of-the-art tracking-by-detection methods feature global appearance optimization but require a sufficiently dense number of high-confidence detections across the whole sequence. However, target detection as such is extremely challenging in such scenes. We therefore considered a sequential Bayesian approach without appearance model adaptation. The Hybrid Joint–Separable Particle Filter (HJS-PF) tracker [32] was specifically developed for occlusion handling and has been applied to tracking in social scenes [63,51,61].

We report HJS-PF tracking results on SALSA using an evaluation protocol similar to that of the Visual Object Tracking Challenge (VOT). The color model for each target was manually extracted from the initial part of the sequence where the target was free of occlusions, prior to tracking. These models were used for the whole sequence and were not reinitialized or adapted during tracking. HJS-PF was initialized for each target with the first available annotation, and tracking was performed at full frame rate (15 Hz) with 320×240 resolution, while evaluation was done every 3 s (or every 45 frames) consistent with the annotations. If the position estimate was over 70 cm from its reference, it was counted as a failure and the tracking of that target was reinitialized at the reference. Otherwise, the distance from the reference was accumulated to compute precision.

In Table 14.4, the average precision (average distance from the references), per-target failure rate (% of failures over 20K annotations), and frames-to-failure count (number of subsequent frames successfully tracked) are reported for (i) the first 30K frames (*Poster*),

Figure 14.3 Tracking results sample on the poster session part of SALSA.

(ii) the remaining 25K frames (*Party*), and (iii) the total 60 min recording. Our implementation tracks the 18 targets using 50 particles per target at about 7 fps on a 3 GHz PC. While overall precision is high considering the large tracking area, low image-resolution and high occlusion rate (cf. last row of table; to our best knowledge, no comparable dataset exists for tracking evaluation), a sensible increase in failure rate is observed for the *party* sequence. Indeed, in FCGs, persons tend to occupy every available space such that they are hardly visible in some of the camera views. Also, targets more often bend their bodies to grab food and beverages, and illumination varies over the scene challenging color-based tracking. However, low failure rate in the *poster* session where targets tend to arrange themselves in a more orderly manner around posters indicates that occlusion handling is effective. Tracking during the *poster* and the *party* scenarios is shown in Fig. 14.3. Based on these results, we identify some key elements necessary for FCG tracking: (i) perform ground tracking with explicit occlusion handling at frame level, (ii) extract discriminative descriptors for each target to resolve identity switches, and (iii) learn the illumination pattern of the scene to adapt target models locally to lighting conditions.

14.5.2 Head and Body Pose Estimation from Visual Data

The estimation of the head and body pose is still an important research topic in the computer vision community. Specifically, when focusing on estimating the positions and head and body orientation of individuals in FCGs monitored by distant surveillance cameras, several challenges arise due to low resolution, clutter, and occlusions. To demonstrate these challenges on SALSA, we considered the recent work of Chen et al. [14], which is one of a few methods that jointly compute both head and body orientation from low resolution images. In a nutshell, this algorithm consists of two phases. First, Histograms of Oriented Gradients (HoG) are computed from head and body bounding boxes obtained from training data. Then, a convex optimization problem that jointly learns two classifiers for head and body pose, respectively, is solved. Importantly, the classifiers are learned simultaneously, imposing consistency on the computed head and body classes so as to reflect human anatomic constraints (i.e., the body orientation naturally limits the range of possible head directions). The approach in [14] leverages

Table 14.5 Head and body pose estimation error (in degrees)

% training data		View 1	View 2	View 3	View 4
10%	Head	45.7 ± 0.6	47.2 ± 0.3	48.4 ± 0.8	49.5 ± 1.2
	Body	49.3 ± 0.5	51.6 ± 0.9	51.2 ± 0.4	54.6 ± 0.8
5%	Head	43.6 ± 0.5	46.2 ± 0.3	46.4 ± 0.8	47.5 ± 0.9
	Body	47.3 ± 0.5	49.4 ± 0.5	49.9 ± 0.5	52.5 ± 0.7
1%	Head	42.2 ± 0.4	45.3 ± 0.3	43.4 ± 0.8	44.9 ± 1.5
	Body	45.4 ± 0.5	47.5 ± 0.8	48.7 ± 0.7	51.7 ± 0.5

information from both annotated and unsupervised data via a manifold term which imposes smoothness on the classifiers, typical of semisupervised learning methods. In our experiments, only labeled data were used for training.

The method proposed in [14] is monocular and considers 8 classes (angular resolution of 45°) for both head and body classification. Therefore, to test it on SALSA, we also considered each camera view separately. In this series of experiments, the target head and body bounding boxes were obtained by manual annotation, and a subset of about 7.5K samples was employed (bounding boxes were not available for targets going out of the field of view). To compute visual features for both head and body, we used the HoG descriptors. In our tests, a small percentage of the frames (1%, 5%, and 10%) were used for training, while the rest were used for testing. For performance evaluation, we used the mean angular error (in degrees) defined in [14] for computing head and body pose estimation accuracy.

Experiments were repeated ten times with random training sets, and corresponding average error and standard deviation are reported in Table 14.5. Despite many occlusions and the presence of clutter, a state-of-the-art pose classification approach achieves satisfactory performance (maximum error of around one class width). However, it is worth noticing that our experiments were performed with homogeneous training and test data, in contrast with the heterogeneous data employed in [14]. We expect a significant decrease in performance when heterogeneous training data are used for pose estimation. Also, errors observed for head pose are considerably smaller than for body pose over all four camera views – this is because body pose classifiers are impeded by severe occlusions in crowded scenes. Precisely for this reason, previous works on F-formation detection from FCGs [18,51,56] have primarily employed head orientation as the primary cue, even though body pose has been widely acknowledged as a more reliable cue for determining interacting persons. We believe that devising a multiview [61] and multimodal [5] approach also employing IR- and Bluetooth-based sensors for body pose estimation would be advantageous as compared to a purely visual analysis, which was one of the primary motives for compiling the SALSA dataset.

Table 14.6 F-formation detection with ground-truth data

Dataset	SALSA						Cocktail Party		
Pose cue	Body			Head					
Measure	Prec.	Rec.	F1	Prec.	Rec.	F1	Prec.	Rec.	F1
HVFF lin [18]	0.59	0.74	0.67	0.56	0.72	0.63	—	—	—
HVFF ent [50]	0.66	0.8	0.73	0.63	0.77	0.69	0.59	0.74	0.66
HVFF ms [51]	0.61	0.76	0.68	0.58	0.73	0.64	0.69	0.74	0.71
GT [56]	0.89	0.70	0.78	0.90	0.75	0.82	0.86	0.82	0.84
GC [52]	0.82	0.85	0.83	0.80	0.85	0.82	—	—	—

14.5.3 F-Formation Detection

Detecting F-formations by visual observing crowded scenes is a challenging task. Several factors such as low video resolution, occlusions, and complexities of human interactions hinder robust and accurate F-formation detection. We first considered five state-of-the-art vision-based approaches for individuating FCGs in SALSA. We adopted (i) Hough voting [18] (HVFF-lin), (ii) its nonlinear variant [50], and (iii) multiscale extensions [51] (denoted as HVFF-ent and HVFF-ms), (iv) the game-theoretic (GT) approach [56], and (v) the graph cut (GC) approach [52], as associated codes are publicly available.[4] These approaches take the target positions and head orientations as input, and compute F-formations independently for each frame. In particular, the Hough-voting methods work by generating a set of virtual samples around each target. These samples are candidate locations for the O-space center. By quantizing the space of all possible locations, aggregating samples in the same cell and finding the local maxima in the discrete accumulation space, the O-space centers and F-formations therefrom are identified. On the other hand, in the graph-cut algorithm, an optimization problem is solved to compute the O-space center coordinates.

We first evaluated the above F-formation detection approaches using ground-truth position and head and body pose annotations, and considered all the annotated frames. F-formation estimation accuracy is evaluated using precision, recall and F1-score as in [18]. In each frame, we consider a group as correctly estimated if at least $T \cdot |G|$ of the members are correctly found, and if no more than $1 - T \cdot |G|$ nonmembers are wrongly identified, where $|G|$ is the cardinality of the F-formation G and $T = 2/3$. Results are reported in Table 14.6. Even the most accurate approach, i.e., the graph-cut method, only achieves an F1-score of about 0.83, clearly demonstrating the need of devising more sophisticated algorithms for detecting F-formations in challenging datasets such as SALSA. Moreover, it is worth noting that our results are consistent with the observations in previous works such as [18], i.e., using the body pose is more advantageous than using head orientation for detecting a group of interacting persons.

---

[4] http://profs.sci.univr.it/~cristanm/ssp/ and http://www.iit.it/it/datasets-and-code/code/gtcg.html.

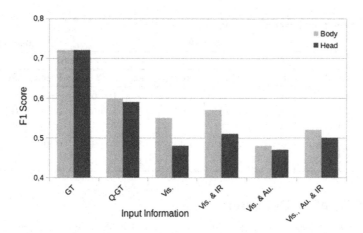

Figure 14.4 F-formation detection results (F1 score) for the graph-cut method [52]. Head and body poses were taken from ground-truth (GT) or quantized ground-truth (Q-GT) or automatically estimated from different combinations of visual (Vis.), infrared (IR), and audio (Au.) data.

In a second series of experiments, we evaluated the graph–cut approach using *automatically* estimated head and body orientations from the multisensor badge data. Specifically, we considered annotations for the target positions, and estimated head and body pose from visual data with the method proposed in [14]. In these experiments, HoG features extracted from head and body crops for the four camera views were concatenated and provided as input to the classifiers. In this series of experiments, we only considered a subset of frames where *all* the targets were in the camera field of view. To train the coupled head–body pose classifier, we used 1% of the available samples as training data. Experiments were repeated ten times, and the average performance is reported. We also integrated information from IR and audio sensors. Audio and IR are sparse observations, whilst visual data are available at every time-stamp. The likelihood of target *n* addressing *m* was estimated from audio and IR data. The maximum likelihood points to the person with whom *m* (the addressee) is more likely to interact. The audio and IR observations correspond to the direction of the addresser (*n*). For integrating multiple angles, we considered their weighted average.

All these methods are based on the concept of transactional segment of an individual, i.e., "the area in front of him/her that can be reached easily, and where hearing and sight are most effective" [17]. The transactional segment is modeled with a Gaussian distribution $\mathcal{N}(\boldsymbol{\mu}, \boldsymbol{\Sigma})$, where $\boldsymbol{\mu} = (x + D\cos\theta, y + D\sin\theta)$, x, y are the target position, θ is the head/body pose and $\boldsymbol{\Sigma} = \text{diag}(\sigma_x, \sigma_y, \sigma_\theta)$. In all our experiments, except for [56], we set $D = 60$, $\sigma_x = \sigma_y = 400$, and $\sigma_\theta = 0.005$. The number of samples generated from each individual's Gaussian distribution, $\mathcal{N}(\boldsymbol{\mu}, \boldsymbol{\Sigma})$, for the Hough voting is fixed to $N = 100$. In the case of [56], $D = 80$ and the Gaussian kernel standard deviation was set to 0.35.

The results of our experiments are reported in Fig. 14.4. Clearly, when the head and body pose are automatically computed from visual analysis, the performance significantly decreases with respect to the use of ground-truth (GT) annotations.[5] Furthermore, by combining information from multiple modalities, we obtain a modest improvement with respect to using only visual data. Specifically, while jointly employing visual and infrared data is advantageous with respect to exclusively employing the visual data, the integration of audio sensors provides minimum benefit. Finally, it is worth noting that a decrease in performance with respect to the ground-truth is also due to the angle quantization process. This can be observed by comparing the four leftmost bars in Fig. 14.4. Therefore, casting head and body pose estimation as a classification task (as typically done in previous works) appears to be insufficient for robustly detecting F-formations. Our experiments suggest that a better strategy entails casting head/body pose estimation into a regression task.

14.6 CONCLUSIONS AND FUTURE WORK

Via extensive experiments, we have demonstrated how SALSA represents a rich but challenging dataset for analysis of FCGs. Vision-based analysis for target tracking, head and body pose estimation and F-formation detection evidenced the shortcomings of state-of-the-art methodologies when posed with cluttered scenes with persisting and extreme occlusions. However, additional sensors available as part of the sociometric badge were found to be helpful in cases where visual analysis was difficult. In particular, the infrared sensor which indicates both the proximity and body pose of the interacting counterpart was found to improve F-formation detection.

Future research directions include: (a) developing new methodologies for robust audio processing in cluttered environments with many dynamic targets, (b) utilizing the Bluetooth and accelerometer data for F-formation detection and personality trait recognition, (c) designing tracking and head/body pose estimation algorithms capable of exploiting multimodal data, and (d) inferring emotional or physiological signals as done in other contexts [2,55]. Given the extensive raw data and accompanying annotations available for analysis and benchmarking, we believe SALSA can spur systematic and intensive research to address the highlighted problems in a multimodal fashion in the near future. Evidently, SALSA would serve as a precious resource for the computer vision, audio processing, social robotics, social signal processing, and affective computing communities among others.

---

[5] Note that the accuracies with GT data reported in Fig. 14.4 are different from those presented in Table 14.6 since only a subset of frames is used in these experiments. In these considered frames, the scene is crowded since all 18 targets are inside the field of view.

REFERENCES

[1] Alameda-Pineda X, Horaud R. A geometric approach to sound source localization from time-delay estimates. IEEE/ACM Trans Audio Speech Lang Process 2014;22(6):1082–95.

[2] Alameda-Pineda X, Ricci E, Yan Y, Sebe N. Recognizing emotions from abstract paintings using non-linear matrix completion. In: CVPR. 2016.

[3] Alameda-Pineda X, Sanchez-Riera J, Wienke J, Franc V, Cech J, Kulkarni K, Deleforge A, Horaud R. Ravel: an annotated corpus for training robots with audiovisual abilities. J Multimodal User Interfaces 2013;7(1–2):79–91.

[4] Alameda-Pineda X, Staiano J, Subramanian R, Batrinca L, Ricci E, Lepri B, Lanz O, Sebe N. SALSA: a novel dataset for multimodal group behavior analysis. IEEE Trans Pattern Anal Mach Intell 2016;38(8):1707–20.

[5] Alameda-Pineda X, Yan Y, Ricci E, Lanz O, Sebe N. Analyzing free-standing conversational groups: a multimodal approach. In: ACM MM. 2015.

[6] Anguera Miro X, Bozonnet S, Evans N, Fredouille C, Friedland G, Vinyals O. Speaker diarization: a review of recent research. IEEE/ACM Trans Audio Speech Lang Process 2012;20(2):356–70.

[7] Badeau R, Plumbley M. Multichannel HR-NMF for modelling convolutive mixtures of non-stationary signals in the time–frequency domain. In: WASPAA. 2013. p. 1–4.

[8] Bazzani L, Cristani M, Tosato D, Farenzena M, Paggetti G, Menegaz G, Murino V. Social interactions by visual focus of attention in a three-dimensional environment. Expert Syst 2013;30(2):115–27.

[9] Ben Shitrit H, Berclaz J, Fleuret F, Fua P. Multi-commodity network flow for tracking multiple people. IEEE Trans Pattern Anal Mach Intell 2014;36(8):1614–27.

[10] Benfold B, Reid I. Unsupervised learning of a scene-specific coarse gaze estimator. In: ICCV. 2011.

[11] Berclaz J, Fleuret F, Turetken E, Fua P. Multiple object tracking using k-shortest paths optimization. IEEE Trans Pattern Anal Mach Intell 2011;33(9):1806–19.

[12] Breitenstein MD, Reichlin F, Leibe B, Koller-Meier E, van Gool L. Online multiperson tracking-by-detection from a single, uncalibrated camera. IEEE Trans Pattern Anal Mach Intell 2011;33(9):1820–33.

[13] Cattuto C, Van den Broeck W, Barrat A, Colizza V, Pinton JF, Vespignani A. Dynamics of person-to-person interactions from distributed RFID sensor networks. PLoS ONE 2010;5(7).

[14] Chen C, Odobez J-M. We are not contortionists: coupled adaptive learning for head and body orientation estimation in surveillance video. In: CVPR. 2012.

[15] Choi W, Chao YW, Pantofaru C, Savarese S. Discovering groups of people in images. In: ECCV. 2014. p. 417–33.

[16] Choudhury T, Pentland A. Sensing and modeling human networks using the sociometer. In: IEEE international symposium on wearable computing. 2003.

[17] Ciolek T. The proxemics lexicon: a first approximation. J Nonverbal Behav 1983;8:55–79.

[18] Cristani M, Bazzani L, Paggetti G, Fossati A, Tosato D, Del Bue A, Menegaz G, Murino V. Social interaction discovery by statistical analysis of F-formations. In: BMVC. 2011.

[19] Eagle N, Pentland A. Reality mining: sensing complex social systems. Personal Ubiquitous Comput 2006;10(4):255–68.

[20] Eichner M, Ferrari V. We are family: joint pose estimation of multiple persons. In: ECCV. 2010.

[21] Gan T, Wong Y, Zhang D, Kankanhalli MS. Temporal encoded F-formation system for social interaction detection. In: ACM multimedia. 2013.

[22] Ge W, Collins RT, Ruback RB. Vision-based analysis of small groups in pedestrian crowds. IEEE Trans Pattern Anal Mach Intell 2012;34(5):1003–16.

[23] Gebru I-D, Alameda-Pineda X, Forbes F, Horaud R. EM algorithms for weighted-data clustering with application to audio-visual scene analysis. IEEE Trans Pattern Anal Mach Intell 2016;38(12):2402–15.

[24] Goffman E. Behavior in public places: notes on the social organization of gatherings. Glencoe (IL): Free Press; 1963.

[25] Hung H, Chittaranjan G. The Idiap WOLF corpus: exploring group behaviour in a competitive role-playing game. In: ACM multimedia. 2010.

[26] Hung H, Englebienne G, Cabrera-Quiros L. Detecting conversing groups with a single worn accelerometer. In: ICMI. 2014.

[27] Hung H, Kröse B. Detecting F-formations as dominant sets. In: ICMI. 2011. p. 213–38.

[28] John O, Srivastava S. The big five trait taxonomy: history, measurement, and theoretical perspectives. In: Handbook of personality: theory and research. 1999. p. 102–38.

[29] Kendon A. Conducting interaction: patterns of behavior in focused encounters, vol. 7. Cambridge: CUP Archive; 1990.

[30] Kounades-Bastian D, Girin L, Alameda-Pineda X, Gannot S, Horaud R. A variational EM algorithm for the separation of time-varying convolutive audio mixtures. IEEE/ACM Trans Audio Speech Lang Process 2016;24(8):1408–23.

[31] Kurtic E, Wells B, Brown GJ, Kempton T, Aker A. A corpus of spontaneous multi-party conversation in Bosnian Serbo-Croatian and British English. In: International conference on language resources and evaluation. 2012. p. 1323–7.

[32] Lanz O. Approximate Bayesian multibody tracking. IEEE Trans Pattern Anal Mach Intell 2006;28(9):1436–49.

[33] LealTaixè L, Pons-Moll G, Rosenhan B. Everybody needs somebody: modeling social and grouping behaviour on a linear programming multiple people tracker. In: ICCV workshop. 2015.

[34] Lepri B, Staiano J, Rigato G, Kalimeri K, Finnerty A, Pianesi F, Sebe N, Pentland A. The sociometric badges corpus: a multilevel behavioral dataset for social behavior in complex organizations. In: IEEE international conference on social computing. 2012.

[35] Lepri B, Subramanian R, Kalimeri K, Staiano J, Pianesi F, Sebe N. Connecting meeting behavior with extraversion – a systematic study. IEEE Trans Affect Comput 2012;3(4):443–55.

[36] Luo C, Chan Mun C. SocialWeaver: collaborative inference of human conversation networks using smartphones. In: ACM conference on embedded networked sensor systems. 2013.

[37] Madan A, Cebrian M, Moturu S, Farrahi K, Pentland A. Sensing the "health state" of a community. IEEE Pervasive Comput 2012;11(4):36–45.

[38] Marin-Jimenez MJ, Zisserman A, Eichner M, Ferrari V. Detecting people looking at each other in videos. Int J Comput Vis 2014;106(3):282–96.

[39] Mathias M, Benenson R, Timofte R, Van Gool L. Handling occlusions with Franken-classifiers. In: ICCV. 2013.

[40] Matic A, Osmani V, Maxhuni A, Mayora O. Multi-modal mobile sensing of social interactions. In: International conference on pervasive computing technologies for healthcare. 2012.

[41] McCowan I, et al. The AMI meeting corpus. In: International conference on methods and techniques in behavioral research. 2005.

[42] Milan A, Schindler K, Roth S. Multi-target tracking by discrete-continuous energy minimization. IEEE Trans Pattern Anal Mach Intell 2015.

[43] Olguín Olguín D, Pentland A. Sensor-based organizational design and engineering. Int J Organ Des Eng 2010;1(1–2):69–97.

[44] Olguín Olguín D, Waber B, Kim T, Mohan A, Ara K, Pentland A. Sensible organizations: technology and methodology for automatically measuring organizational behavior. IEEE Trans Syst Man Cybern, Part B, Cybern 2009;39(1):43–55.

[45] Patron-Perez A, Marszalek M, Reid I, Zisserman A. Structured learning of human interactions in TV shows. IEEE Trans Pattern Anal Mach Intell 2012;34(12):2441–53.

[46] Pellegrini S, Ess A, Schindler K, van Gool L. You'll never walk alone: modeling social behavior for multi-target tracking. In: ICCV. 2009.

[47] Perugini M, Di Blas L. The Big Five marker scales (BFMS) and the Italian AB5C taxonomy: analyses from an emic–etic perspective. Boston: Hogrefe & Huber Publishers; 2002.

[48] Rajagopal AK, Subramanian R, Ricci E, Vieriu RL, Lanz O, Sebe N. Exploring transfer learning approaches for head pose classification from multi-view surveillance images. Int J Comput Vis 2014;109(1–2):146–67.

[49] Ricci E, Varadarajan J, Subramanian R, Rota Bulò S, Ahuja N, Lanz O. Uncovering interactions and interactors: joint estimation of head, body orientation and F-formations from surveillance video. In: ICCV. 2015.

[50] Setti F, Hung H, Cristani M. Group detection in still images by F-formation modeling: a comparative study. In: WIAMIS. 2013.

[51] Setti F, Lanz O, Ferrario R, Murino V, Cristani M. Multi-scale F-formation discovery for group detection. In: ICIP. 2013.

[52] Setti F, Russell C, Bassetti C, Cristani M. F-formation detection: individuating free-standing conversational groups in images. PLoS ONE 2015;10(5).

[53] Smith K, Ba SO, Odobez J-M, Gatica-Perez D. Tracking the visual focus of attention for a varying number of wandering people. IEEE Trans Pattern Anal Mach Intell 2008;30(7):1212–29.

[54] Stiefelhagen R. Tracking focus of attention in meetings. In: ICMI. 2002.

[55] Tulyakov S, Alameda-Pineda X, Ricci E, Yin L, Cohn JF, Sebe N. Self-adaptive matrix completion for heart rate estimation from face videos under realistic conditions. In: CVPR. 2016.

[56] Vascon S, Mequanint EZ, Cristani M, Hung H, Pelillo M, Murino V. A game theoretic probabilistic approach for detecting conversational groups. In: ACCV. 2014. p. 658–756.

[57] Vinciarelli A, Dielmann A, Favre S, Salamin H. Canal9: a database of political debates for analysis of social interactions. In: International conference on affective computing and intelligent interaction. 2009.

[58] Vinciarelli A, Pantic M, Bourlard H. Social signal processing: survey of an emerging domain. Image Vis Comput 2009;27(12):1743–59.

[59] Wojek C, Walk S, Roth S, Schiele B. Monocular 3D scene understanding with explicit occlusion reasoning. In: CVPR. 2011.

[60] Wyatt D, Choudhury T, Bilmes J, Kitts JA. Inferring colocation and conversation networks from privacy-sensitive audio with implications for computational social science. ACM Trans Intell Syst Technol 2011;2(1):7:1–41.

[61] Yan Y, Ricci E, Subramanian R, Liu G, Lanz O, Sebe N. A multi-task learning framework for head pose estimation under target motion. IEEE Trans Pattern Anal Mach Intell 2016;38(6):1070–83.

[62] Yang Y, Baker S, Kannan A, Ramanan D. Recognizing proxemics in personal photos. In: CVPR. 2012.

[63] Zen G, Lepri B, Ricci E, Lanz O. Space speaks: towards socially and personality aware visual surveillance. In: ACM international workshop on multimodal pervasive video analysis. 2010.

CHAPTER 15

Zero-Shot Crowd Behavior Recognition

Xun Xu, Shaogang Gong, Timothy M. Hospedales
Queen Mary University of London, London, United Kingdom

Contents

15.1 INTRODUCTION

Crowd behavior analysis is important in video surveillance for public security and safety. It has drawn increasing attention in computer vision research over the past decade [51, 47,50,36,42,41]. Most existing methods employ a video analysis processing pipeline

Group and Crowd Behavior for Computer Vision
DOI: 10.1016/B978-0-12-809276-7.00018-7

that includes: crowd scene representation [51,47,50,36], definition and annotation of crowd behavioral attributes for detection and classification, and learning discriminative recognition models from labeled data [42,41]. However, this conventional pipeline is limited for scaling up to recognizing ever increasing number of behavior types of interest, particularly for recognizing crowd behaviors of no training examples in a new environment. Firstly, conventional methods rely on exhaustively annotating examples of every crowd attribute of interest [41]. This is often neither plausible nor scalable due to the complexity and the cost of annotating crowd *videos* which requires spatiotemporal localization. Secondly, many crowd attributes may all appear simultaneously in a single video instance, e.g., *"outdoor"*, *"parade"*, and *"fight"*. To achieve *multilabel* annotation consistently, it is significantly more challenging and costly than conventional single-label multiclass annotation. Moreover, the most interesting crowd behaviors often occur rarely, or have never occurred previously in a given scene. For example, crowd attributes such as *"mob"*, *"police"*, *"fight"*, and *"disaster"* are rare in the *WWW* crowd video dataset, both relative to others and in absolute numbers (see Fig. 15.1). Given that such attributes have few or no training samples, it is hard to learn a model capable of detecting and recognizing them using the conventional supervised learning based crowd analysis approach.

In this chapter, we investigate and develop methods for zero-shot learning (ZSL) [23] based crowd behavior recognition. We want to learn a generalizable model on well annotated common crowd attributes. Once learned, the model can then be deployed to recognize novel (unseen) crowd behaviors or attributes of interest without any annotated training samples. The ZSL approach is mostly exploited for object image recognition: A regressor [43] or classifier [23] is commonly learned on known categories to map an image's visual feature to the continuous semantic representation of corresponding category or the discrete human-labeled semantic attributes. Then it is deployed to project unlabeled images into the same semantic space for recognizing previously unseen object categories [23,43,10,1]. There have also been recent attempts on ZSL recognition of single-label human actions in video instances [53,2] where similar pipeline is adopted. However, for ZSL crowd behavior recognition, there are two open challenges. First, crowd videos contain significantly more complex and cluttered scenes, making accurate and consistent interpretation of crowd behavioral attributes in the absence of training data very challenging. Second, crowd scene videos are inherently multilabeled. That is, there are almost always multiple attributes concurrently existing in each crowd video instance. The most interesting ones are often related to other noninteresting attributes. Thus we wish to infer these interesting attributes/behaviors from the detection of noninteresting but more readily available attributes. However, this has not been sufficiently studied in crowd behavior recognition, not to mention in the context of zero-shot learning.

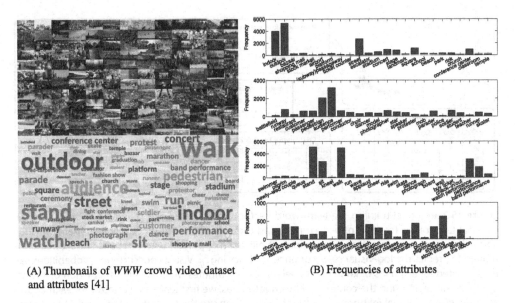

(A) Thumbnails of *WWW* crowd video dataset and attributes [41]

(B) Frequencies of attributes

Figure 15.1 A thumbnail visualization and a summary on the popularities of all 94 attributes in the *WWW* crowd video dataset [41].

It has been shown that in a *fully supervised setting*, exploring cooccurrence of multi-labels in a common context can improve the recognition of each individual label [55, 16,26]. For example, the behavioral attribute *"protest"* [41] is more likely to occur in *"outdoor"* rather than *"indoor"*. Therefore, recognizing the indoor/outdoor attribute in video can help to predict more accurately the *"protest"* behavior. However, it is not only unclear how, but also nontrivial, to extend this idea to the ZSL setting. For instance, predicting a previously unseen behavior *"violence"* in a different domain [19] would be much harder than the prediction of *"protest"*. As it is unseen, it is impossible to leverage the cooccurrence here as we have no *a priori* annotated data to learn their cooccurring context. The problem addressed in this chapter is on how to explore contextual cooccurrence among multiple known crowd behavioral attributes in order to facilitate the prediction of an unseen behavioral attribute, likely in a different domain.

More precisely, in this chapter we develop a zero-shot multilabel attribute contextual prediction model (Fig. 15.2). We make the assumption that the detection of known attributes helps the recognition of unknown ones. For instance, a putative unknown attribute such as *"violence"* may be related to known attributes *"outdoor"*, *"fight"*, *"mob"*, and *"police"*, among others. Therefore, high confidence in these attributes would support the existence of *"violence."* Specifically, our model first learns a probabilistic P-way classifier on P known attributes, e.g., $p(\text{"outdoor"}|\mathbf{x})$. Then we estimate the probability of each novel (unseen) attribute conditioned on the confidence of P known attributes, e.g., $p(\text{"violence"}|\text{"outdoor"})$. Recall that due to *"violence"* in this example

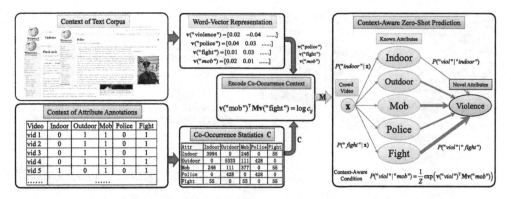

Figure 15.2 In model training, we learn word-vector representations of training attributes from an external text corpus (context of text corpus), and their visual cooccurrence from the training video annotations (context of attribute annotations). A bilinear mapping **M** between pairs of word vectors is trained to predict the log-visual co-occurrence statistics $\log c_{ij}$. Visual cooccurrence probabilities can be estimated for any pairs of known or novel (unseen) attributes. To enable the prediction of a novel attribute *"violence"* using the context of known attributes, we first learn a recognizer for each known attribute given its visual features, e.g., $p("mob"|\mathbf{x})$; we then use the trained context model to estimate the conditional probability $P("violence"|"mob")$ between novel and known attributes.

being a novel attribute, this conditional probability cannot be estimated directly by tabulation of annotation statistics. To model this conditional, we consider two contextual learning approaches. The first approach relies on the semantic relatedness between the two attributes. For instance, if *"fight"* is semantically related to *"violence"*, then we would assume a high conditional probability $p("violence"|"fight")$. Crucially, such semantic relations can be learned in the absence of annotated video data. This is achieved by using large text corpora [14] and language models [29,33]. However, this text-only based approach has the limitation that linguistic relatedness may not correspond reliably to the visual contextual cooccurrence that we wish to exploit. For example, the word *"outdoor"* has high linguistic semantic relatedness, e.g., measured by a cosine similarity, with *"indoor"*, whilst they would never cooccur in video annotations. Therefore, our second approach to conditional probability estimation is based on learning to map from *pairwise* linguistic semantic relatedness to visual cooccurrence. Specifically, on the known training attributes, we train a bilinear mapping to map *the pair of* training word-vectors (e.g., $\mathbf{v}("fight")$ and $\mathbf{v}("mob")$) to the training attributes' cooccurrence. This bilinear mapping can then be used to better predict the conditional probability between known and novel/unseen attributes. This is analogous to the standard ZSL idea of learning a visual-semantic mapping from a set of single attributes and reusing this mapping across different unseen attributes. Here, we focus instead on a set of attribute-pairs to learn cooccurrence mapping, and reusing this pairwise mapping across new attribute pairs.

As a proof-of-concept case study, we consider the task of violent behavior (event) detection in videos. This task has received increasing interest in recent years [19], but it is challenging due to the difficulty of obtaining violent event videos for training reliable recognizers. In this chapter, we demonstrate our approach by training our model on an independent large *WWW* crowd-attribute video dataset, which does not contain "*violence*" as a known attribute, and then apply the model to violent event detection on the *Violent Flow* video dataset [19].

In summary, we make the following contributions in this chapter: (1) For the first time, we investigate zero-shot learning for crowd behavior recognition to overcome the costly and semantically ambiguous video annotation of multilabels. (2) We propose a contextual learning strategy which enhances novel attribute recognition through context prediction by estimating attribute-context cooccurrence with a bilinear model. (3) A proof-of-concept case study is presented to demonstrate the viability of transferring zero-shot recognition of violent event cross-domain with very promising performance.

15.2 RELATED WORK

15.2.1 Crowd Analysis

Crowd analysis is one of the central topics in computer vision research for surveillance [17]. There are a variety of tasks including: (1) crowd density estimation and person counting [7,27], (2) crowd tracking [3,36], and (3) crowd behavior recognition [4,50,41]. There are several challenges in crowd behavior analysis. First of all, one requires both informative and robust visual features from crowd videos. Although simple optical flow [52,36,40], tracklets [57,58], or a combination of motion and static features [25] have been adopted. None of them is both informative and robust. More desirable scene-level features can be further constructed from these low-level features, using probabilistic topic models [50,52] or Gaussian mixtures [40]. However, these mid-level representations are mostly scene-specific, with a few exceptions such as [52] which models multiple scenes to learn a scene-independent representation. Second, for recognition in different scenes, existing methods rely heavily upon the assumption of the availability of sufficient observations (a large number of repetitions with variations) from these scenes in order to either learn behavior models from scratch [50,25,40], or inherit models from related scenes [52]. To generalize models across scenes, studies have proposed scene-invariant crowd/group descriptors inspired by socio-psychological and biological research [42], and more recently from deep learning mined crowd features [41]. In addition to these purpose-built crowd features, dense trajectory features [49] capturing both dynamic (motion boundary) and static textural information have also been adopted for crowd analysis [41]. For learning a scene-invariant model, the method of [41] requires extensive manual annotation of crowd attributes: The *WWW* crowd video dataset [41] has 94 attributes captured by over 10,000 annotated crowd videos, where each crowd video is annotated with multiple attributes. The effort required for annotating these videos is huge. This poses significant challenge to scale up the annotation of any larger video dataset from diverse domains. Third, often the most interesting crowd behavior is also novel in a given scene/domain. That is, the particular behavioral attribute has not been seen previously in that domain. To address these challenges, in this study we explore a different approach to crowd behavior recognition, by which crowd attribute context is learned from a large body of text descriptions rather than relying on exhaustive visual annotations, and this semantic contextual knowledge is exploited for zero-shot recognition of novel crowd behavioral attributes without labeled training samples.

15.2.2 Zero-Shot Learning

Zero-shot learning (ZSL) addresses the problem of constructing recognizers for novel categories without labeled training data (unseen) [23]. ZSL is made possible by leveraging an intermediate semantic space that bridges visual features and class labels (semantics). In general, the class labels can be obtained by manually labeled attributes [23,11], word-vector embeddings [43,53], structured word databases such as the WordNet [37, 14], and cooccurrence statistics from external sources [28].

15.2.2.1 Attributes

Attributes are manually defined binary labels of mid-level concepts [23] which can be used to define high-level classes, and thus bridge known and unknown classes. Traditional supervised classifiers can be trained to predict attributes rather than categories. In the testing phase, recognizers for new classes can then be defined based on novel classes' attributes, e.g., Direct Attribute Prediction (DAP) [23], or relations to known classes by the attributes, e.g., Indirect Attribute Prediction (IAP) [23]. This intuitive attribute based strategy inspired extensive research into ZSL. However, attributes themselves are manually annotated and thus suffer from: (i) the difficulty of determining an appropriate ontology of attributes; (ii) prohibitive annotation cost, in particular for videos due to their spatiotemporal nature; and (iii) labeling each video with a large vocabulary of attributes is particularly costly and ambiguous.

Note that attributes in the context of a ZSL *semantic representation* are different from the attributes we aim to predict in this chapter. In the attribute-ZSL case, all attributes are predefined and annotated in order to train supervised classifiers to generate a representation that bridges known and unknown high-level classes for *multiclass* ZSL prediction. In our case, we want to predict multiple crowd attributes for each video. That is, our final goal is *multilabel* ZSL prediction, as some of these attributes are zero-shot, i.e., not predefined or annotated in training data.

15.2.2.2 WordNet

As an alternative to attributes, WordNet [9] is a large English lexical database which organizes words in groups (aka synsets). WordNet is notably exploited for the graph structure which provides a direct relatedness measurement between classes as a path length between concepts [38,14]. The IAP model can be implemented without attribute annotation by replacing the novel to known class relation by WordNet induced relation. However, due to the absence of explicit representation for each individual word, the WordNet semantics are less likely to generalize to ZSL models with alternative training losses (e.g., ranking loss and regression loss) which require explicit embedding of words.

15.2.2.3 Cooccurrence

Studies have also explored external sources for measuring the relation between known and novel classes. In particular, web hit count has been considered as a source to induce a cooccurrence based representation [38,28]. Intuitively, two labels/concepts are treated closely related if they often cooccur in search engine results. As with the WordNet based approaches, cooccurrence models are not able to produce explicit representations for classes therefore are not compatible with learning alternative losses.

15.2.2.4 Word-Vector

The word-vector representation [29,43] generated by unsupervised learning on text corpora has emerged as a promising representation for ZSL in that: (i) As the product of unsupervised learning on existing text corpora, it avoids manual annotation bottlenecks; (ii) Semantic similarity between words/phrases can be measured as cosine distance in the word-vector space thus enables probabilistic views of zero-shot learning, e.g., DAP [23] and semantic inter-relations [14], and training with alternative models, e.g., ranking loss [1,10] and regression loss [43,53].

15.2.3 Multilabel Learning

Due to the multiple aspects of crowd behavior to be detected/recognized, videos are often annotated with more than one attribute. The multiattribute nature of crowd video, makes crowd behavior understanding a *multilabel learning* (MLL) [55] problem. MLL [55] is the task of assigning a single instance simultaneously to multiple categories. MLL can be decomposed into a set of independent single-label problems to avoid the complication of label correlation [54,5]. Although this is computationally efficient, ignoring label correlation produces suboptimal recognition. Directly tackling the joint multilabel problem through considering all possible label combinations is intractable, as the size of the output space and the required training data grow exponentially w.r.t. the number of unique labels [45]. As a compromise, tractable solutions to correlated multilabel prediction typically involve considering *pairwise* label correlations [13,35,46], e.g., using conditional random fields (CRF)s. However, all existing methods require to learn these pairwise label correlations in advance from the statistics of large labeled datasets. In this chapter, we solve the challenge of multilabel prediction for labels without any existing annotated datasets from which to extract cooccurrence statistics.

15.2.4 Multilabel Zero-Shot Learning

Although zero-shot learning is now quite a well studied topic, only a few studies have considered multilabel zero-shot learning [12,28]. Joint multilabel prediction is challenging because conventional multilabel models require precomputing the label cooccurrence statistics, which is not available in the ZSL setting. The study given by [12] proposed a Direct Multilabel zero-shot Prediction (DMP) model. This method synthesizes a power-set of potential testing label vectors so that visual features projected into this space can be matched against every possible combination of testing labels with simple NN matching. This is analogous to directly considering the jointly multilabel problem, which is intractable due to the size of the label power-set growing exponentially (2^n) with the number of labels being considered. An alternative study was provided by [28]. Although applicable to the multilabel setting, this method used cooccurrence statistics as the semantic bridge between visual features and class names, rather than jointly predicting multiple labels that can disambiguate each other. A related problem is to jointly predict multiple attributes when attributes are used as the semantic embedding for ZSL [18]. In this case, the correlations of mid-level attributes, which are multilabeled, are exploited in order to improve single-label ZSL, rather than the interclass correlation being exploited to improve multilabel ZSL.

15.3 METHODOLOGY

We introduce in this section a method for recognizing novel crowd behavioral attributes by exploring the context from other recognizable (known) attributes. In Section 15.3.1, we introduce a general procedure for predicting novel behavioral attributes based on their relation to known attributes. This is formulated as a probabilistic graphic model adapted from [23] and [15]. We then give the details in Section 15.3.2 on how to learn a behavior predictor that estimates the relations between known and novel attributes by inferring from text corpus and cooccurrence statistics of known attribute annotations.

Let us first give an overview of the notations used in this chapter in Table 15.1. Formally, we have training dataset $\mathscr{T}^S = \{\mathbf{X}^S, \mathbf{Y}^S, \mathbf{V}^S\}$ associated with P known attributes and testing dataset $\mathscr{T}^T = \{\mathbf{X}^T, \mathbf{Y}^T, \mathbf{V}^T\}$ associated with Q novel/unseen attributes. We denote the visual feature for training and testing videos as $\mathbf{X}^S = [\mathbf{x}_1, \ldots, \mathbf{x}_{N_S}] \in \mathbb{R}^{D_x \times N_S}$ and $\mathbf{X}^T = [\mathbf{x}_1, \ldots, \mathbf{x}_{N_T}] \in \mathbb{R}^{D_x \times N_T}$, multiple binary labels for training and testing videos as $\mathbf{Y}^S = [\tilde{\mathbf{y}}_1, \ldots, \tilde{\mathbf{y}}_{N_S}] \in \{0, 1\}^{P \times N_S}$ and $\mathbf{Y}^T = [\mathbf{y}^*_1, \ldots, \mathbf{y}^*_{N_T}] \in \{0, 1\}^{Q \times N_T}$,

Table 15.1 Notation summary

Notation	Description
N_S; N_T	Number of training/source instances; testing/target instances
D_x; D_v	Dimension of visual feature; of word–vector embedding
P; Q	Number of training/source attributes; testing/target attributes
$\mathbf{X} \in \mathbb{R}^{D_x \times N}$; $\mathbf{x}$	Visual feature matrix for N instances; column representing one instance
$\mathbf{Y} \in \{0, 1\}^{P \times N}$; $\mathbf{y}$	Binary labels for N instances with P (or Q) labels; column representing one instance
$\mathbf{V} \in \mathbb{R}^{D_v \times P}$; $\mathbf{v}$	Word–vector embedding for P (or Q) attributes; column representing embedding for one attribute

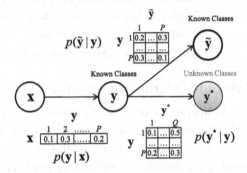

Figure 15.3 A probabilistic graphical representation of a context-aware multilabel zero-shot prediction model.

and the continuous semantic embedding (word–vector) for training and testing attributes as $\mathbf{V}^S = [\mathbf{v}_1, \dots, \mathbf{v}_P] \in \mathbb{R}^{D_v \times P}$ and $\mathbf{V}^T = [\mathbf{v}_1, \dots, \mathbf{v}_Q] \in \mathbb{R}^{D_v \times Q}$. Note that according to the zero–shot assumption, the training and testing attributes are disjoint, i.e., $\forall p \in \{1, \dots, P\}, q \in \{1, \dots, Q\} : \mathbf{v}_p \in \mathbf{V}^S, \mathbf{v}_q \in \mathbf{V}^T, \mathbf{v}_p \neq \mathbf{v}_q$.

15.3.1 Probabilistic Zero-Shot Prediction

To predict novel attributes by reasoning about the relations between known and novel attributes, we formulate this reasoning process as a probabilistic graph (see Fig. 15.3).

Given any testing video $\mathbf{x}$, we wish to assign it with one or many of the P known attributes or Q novel attributes. This problem is equivalent to inferring a set of conditional probabilities $p(\mathbf{y}^*|\mathbf{x}) = \{p(y_q^*|\mathbf{x})\}_{q=1,\dots,Q}$ and/or $p(\tilde{\mathbf{y}}|\mathbf{x}) = \{p(\tilde{y}_p|\mathbf{x})\}_{p=1,\dots,P}$. To achieve this, given the video instance $\mathbf{x}$, we first infer the likelihood of it being one of the P known attributes as $p(\mathbf{y}|\mathbf{x}) = \{p(y_p|\mathbf{x})\}_{p=1,\dots,P}$. Then, given the relation between known and novel/known attributes as conditional probability $P(\mathbf{y}^*|\mathbf{y})$ or $P(\tilde{\mathbf{y}}|\mathbf{y})$, we formulate the conditional probability similar to Indirect Attribute Prediction (IAP) [23,14] as follows:

$$p(\gamma_q^*|\mathbf{x}) = \sum_{p=1}^{P} p(\gamma_q^*|\gamma_p)p(\gamma_p|\mathbf{x}),$$

$$p(\tilde{\gamma}_{\tilde{p}}|\mathbf{x}) = \sum_{p=1}^{P} p(\tilde{\gamma}_{\tilde{p}}|\gamma_p)p(\gamma_p|\mathbf{x}).$$

(15.1)

The zero-shot learning task is to infer the probabilities $\{p(\gamma_q^*|\mathbf{x})\}_{p=1,...,P}$ for unseen labels $\{\gamma_q^*\}$. We estimate the multinomial conditional probability of known attributes $p(\gamma_p|\mathbf{x})$ based on the output of a probabilistic P-way classifier, e.g., SVM or Softmax Regression with probability output. Then the key to the success of zero-shot prediction is to estimate the known to novel contextual attribute relation as conditional probabilities $\{p(\gamma_q^*|\gamma_p)\}$. We introduce two approaches to estimate this contextual relation.

15.3.2 Modeling Attribute Relation from Context

In essence, our approach to the prediction of novel attributes depends on the prediction of known attributes and then predicting the novel attributes based on the confidence of each known attribute. The key to the success of this zero-shot prediction is therefore appropriately estimating the conditional probability of novel attribute given known attributes. We first consider a more straightforward way to model this conditional by exploiting the relation encoded by a *text* corpus [14]. We then extend this idea to predict the expected *visual* cooccurrence between novel and known attributes without labeled samples of the novel classes.

15.3.2.1 Learning Attribute Relatedness from Text Corpora

The first approach builds on semantic word embedding [14]. The semantic embedding represents each English word as a continuous vector $\mathbf{v}$ by training a skip-gram neural network on a large text corpus [29]. The objective of this neural network is to predict the adjacent c words to the current word w_t as follows:

$$\frac{1}{T}\sum_{t=1}^{T}\sum_{-c\leq j\leq c, j\neq 0} p(w_{t+j}|w_t).$$

(15.2)

The conditional probability is modeled by a softmax function, a normalized probability distribution, based on each word's representation as a continuous vector

$$p(w_{t+j}|w_t) = \frac{\exp(\mathbf{v}_{t+j}^\top \mathbf{v}_t)}{\sum_{j=1}^{W}\exp(\mathbf{v}_{t+j}^\top \mathbf{v}_t)}.$$

(15.3)

By maximizing the above objective function, the learned word-vectors $\mathbf{V} = \{\mathbf{v}\}$ capture contextual cooccurrence in the text corpora so that frequently cooccurring words result in high cumulative log-probability in Eq. (15.2). We apply the softmax function to model conditional attribute probability as Eq. (15.4) where γ is a temperature parameter, i.e.,

$$p(\gamma_q^*|\gamma_p) = \frac{\exp(\frac{1}{\gamma}\mathbf{v}_q^\top \mathbf{v}_p^S)}{\sum_{p=1}^{P} \exp(\frac{1}{\gamma}\mathbf{v}_q^\top \mathbf{v}_p^S)}. \tag{15.4}$$

This can be understood intuitively from the following example: An attribute "*Shopping*" has high affinity with attribute "*ShoppingMall*" in word-vector inner product because they cooccur in the text corpus. Our assumption is that the existence of known video attribute "*Shopping*" would support the prediction of unseen attribute "*ShoppingMall*".

15.3.2.2 Context Learning from Visual Cooccurrence

Although attribute relations can be discovered from text context as described above, these relations may *not* ideally suit crowd attribute prediction in videos. For example, the inner product of $vec("Indoor")$ and $vec("Outdoor")$ is 0.7104 which is ranked the first w.r.t. "Indoor" among 93 attributes in the *WWW* crowd video dataset. As a result, the estimated conditional probability $p(\tilde{\gamma}_{Indoor}|\gamma_{Outdoor})$ is the highest among all $\{p(\tilde{\gamma}_{Indoor}|\gamma_p)\}_{p=1,...,P}$. However, whilst these two attributes are similar because they occur nearby in the text semantical context, it is counterintuitive for visual cooccurrence as a video is very unlikely to be *both* indoor and outdoor. Therefore in visual context, their conditional probability should be small rather than large.

To address this problem, instead of directly parameterizing the conditional probability using word-vectors, we use pairs of word vectors to *predict* the actual visual cooccurrence. More precisely, we train a word-vector→cooccurrence predictor based on an auxiliary set of known attributes annotated on videos, for which both word-vectors and annotations are known. We then redeploy this learned predictor for zero-shot recognition on novel attributes. Formally, given binary multilabel annotations $\mathbf{Y}^S$ on training video data, we define the contextual attribute occurrence as $\mathbf{C} = \mathbf{Y}^S \mathbf{Y}^{ST}$. The occurrence of the jth attribute in the context of the ith attribute is thus c_{ij} of the $\mathbf{C}$. The prevalence of the ith attribute is defined as $c_i = \sum_j c_{ij}$. The normalized cooccurrence thus defines the conditional probability as

$$p(\tilde{\gamma}_j|\tilde{\gamma}_i) = \frac{c_{ij}}{c_i}. \tag{15.5}$$

The conditional probability can only be estimated based on visual cooccurrence in the case of training attributes with annotations $\mathbf{Y}^S$. To estimate the conditional probability for testing data of novel attributes without annotations $\mathbf{Y}^T$, we consider *predicting* the expected cooccurrence based on a bilinear mapping $\mathbf{M}$ from the pair of word-vectors.

Specifically, we approximate the unnormalized cooccurrence as $\exp(\mathbf{v}_i^\top \mathbf{M} \mathbf{v}_j) = c_{ij}$. To estimate $\mathbf{M}$, we optimize the regularized linear regression problem:

$$J = \sum_i^P \sum_j^P w(c_{ij}) \left(\mathbf{v}_i^\top \mathbf{M} \mathbf{v}_j - \log c_{ij} \right)^2 + \lambda \|\mathbf{M}\|_F^2, \qquad (15.6)$$

where λ is the regularization strength, and a weight function $w(c_{ij})$ is applied to the regression loss function above in order to penalize rarely occurring cooccurrence statistics. We choose the weight function according to [28], which is

$$w(c_{ij}) = \left(\frac{c_{ij}}{C_{max}} \right)^{(\alpha \cdot \mathbb{1}(c_{ij} \le C_{max}))}, \qquad (15.7)$$

where C_{max} is a threshold of cooccurrence, α controls the increasing rate of the weight function, and the $\mathbb{1}$ is an indicator function. This bilinear mapping is related to the model in [33], but differs in that: (i) The input of the mapping is the word-vector representations $\mathbf{v}$ learned from the skip-gram model [29] in order to generalize to novel attributes where no cooccurrence statistics are available. (ii) The mapping is trained to account for *visual* compatibility, e.g., "*Outdoor*" is unlikely to cooccur with "*Indoor*" in a visual context, although the terms are closely related in their representations learned from the text corpora alone. The bilinear mapping can be seamlessly integrated with the softmax conditional probability as

$$p(y_q^*|y_p) = \frac{\exp(\mathbf{v}_q^\top \mathbf{M} \mathbf{v}_p)}{\sum_p \exp(\mathbf{v}_q^\top \mathbf{M} \mathbf{v}_p)}. \qquad (15.8)$$

Note that by setting $\mathbf{M} = \mathbf{I}$, this conditional probability degenerates to the conventional word-vector based estimation in Eq. (15.4). The regression to predict visual co-occurrence from word-vectors (Eq. (15.6)) can be efficiently solved by gradient descent using the following gradient:

$$\nabla \mathbf{M} = \sum_{i=1}^P \sum_{j=1}^P f(c_{ij}) \left(2\mathbf{v}_i \mathbf{v}_i^\top \mathbf{M} \mathbf{v}_j \mathbf{v}_j^\top - 2 \log c_{ij} \mathbf{v}_i \mathbf{v}_j^\top \right) + 2\lambda \mathbf{M}. \qquad (15.9)$$

15.4 EXPERIMENTS

We evaluate our multilabel crowd behavior recognition model on the large *WWW* crowd video dataset [41]. We analyze each component's contribution to the overall multilabel ZSL performance. Moreover, we present a proof-of-concept case study for performing transfer zero-shot recognition of violent behavior in the *Violence Flow* video dataset [19].

15.4.1 Zero-Shot Multilabel Behavior Inference

15.4.1.1 Experimental Settings

Dataset

The *WWW* crowd video dataset is specifically proposed for studying scene-independent attribute prediction for crowd scene analysis. It consists of over 10,000 videos collected from online resources of 8257 unique scenes. The crowd attributes are designed to answer the following questions: "Where is the crowd?", "Who is in the crowd?", and "Why is the crowd here?". All videos are manually annotated with 94 attributes with 6 positive attributes per video on average. Fig. 15.4 shows a collection of 94 examples with each example illustrating each attribute in the *WWW* crowd video dataset.

Data Split

We validated the ability to utilize known attributes for recognizing novel attributes in the absence of training samples on the *WWW* dataset. To that end, we divided the 94 attributes into 85 for training (known) and 9 for testing (novel). This was repeated for 50 random splits. In every split, any video which has no positive label from the 9 novel attributes was used for training and the rest for testing. The distributions of the number of multiattributes (labels) per video over all videos and over the testing videos are shown in Fig. 15.5A–B, respectively. Fig. 15.5C also shows the distribution of the number of testing videos over the 50 random splits. In most splits, the number of testing videos is in the range of 3000 to 6000. The training to testing video number ratio is between 2:1 and 1:1. This low training–testing ratio makes for a challenging zero-shot prediction setting.

Visual Features

Motion information can play an important role in crowd scene analysis. To capture crowd dynamics, we extracted the improved dense trajectory features [49] and performed Fisher vector encoding [34] on these features, generating a 50,688-dimensional feature vector to represent each video.

Evaluation Metrics

We evaluated the performance of multilabel prediction using five different metrics [55]. These multilabel prediction metrics fall into two groups: example-based metric and label-based metric. Example-based metrics evaluate the performance per video instance and then average over all instances to give the final metric. Label-based metrics evaluate the performance per label category and return the average over all label categories as the final metric. The five multilabel prediction performance metrics are:

- **AUC** – The Area Under the ROC Curve. AUC evaluates binary classification performance. It is invariant to the positive/negative ratio of each testing label. Random

(A) 27 attributes by "Where"

(B) 24 attributes by "Who"

(C) 44 attributes by "Why"

Figure 15.4 Examples of all attributes in the *WWW* crowd video dataset [41].

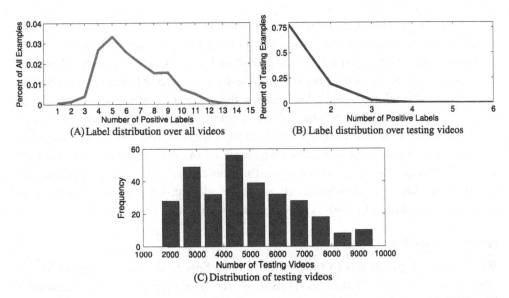

Figure 15.5 Statistics of the dataset split for our experiments on the *WWW* dataset. (A)–(B) The distributions of multilabel per video over all the videos and over the testing videos respectively. (C) The distribution of the number of testing videos over all 50 random splits.

guess leads to AUC of 0.5. For multilabel prediction, we measure the AUC for each testing label and average the AUC over all 50 splits to yield the AUC per category. The final mean AUC is reported as the mean over all label categories.

- **Label-based AP** – Label-based Average Precision. We measure the average precision for each attribute as the average fraction of relevant videos ranked higher than a threshold. The random guess baseline for label-based AP is determined by the prevalence of positive videos.

- **Example-based AP** – Example-based Average Precision. We measure the average precision for each video as the average fraction of relevant label prediction ranked higher than a threshold. Example-based AP focuses on the rank of attributes within each instance rather than rank of examples for each label as for label-based AP.

- **Hamming Loss** – Hamming Loss measures the percentage of incorrect predictions from ground-truth labels. Optimal hamming loss is 0, indicating perfect prediction. Due to the nature of Hamming loss, the distance of [000] and [110] w.r.t. [010] are equal. Thus it does not differentiate overestimation from underestimation. Hamming loss is a label-based metric. The final mean is reported as the average over all instances.

- **Ranking Loss** – Ranking Loss measures, for every instance, the percentage of negative labels ranked higher than positive labels among all possible positive–negative label pairs. Similar to example-based AP, the ranking loss is example-based metric focusing on pushing positive labels ahead of negative labels for each instance.

Both AUC and label-based AP are label-based metrics, whilst example-based AP, Hamming Loss, and Ranking Loss are example-based metrics. Moreover, as both Hamming Loss and Ranking Loss are loss metrics, the lower their values, the better. To the contrary, the higher the AUC and AP values, the better. In a typical surveillance application of crowd behavior recognition in videos, we are interested in detecting video instances of a particular attribute that triggers an alarm event, e.g., searching for video instances with the *"fighting"* attribute. In this context, label-based performance metrics such as AUC and label-based AP are more relevant. Overall, we present model performance evaluated by both types of evaluation metrics.

Parameter Selection

We have several parameters to tune-in our model. Specifically, for training SVM classifiers for known classes/attributes $\{p(y|\mathbf{x})\}$, we set the slack parameter to a constant 1. The ridge regression coefficient λ in Eq. (15.6) is essential to avoid over-fitting and numerical problems. It is empirically set to small nonzero value. We choose $\lambda = 0.001$ in our experiments. For the temperature parameter γ in Eq. (15.4), we cross-validated and found best value to be around 0.1. In addition, we used a word-vector dictionary pretrained on Google News dataset [29] with 100 billion words where word-vectors were trained with 300 dimensions ($D_v = 300$) and context size of 5 ($c = 5$).

15.4.1.2 *Comparative Evaluation*

In this first experiment, we evaluated zero-shot multilabel prediction on *WWW* crowd video dataset. We compared our context-aware multilabel ZSL models, both purely text-based and visual cooccurrence based, against four contemporary and state-of-the-art zero-shot learning models.

Sate-of-the-Art ZSL Models

1. Word-Vector Embedding (**WVE**) [53]. The WVE model constructs a vector representation $\mathbf{z}_{tr} = g(y_{tr})$ for each training instance according to its category name y_{tr} via word-vector embedding $g(\cdot)$ and then learns a support vector regression $f(\cdot)$ to map the visual feature $\mathbf{x}_{tr}$. For testing instance $\mathbf{x}_{te}$, it is first mapped into the semantic embedding space via the regressor $f(\mathbf{x}_{te})$. Novel category $y_{te} \in \mathcal{Y}_{te} = \{1, \ldots, Q\}$ is then mapped into the embedding space via $g(y_{te})$. Nearest neighbor matching is applied to match $\vec{x}_{te}$ with category y^* using the L_2 distance:

$$y^* = \arg \min_{y_{te} \in \mathcal{Y}_{te}} \|f(\mathbf{x}_{te}) - g(y_{te})\|_2^2. \tag{15.10}$$

We do not assume having access to the whole testing data distribution, so we do not exploit transductive self-training and data augmentation post processing, unlike in the cases of [53,2].

2. Embarrassingly Simple Zero-Shot Learning (**ESZSL**) [39]. The ESZSL model considers ZSL as training an L_2 loss classifier. Specifically, given known categories' binary labels **Y** and word-vector embedding $\mathbf{V}_{tr}$, we minimize the L_2 classification loss as

$$\min_{\mathbf{M}} \sum_{i=1}^{N} \|\mathbf{x}_i^\top \mathbf{M} \mathbf{V}_{tr} - \mathbf{y}_i\|_2^2 + \Omega(\mathbf{M}; \mathbf{V}_{tr}, \mathbf{X}) \qquad (15.11)$$

where $\Omega(\mathbf{M}; \mathbf{V}_{tr}, \mathbf{X})$ is a regularizer defined as

$$\Omega(\mathbf{M}; \mathbf{V}_{tr}, \mathbf{X}) = \lambda_1 \|\mathbf{M} \mathbf{V}_{tr}\|_F^2 + \lambda_2 \|\mathbf{X}^\top \mathbf{M}\|_F^2 + \lambda_3 \|\mathbf{M}\|_F^2. \qquad (15.12)$$

Novel categories are predicted by

$$\mathbf{y}^* = \mathbf{x}_{te}^\top \mathbf{M} \mathbf{V}_{te}. \qquad (15.13)$$

3. Extended DAP (**ExDAP**) [12]. ExDAP was specifically proposed for multilabel zero-shot learning [12]. This is an extension of single-label regression models to multilabel. Specifically, given training instances $\mathbf{x}_i$, associated multiple binary labels $\mathbf{y}_i$, and word-vector embedding of known labels $\mathbf{V}_{tr}$, we minimize the L_2 regression loss for learning a regressor **M**:

$$\min_{\mathbf{M}} \sum_{i=1}^{N} \|\mathbf{x}_i^\top \mathbf{M} - \mathbf{V}_{tr} \mathbf{y}_i\|_2^2 + \lambda \|\mathbf{M}\|_2^2. \qquad (15.14)$$

For zero-shot prediction, we minimize the same loss, but w.r.t. the binary label vector **y** with L_2 regularization:

$$\mathbf{y}^* = \arg\min_{\mathbf{y}^* \in \mathbb{R}} \|\mathbf{x}_{te}^\top \mathbf{M} - \mathbf{V}_{te} \mathbf{y}^*\|_2^2 + \lambda \|\mathbf{y}^*\|_2^2. \qquad (15.15)$$

A closed-form solution exists for prediction:

$$\mathbf{y}^* = \left(\mathbf{V}_{te}^\top \mathbf{V}_{te} + \lambda \mathbf{I}\right)^{-1} \mathbf{V}_{te}^\top \mathbf{x}_{te}^\top \mathbf{M}. \qquad (15.16)$$

4. Direct Multilabel Prediction (**DMP**) [12]. DMP was proposed to exploit the correlation between testing labels so to benefit the multilabel prediction. It shares the same training procedure with ExDAP in Eq. (15.14). For zero-shot prediction, given testing categories $\mathcal{Y}_{te}$ we first synthesize a power-set of all labels $\mathcal{P}(\mathcal{Y}_{te})$. The

multilabel prediction $\mathbf{y}^*$ is then determined by nearest neighbor matching of visual instances mapped into word-vector embedding $\mathbf{x}_{te}^\top \mathbf{M}$ against the synthesized power-set:

$$\mathbf{y}^* = \arg \min_{\mathbf{y}^* \in \mathcal{P}(\mathcal{Y}_{te})} \|\mathbf{x}_{te}^\top \mathbf{M} - \mathbf{V}_{te}\mathbf{y}^*\|_2^2. \tag{15.17}$$

Context-Aware Multilabel ZSL Models

1. Text Context-Aware ZSL (**TexCAZSL**). In our text corpus context-aware model introduced in Section 15.3.2.1, only word-vectors learned from text corpora [29] are used to model the relation between known and novel attributes $p(\mathbf{y}^*|\mathbf{y})$, as defined by Eq. (15.4). We implemented the video instance to known attributes probabilities $p(y_p|\mathbf{x})$ as P linear SVM classifiers with normalized probability outputs [6]. Novel attribute prediction $p(y_q^*|\mathbf{x})$ is computed by marginalizing over the known attributes defined by Eq. (15.1).

2. Visual Cooccurrence Context-Aware ZSL (**CoCAZSL**). We further implemented a visual cooccurrence context-aware model built on top of the **TexCAZSL** model. This is done by predicting the expected cooccurrence context using bilinear mapping $\mathbf{M}$, as introduced in Section 15.3.2.2. The known to novel attribute relation is thus modeled by a weighted inner-product between the word-vectors of known and novel attributes given by Eq. (15.8). Novel attribute prediction $p(y_q^*|\mathbf{x})$ is computed in the same way as for **TexCAZSL**, defined by Eq. (15.1).

Quantitative Comparison

Table 15.2 shows the comparative results of our models against four state-of-the-art ZSL models and the baseline of "Random Guess", using all five evaluation metrics. Three observations can be made from these results: (1) All zero-shot learning models can substantially outperform random guessing, suggesting that zero-shot crowd attribute prediction is valid. This should inspire more research into zero-shot crowd behavior analysis in the future. (2) It is evident that our context-aware models improve on existing ZSL methods when measured by the label-based AUC and AP metrics. As discussed early under evaluation metrics, for typical surveillance tasks, label-based metrics provide a good measurement on detecting novel alarm events in the mist of many other contextual attributes in crowd scenes. (3) It is also evident that our context-aware models perform comparably to the alternative ZSL models under the example-based evaluation metrics, with the exception that DMP [12] performs extraordinarily well on Hamming Loss but poorly on Ranking Loss. This is due to the direct minimization of Hamming Loss between synthesized power-set and embedded video in DMP. However, since the relative order between attributes are ignored in DMP, low performance in ranking loss as well as other label-based metrics is expected.

Table 15.2 Comparison of zero-shot multilabel attribute prediction on the *WWW* crowd video dataset. The ↑ and ↓ symbols indicate whether a higher metric's value is better or worse

Feature	Model	Label-based			Example-based	
		AUC ↑	AP ↑	AP ↑	Hamming Loss ↓	Ranking Loss ↓
—	Random Guess	0.50	0.14	0.31	0.50	—
ITF	WVE [53]	0.65	0.24	0.52	0.45	0.32
ITF	ESZSL [39]	0.63	0.22	0.53	0.46	0.32
ITF	ExDAP [12]	0.62	0.21	0.52	0.45	0.32
ITF	DMP [12]	0.59	0.20	0.45	**0.30**	0.70
ITF	TexCAZSL	0.65	0.24	0.52	0.43	0.32
ITF	CoCAZSL	**0.69**	**0.27**	**0.53**	0.42	**0.31**

Qualitative Analysis

We next give some qualitative examples of zero-shot attribute predictions in Fig. 15.6. To get a sense of how well the attributes are detected in the context of label-based AP, we present the AP number with each attribute. Firstly, we give examples of detecting videos matching some randomly chosen attributes (label-centric evaluation). By designating an attribute to detect, we list the crowd videos sorted in the descending order of probability $p(\gamma^*|\mathbf{x})$. In general, we observe good performance in ranking crowd videos according to the attribute to be detected. The false detections are attributed to the extremely ambiguous visual cues. For example, the third video in "*fight*", fifth video in "*police*", and second video in "*parade*" are very hard to interpret.

In addition to detecting each individual attribute, we also present some examples of simultaneously predicting multiple attributes in Fig. 15.7 (example-centric evaluation). For each video we give the prediction score for all testing attributes as $\{p(\gamma_q^*|\mathbf{x})\}_{q=1,...,Q}$. For the ease of visualization, we omit the attribute with least score. We present the example-based ranking loss number along with each video to give a sense of how the quantitative evaluation metric relates to the qualitative results. In general, ranking loss less than 0.1 would yield very good multilabel prediction as all labels would be placed among the top 3 out of 9 labels to be predicted. Whilst ranking loss around 0.3 (roughly the average performance of our CoCAZSL model, see Table 15.2) would still give reasonable predictions by placing positive labels in the top 5 out of 9.

15.4.2 Transfer Zero-Shot Recognition in Violence Detection

Recognizing violence in surveillance scenario has an important role in safety and security [19,17]. However, due to the sparse nature of violent events in day-to-day surveillance scenes, it is desirable to exploit zero-shot recognition to detect violent events without human annotated training videos. Therefore, we explore a proof of concept case study of transfer zero-shot violence detection. We learn to recognize labeled

Figure 15.6 Illustration of crowd videos ranked in accordance with prediction scores (marginalized conditional probability) w.r.t. each attribute.

attributes in *WWW* dataset [41] and then transfer the model to detect violence event in Violence Flow dataset [19]. This is zero-shot because we use no annotated examples of violence to train, and violence does not occur in the label set of WWW. It is contextual because the violence recognition is based on the predicted visual cooccurrence between each known attribute in WWW and the novel violence attribute. For example, "*mob*" and "*police*" attributes known from WWW may support the violence attribute in the new dataset.

15.4.2.1 Experiment Settings

Dataset

The Violence Flow dataset [19] was proposed to facilitate the study into classifying violent events in crowded scenes. 246 videos in total are collected from online video repositories (e.g., YouTube) with 3.6 seconds length on average. Half of the 246 video

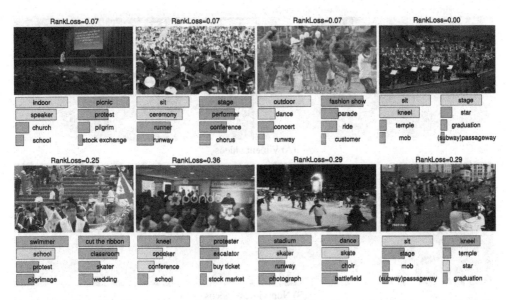

Figure 15.7 Examples of zero-shot multilabel attribute prediction. Bars under each image indicate the normalized score for testing attributes. Blue and pink bars indicate positive and negative ground-truth labels, respectively. (For interpretation of the references to color in this figure legend, the reader is referred to the web version of this chapter.)

are with positive violence content and the another half are with nonviolent crowd contents. We illustrate example frames of both violent and nonviolent videos in Fig. 15.8.

Data Split

A standard fully supervised 5-fold cross validation split was proposed by [19]. The standard split partitions the whole dataset into 5 splits each of which is evenly divided into positive and negative videos. For each testing split, the other 4 splits are used as the training set and the left-out one is the testing set. Results are reported as both the mean classification accuracy over 5 splits plus standard deviation and the area under the ROC curve (AUC).

Beyond the standard cross validation split we create a new zero-shot experimental design. Our zero-shot split learns attribute detection models on all 94 attributes from *WWW* dataset and then tests on the same testing set as the standard 5 splits in [19]. We note that there are 123 overlapped videos between WWW and Violence Flow. To make fair comparison, we exclude these overlapped videos from constructing the training data for 94 attributes. In this way zero-shot prediction performance can be directly compared with supervised prediction performance using AUC metric. We define the event/attribute to be detected as the word "*violence*".

(A) Violent videos

(B) Non-violent videos

Figure 15.8 Example frames of violence flow dataset [19].

Zero-Shot Recognition Models

We explore the transfer zero-shot violence recognition by comparing the same set of zero-shot learning models as in Section 15.4.1.2: competitors are WVE, ESZSL, and ExDAP; and our TexCAZSL and CoCAZSL.

Fully Supervised Model

To put zero-shot recognition performance in context, we also report fully supervised models' performance. These models are evaluated on the 5-fold cross-validation split and the average accuracy and AUC are reported. Specifically, we report the best performance of [19], that is, linear SVM with Violent Flows (ViF) descriptor, and our fully supervised baseline, i.e., linear SVM with Improved Trajectory Feature (ITF).

15.4.2.2 Results and Analysis

The results of both transfer zero-shot and supervised violence prediction are summarized in Table 15.3. We make the following observations: Our context-aware models perform consistently better than alternative zero-shot models, suggesting that context does facilitate zero-shot recognition. Surprisingly, our zero-shot models moreover perform very competitively compared to the fully supervised models. Our **CoCAZSL** (albeit with better ITF feature) beats the fully supervised Linear SVM with ViF feature in AUC metric (87.22 vs. 85.00). The context-aware model is also close to the fully

Table 15.3 Evaluation of violence prediction in Violence Flow dataset: zero-shot versus fully supervised prediction (%)

Model	Split	Feature	Accuracy	AUC
WVE [53]	Zero-Shot	ITF	64.27 ± 5.06	64.25
ESZSL [39]	Zero-Shot	ITF	61.30 ± 8.28	61.76
ExDAP [12]	Zero-Shot	ITF	54.47 ± 7.37	52.31
TexCAZSL	Zero-Shot	ITF	67.07 ± 3.87	69.95
CoCAZSL	Zero-Shot	ITF	**80.52 ± 4.67**	**87.22**
Linear SVM	5-fold CV	ITF	94.72 ± 4.85	98.72
Linear SVM [18]	5-fold CV	ViF	81.30 ± 0.21	85.00

supervised model with the same ITF feature (87.22 vs. 98.72). This is in contrast to the common result in the literature were zero-shot recognition "works", but does so much worse than fully supervised learning. The promising performance is partly due to modeling the cooccurrence on large known crowd attributes help the correct prediction of known to novel attribute relation prediction. Overall the result shows that by transferring our attribute recognition model trained for a wide set of 94 attributes on a large 10,000 video dataset it is possible to perform effective zero-shot recognition of a novel behavior type in a new dataset.

15.5 FURTHER ANALYSIS

In this section we provide further analysis on the importance of the visual feature used, and also give more insight into how our contextual zero-shot multilabel prediction works by visualizing the learned label-relations.

15.5.1 Feature Analysis

We first evaluate different static and motion features on the standard supervised attribute prediction task. Both hand-crafted and deeply learned features are reported for comparison.

15.5.1.1 Static Features

We report the both the hand-crafted and deeply learned static feature from [41] including Static Feature (SFH) and Deeply Learned Static Feature (DLSF). SFH captures general image content by extracting Dense SIFT [24], GIST [31] and HOG [8]. Color histogram in HSV space is further computed to capture global information, and LBP [56] is extracted to quantify local texture. Bag-of-words encoding is used to create comparable features, leading to a 1536-dimensional static feature. DLSF is initialized using a

Table 15.4 Comparison between different visual features for attribute prediction

Alternative features	Mean AUC
SFH [41]	0.81
DLSF [41]	0.87
DenseTrack [41]	0.63
DLMF [41]	0.68
SFH+DenseTrack [41]	0.82
DLSF+DLMF [41]	0.88
Our features	
Improved Trajectory Feature (ITF)	0.91

pre-trained model for ImageNet detection task [32] and then fine-tuned on the WWW attribute recognition task with cross-entropy loss.

15.5.1.2 Motion Features

We report both the hand-crafted and deeply learned motion features from [41] including DenseTrack [48], spatiotemporal motion patterns (STMP) [21], and Deeply Learned Motion Feature (DLMF) [41]. Apart from the reported evaluations, we compare them with the improved trajectory feature (ITF) [49] with Fisher vector encoding. Though ITF is constructed in the same way as DenseTrack reported in [41], we make a difference in that the visual codebook is trained on a collection of human action datasets (HMDB51 [22], UCF101 [44], Olympic Sports [30], and CCV [20]).

15.5.1.3 Analysis

Performance on the standard WWW split [41] for static and motion features is reported in Table 15.4. We can clearly observe that the improved trajectory feature is consistently better than all alternative static and motion features. Surprisingly, ITF is even able to beat deep features (DLSF and DLMF). We attribute this to ITF's ability to capture both motion information by motion boundary histogram (MBH) and histogram of flow (HoF) descriptors and texture information by Histogram of Gradient (HoG) descriptor.

More interestingly, we demonstrate that motion feature encoding model (Fisher vector) learned from action datasets can benefit the crowd behavior analysis. Due to the vast availability of action and event datasets and limited crowd behavior data, a natural extension work is to discover if deep motion model pretrained on action or event dataset can help crowd analysis.

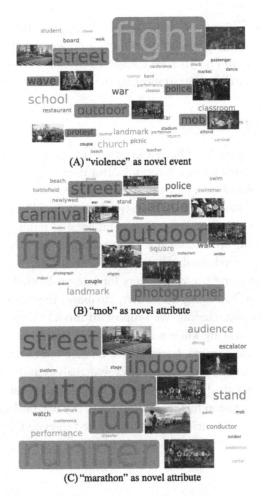

(A) "violence" as novel event

(B) "mob" as novel attribute

(C) "marathon" as novel attribute

Figure 15.9 Importance of known attributes w.r.t. novel event/attributes. The fontsize of each attribute is proportional to the conditional probability, e.g., $p(\text{"violence"}|y_p)$.

15.5.2 Qualitative Illustration of Contextual Cooccurrence Prediction

Recall that the key step in our method's approach to zero-shot prediction is to estimate the visual cooccurrence (between known attributes and held out zero-shot attributes) based on the textually derived word-vectors of each attributes. To illustrate what is learned, we visualize the predicted importance of 94 attributes from WWW in terms of supporting the detection of the held out attribute "*violence.*" The results are presented as a word cloud in Fig. 15.9, where the size of each word/attribute p is proportional to the conditional probability, e.g., $p(\text{"violence"} \,|\, \gamma_p)$. As we see from Fig. 15.9A, attribute "*fight*" is the most prominent attribute supporting the detection of "*violence*". Besides this, actions like *"street"*, *"outdoor"*, and *"wave"* all support the existence of *"violence"*, while *"disaster"* and *"dining"*, among others, do not. We also illustrate the support of "mob" and "marathon" in Fig. 15.9B–C, respectively. All these give us very reasonable importance of known attributes in supporting the recognition of novel attributes.

15.6 CONCLUSIONS

Crowd behavior analysis has long been a key topic in computer vision research. Supervised approaches have been proposed recently. But these require exhaustively obtaining and annotating examples of each semantic attribute, preventing this strategy from scaling up to ever expanding dataset sizes and variety of attributes. Therefore, it is worthwhile to develop recognizers that require little or no annotated training examples for the attribute/event of interest. We address this by proposing a zero-shot learning strategy in which recognizers for novel attributes are built without corresponding training data. This is achieved by learning the recognizers for known labeled attributes. For testing data, the confidence of belonging to known attributes then supports the recognition of novel ones via attribute relation. We propose to model this relation from the cooccurrence context provided by known attributes and word-vector embeddings of the attribute names from text corpora. Experiments on zero-shot multilabel crowd attribute prediction prove the feasibility of zero-shot crowd analysis and demonstrate the effectiveness of learning contextual cooccurrence. A proof-of-concept case study on transfer zero-shot violence recognition further demonstrates the practical value of our zero-shot learning approach, and its superior efficacy compared to even fully supervised learning approaches.

REFERENCES

[1] Akata Zeynep, Reed Scott, Walter Daniel, Lee Honglak, Schiele Bernt. Evaluation of output embeddings for fine-grained image classification. In: CVPR. 2015.
[2] Alexiou Ioannis, Xiang Tao, Gong Shaogang. Exploring synonyms as context in zero-shot action recognition. In: ICIP. 2016.
[3] Ali Saad, Shah Mubarak. Floor fields for tracking in high density crowd scenes. In: ECCV. 2008.
[4] Andrade Ernesto L, Blunsden Scott, Fisher Robert B. Modelling crowd scenes for event detection. In: ICPR. 2006.
[5] Boutell Matthew R, Luo Jiebo, Shen Xipeng, Brown Christopher M. Learning multi-label scene classification. Pattern Recognit 2004. http://dx.doi.org/10.1016/j.patcog.2004.03.009.
[6] Chang Chih-Chung, Lin Chih-Jen. LIBSVM: a library for support vector machines. ACM Trans Intell Syst Technol 2011. http://dx.doi.org/10.1145/1961189.1961199. Software available from http://www.csie.ntu.edu.tw/~cjlin/libsvm.
[7] Chen Ke, Loy Chen Change, Gong Shaogang, Xiang Tao. Feature mining for localised crowd counting. In: BMVC. 2012.
[8] Dalal Navneet, Triggs Bill. Histograms of oriented gradients for human detection. In: CVPR. 2005.

[9] Fellbaum Christiane. WordNet and wordnets. In: Brown Keith, et al., editors. Encyclopedia of language and linguistics. 2nd edition. Oxford: Elsevier; 2005. p. 665–70.

[10] Frome Andrea, Corrado Greg S, Shlens Jon, Bengio Samy, Dean Jeff, Mikolov Tomas, et al. Devise: a deep visual-semantic embedding model. In: NIPS. 2013.

[11] Fu Yanwei, Hospedales Timothy M, Xiang Tao, Gong Shaogang. Transductive multi-view zero-shot learning. IEEE Trans Pattern Anal Mach Intell 2015. http://dx.doi.org/10.1109/TPAMI.2015.2408354.

[12] Fu Yanwei, Yang Yongxin, Hospedales Timothy M, Xiang Tao, Gong Shaogang. Transductive multi-label zero-shot learning. In: BMVC. 2014.

[13] Fürnkranz Johannes, Hüllermeier Eyke, Mencía Eneldo Loza, Brinker Klaus. Multilabel classification via calibrated label ranking. Mach Learn 2008. http://dx.doi.org/10.1007/s10994-008-5064-8.

[14] Gan Chuang, Lin Ming, Yang Yi, Zhuang Yueting, Hauptmann Alexander G. Exploring semantic inter-class relationships (SIR) for zero-shot action recognition. In: AAAI. 2015.

[15] Gan Chuang, Wang Naiyan, Yang Yi, Yeung Dit-Yan, Hauptmann Alex G. DevNet: a deep event network for multimedia event detection and evidence recounting. In: CVPR. 2015.

[16] Ghamrawi Nadia, McCallum Andrew. Collective multi-label classification. In: ACM CIKM. 2005.

[17] Gong Shaogang, Loy Chen Change, Xiang Tao. Security and surveillance. In: Moeslund, Hilton, Kruger, Sigal, editors. Visual analysis of humans. Springer; 2011. p. 455–72.

[18] Hariharan Bharath, Vishwanathan SVN, Varma Manik. Efficient max-margin multi-label classification with applications to zero-shot learning. Mach Learn 2012. http://dx.doi.org/10.1007/s10994-012-5291-x.

[19] Hassner Tal, Itcher Yossi, Kliper-Gross Orit. Violent flows: real-time detection of violent crowd behavior. In: CVPR workshop. 2012.

[20] Jiang Yu-Gang, Ye Guangnan, Chang Shih-Fu, Ellis Daniel PW, Loui Alexander C. Consumer video understanding: a benchmark database and an evaluation of human and machine performance. In: ICMR. 2011.

[21] Kratz Louis, Nishino Ko. Anomaly detection in extremely crowded scenes using spatio-temporal motion pattern models. In: CVPR. 2009.

[22] Kuehne H, Jhuang H, Garrote E, Poggio T, Serre T. HMDB: a large video database for human motion recognition. In: ICCV. 2011.

[23] Lampert CH, Nickisch H, Harmeling S. Learning to detect unseen object classes by between-class attribute transfer. In: CVPR. 2009.

[24] Lazebnik Svetlana, Schmid Cordelia, Ponce Jean. Beyond bags of features: spatial pyramid matching for recognizing natural scene categories. In: CVPR. 2006.

[25] Li Jian, Gong Shaogang, Xiang Tao. Learning behavioural context. Int J Comput Vis 2012. http://dx.doi.org/10.1007/s11263-011-0487-2.

[26] Li Xin, Zhao Feipeng, Guo Yuhong. Multi-label image classification with a probabilistic label enhancement model. In: UAI. 2014.

[27] Loy Chen Change, Chen Ke, Gong Shaogang, Xiang Tao. Crowd counting and profiling: methodology and evaluation. In: Ali, Nishino, Manocha, Shah, editors. Modeling, simulation and visual analysis of crowds. Springer; 2013.

[28] Mensink Thomas, Gavves Efstratios, Snoek Cees GM. COSTA: co-occurrence statistics for zero-shot classification. In: CVPR. 2014.

[29] Mikolov Tomas, Sutskever Ilya, Chen Kai, Corrado Greg, Dean Jeffrey. Distributed representations of words and phrases and their compositionality. In: NIPS. 2013.

[30] Niebles Juan Carlos, Chen Chih Wei, Fei-Fei Li. Modeling temporal structure of decomposable motion segments for activity classification. In: ECCV. 2010.

[31] Oliva Aude, Torralba Antonio. Modeling the shape of the scene: a holistic representation of the spatial envelope. Int J Comput Vis 2001. http://dx.doi.org/10.1023/A:1011139631724.

[32] Ouyang Wanli, Luo Ping, Zeng Xingyu, Qiu Shi, Tian Yonglong, Li Hongsheng, Yang Shuo, Wang Zhe, Xiong Yuanjun, Qian Chen, et al. DeepID-Net: multi-stage and deformable deep convolutional neural networks for object detection. Available from arXiv:1409.3505, 2014.

[33] Pennington Jeffrey, Socher Richard, Manning Christopher D. Glove: global vectors for word representation. In: EMNLP. 2014.

[34] Perronnin Florent, Sánchez Jorge, Mensink Thomas. Improving the Fisher kernel for large-scale image classification. In: ECCV. 2010.

[35] Qi Guo-Jun, Hua Xian-Sheng, Rui Yong, Tang Jinhui, Mei Tao, Zhang Hong-Jiang. Correlative multi-label video annotation. In: ACM multimedia. 2007.

[36] Rodriguez Mikel, Sivic Josef, Laptev Ivan, Audibert Jean-Yves. Data-driven crowd analysis in videos. In: ICCV. 2011.

[37] Rohrbach Marcus, Stark Michael, Schiele Bernt. Evaluating knowledge transfer and zero-shot learning in a large-scale setting. In: CVPR. 2011.

[38] Rohrbach Marcus, Stark Michael, Szarvas György, Gurevych Iryna, Schiele Bernt. What helps where – and why? Semantic relatedness for knowledge transfer. In: CVPR. 2010.

[39] Romera-Paredes Bernardino. An embarrassingly simple approach to zero-shot learning. In: ICML. 2015.

[40] Saleemi Imran, Hartung Lance, Shah Mubarak. Scene understanding by statistical modeling of motion patterns. In: CVPR. 2010.

[41] Shao Jing, Kang Kai, Loy Chen Change, Wang Xiaogang. Deeply learned attributes for crowded scene understanding. In: CVPR. 2015.

[42] Shao Jing, Loy Chen, Wang Xiaogang. Scene-independent group profiling in crowd. In: CVPR. 2014.

[43] Socher Richard, Ganjoo Milind. Zero-shot learning through cross-modal transfer. In: NIPS. 2013.

[44] Soomro Khurram, Zamir AR, Shah Mubarak. UCF101: a dataset of 101 human actions classes from videos in the wild. Available from arXiv:1212.0402, 2012.

[45] Tsoumakas Grigorios, Vlahavas Ioannis. Random k-labelsets: an ensemble method for multilabel classification. In: ECML. 2007.

[46] Ueda Naonori, Saito Kazumi. Parametric mixture models for multi-labeled text. In: NIPS. 2002.

[47] Varadarajan Jagannadan, Emonet Rémi, Odobez Jean-Marc. A sequential topic model for mining recurrent activities from long term video logs. Int J Comput Vis 2013. http://dx.doi.org/10.1007/s11263-012-0596-6.

[48] Wang Heng, Kläser Alexander, Schmid Cordelia, Liu Cheng-Lin. Action recognition by dense trajectories. In: CVPR. 2011.

[49] Wang Heng, Oneata Dan, Verbeek Jakob, Schmid Cordelia. A robust and efficient video representation for action recognition. Int J Comput Vis 2015. http://dx.doi.org/10.1007/s11263-015-0846-5.

[50] Wang Xiaogang, Ma Xiaoxu, Grimson W Eric L. Unsupervised activity perception in crowded and complicated scenes using hierarchical Bayesian models. IEEE Trans Pattern Anal Mach Intell 2009. http://dx.doi.org/10.1109/TPAMI.2008.87.

[51] Xu X, Hospedales TM, Gong S. Discovery of shared semantic spaces for multi-scene video query and summarization. IEEE Trans Circuits Syst Video Technol 2016. http://dx.doi.org/10.1109/TCSVT.2016.2532719.

[52] Xu Xun, Gong Shaogang, Hospedales Timothy. Cross-domain traffic scene understanding by motion model transfer. In: Proceedings of the 4th ACM/IEEE international workshop on ARTEMIS. 2013.

[53] Xu Xun, Hospedales Timothy, Gong Shaogang. Semantic embedding space for zero-shot action recognition. In: ICIP. 2015.

[54] Zhang Min-Ling, Zhou Zhi-Hua. ML-KNN: a lazy learning approach to multi-label learning. Pattern Recognit 2007. http://dx.doi.org/10.1016/j.patcog.2006.12.019.

[55] Zhang Min-Ling, Zhou Zhi-Hua. A review on multi-label learning algorithms. IEEE Trans Knowl Data Eng 2014. http://dx.doi.org/10.1109/TKDE.2013.39.

[56] Zhao Guoying, Ahonen Timo, Matas Jiří, Pietikäinen Matti. Rotation-invariant image and video description with local binary pattern features. IEEE Trans Image Process 2012. http://dx.doi.org/10.1109/TIP.2011.2175739.

[57] Zhao Xuemei, Gong Dian, Medioni Gérard. Tracking using motion patterns for very crowded scenes. In: ECCV. 2012.

[58] Zhou Bolei, Tang Xiaoou, Wang Xiaogang. Coherent filtering: detecting coherent motions from crowd clutters. In: ECCV. 2012.

CHAPTER 16

The GRODE Metrics
Exploring the Performance of Group Detection Approaches

Francesco Setti, Marco Cristani
Department of Computer Science, University of Verona, Verona, Italy
Institute of Cognitive Science and Technology, Italian National Research Council, Trento, Italy

Contents

16.1 INTRODUCTION

The analysis of the grouping dynamics of interacting entities is a long-running problem in many fields of pattern recognition, with applications spanning from biology [1] to system management [2] and cyber-security [3]. When the entities are individuals, the problem is commonly dubbed *group detection* [4]; in the last years, detection of groups of people has become a relevant issue for many computer vision fields, especially in those areas where the understanding of social activities for high-level profiling and decision making play a crucial role, like video surveillance, social robotics, social signal processing, to quote a few.

On the one side, some approaches are focused on people dynamics, and are built over object tracking frameworks, enriching the usual per-person estimation with labels of group membership: these are added after the tracking in a postprocessing step [5–8], or directly embedded in the state-space of the filtering mechanism [9–12]. On the other side, many more approaches discover groups without temporal reasoning, directly on still images [13–20,4,21,22].

Group and Crowd Behavior for Computer Vision
DOI: 10.1016/B978-0-12-809276-7.00019-9

This growing literature is still in its infancy, with many aspects related to the notion of "group of people" that should be addressed in a more systematic fashion. As an emblematic example, the mere definition of *what is a group* is not clearly stated and widely shared in the computer vision community: in most of the cases, the individuation of a group lies on intuitive observations of proxemics, that is, mutual proximity and common dynamics (especially in the tracking approaches [6,7]), while in some other cases anthropological [23], ethnomethodological [24], sociological [25], and social psychology domains [26] have been consulted for finding a more founded definition. This is the case of the notions of *F-formation* (people free to move that spontaneously decide to jointly interact) [26] or *gathering* (people that are in the same location at the same time not exclusively for a given shared reason) [4]. In particular, the notion of F-formation received increased attention, having a solid computational translation that allowed researchers to design various algorithms to detect it; in particular, F-formations could be detected by minimization problems solved by Hough voting strategies [13,15,19,20], graph theory methods [16,4,21,22], or directly by classification approaches [14,17,18].

All the different definitions of a group existing in the literature generated a consequent problem, namely, the lack of a widely accepted way to quantitatively assess the goodness of a group detection approach: in fact, each researcher usually crafts a specific metric, well suited to measure the characteristics expressed by his own definition of group. In practice, this missed common agreement generated a considerable set of metrics which are specific for the scenario-at-hand, making a fair comparison among different approaches hard.

The most straightforward measures of detection quality are inherited from the multiobject tracking literature, that is, the CLEAR MOT metrics [27], where bounding boxes around single individuals have been replaced by the convex hull enveloping all the people inside the group [9,7]. Similar metrics are designed considering the object detection approaches [14,17,18]. In these cases, groups are intended as atomic entities, so that the errors of including more people in a group or losing some individuals in a formation are not explicitly modeled.

On the contrary, the metrics proposed in [15] deal with this problem, introducing also the concept of *tolerance*: in such a formulation, a group is considered as correctly estimated even if some individuals are missing or erroneously included. In this work, the idea of tolerance will be kept and developed for designing different GRODE indexes of quality.

Despite the fact that these measures focus on many interesting aspects of group detection, they seem to forget that groups can be different in terms of cardinality, ranging from pairs of people to sets of 11–12 people (after that it is reasonable to talk about crowds [28], which is not the subject of this study). As is intuitive by looking at Fig. 16.1, larger groups of people are harder to completely individuate than smaller

(A)   (B)   (C)

Figure 16.1 Examples of groups configurations in daily situations.

formations: this fact is anyway neglected by the current metrics that evaluate instead all the detections equally important in the computation of the final scores.

In this chapter we aim at defining a general set of metrics, dubbed here GROup DEtection (GRODE) metrics, where the presence of groups of different size is of primary importance. In particular, the idea is to capture the behavior of a group detection approach with respect to specific group cardinalities, individuating the group settings where a given algorithm performs better than another. In fact, an interesting phenomenon that can be captured is the tendency of oversegmenting or undersegmenting a group, and specifically if this tendency is polarized with a significant bias on particular formation cardinalities (that is, the tendency of systematically capture/break a group of G elements as one or more groups of g individuals). This sort of report will be of great importance for: (i) deeply understanding the nature of a group detection approach; and (ii) selectively fixing the issues and ameliorating the approach.

Here we extend the metrics presented in [29], integrating them with other indexes and offering a wider and deeper analysis on their meaning as well as similarities and differences with respect to the other measures adopted in the literature. The proposed metrics have been applied to most of the group detection approaches whose code is publicly available in the literature, and to different datasets; this is to ensure a comprehensive, unbiased analysis of the recent state-of-the-art in group detection. The results are enlightening, in the sense that many of the above discussed aspects unveil some unexplored characteristics of the approaches, helping the researcher in understanding which methods are more suited to his/her requirements.

Notably, GRODE metrics reveal that almost all the approaches have the tendency of oversegmenting groups (that is, detecting split versions of the groups), and that the bias is on groups of 2–3 persons. Another interesting fact is an apparent balance between group precision and recall in all the considered approaches.

The code for the GRODE metrics is publicly available on http://vips.sci.univr.it/research/fformation/, to promote its usage in the community. In addition, a repository is actually online, in order to record for each dataset the performance (expressed obviously in GRODE metrics) obtained by the different approaches.

16.2 METRICS IN THE LITERATURE

Over the last years, many and diverse metrics have been proposed in the literature, together with the group detection methods. Most of these metrics are actually extensions/customizations of metrics already used in different fields of computer vision [17, 14], pattern recognition [18], and natural language processing [30]. The main effort in proposing an original metric for this particular topic has been done in [15] and then extended in [4].

As for the metrics inherited from the tracking/detection fields, Manfredi et al. [17] propose two metrics focusing on the image level. In fact, their approach exploits as constitutive element 2D patches, extracted by analyzing motion and temporal information, that are subsequently classified as being a group or not. The detection measures are computed by considering as the main concept the intersection of the group detection convex hulls and the ground truth regions. In particular, they define the overlapping ratio (O_r) as

$$O_r = \frac{\sum_{p=1}^{P} M_{out}(p) \cap M_{gt}(p)}{\sum_{p=1}^{P} M_{out}(p) \cup M_{gt}(p)} \tag{16.1}$$

where p is a generic pixel and P is the total number of pixels in the image, M_{out} is a mask output by the detector algorithm that is 1 if the pixel belongs to a detected group and 0 otherwise, and M_{gt} is the same for ground truth. They also account precision and recall metrics at the object level (P_O and R_O) by considering a group as correctly identified if the spatial overlapping with the ground truth is higher than 50% of the intersection over union. While the concept of pixel-wise association of detected and ground truth groups is well funded in the detection literature, when applied it has a major drawback of being very sensitive to camera orientation. For instance, considering standard video surveillance views, as in Fig. 16.1, and top views of the same scene, results could be very different. At the same time, losing some components of the group could lead to different results depending on its location with respect to the camera. Moreover, working on the image plane, it can be hard to define a unique score for methods based on multicamera systems. Lastly, the metric is very sensitive to occlusions, since the same detector mask M_{out} could intersect many ground truth groups M_{gt}, and vice versa (e.g., Fig. 16.1B).

In the work of Choi et al. [14], the key elements are the people detections that are organized into structured groups by inferring their 3D layout in the scene by an iterative algorithm. In any case, the evaluation metric is still at the image level as the one of [17], but has some interesting characteristics which differentiate it. In fact, other than considering the intersection over the union ratio at the image level, they also force a detected group to be associated with at most one ground truth group, and vice versa. On top of it, the authors use standard precision, recall, and F_1 measures. To avoid problems related to misdetections of single individuals in the scene, the metric simply ignores them. This last point is, in our opinion, critical; in fact, crowded scenes are more difficult to correctly interpret compared with uncrowded ones, and thus a good metric should take into account this fact by including also the singletons.

Rota et al. [18] propose a classification based approach where the starting point is the knowledge of the ground truth positions of people in the scene, in the world plane. On this premise, dynamic features are considered in the frequency domain, in order to capture pairwise interactions among individuals. As a consequence of the setting, the metrics employed are based on the correct identification of each link between pairs of individuals; actually, this is a common choice for clustering problems and is commonly referred as *pairwise loss*. They consider standard precision, recall, and accuracy measures where a true positive is a pair of individuals belonging to the same group both for the estimates and ground truth, a false negative is a pair of individuals belonging to different groups both in estimates and ground truth, false positives are pairs of individuals grouped together by the algorithm but in different groups in ground truth, and vice versa for false negatives. The pairwise loss function has the main drawback to be imprecise when dealing with large crowds, due to the quadratic number of connections generated between members.

To overcome this problem, Solera et al. [30] propose in a similar experimental setting (that is, knowing the trajectories of the individuals in the scene on the floor) to extend the *MITRE loss* function [31], commonly used in NLP for the coreference problem, to handle groups; in particular, the MITRE loss is not able to handle singletons. The *GROUP–MITRE loss* is obtained by adding, for each individual, a fake counterpart to which only singletons are connected. Still, problems arise for high numbers of different groups, resulting in an unbalanced number of positive and negative links.

In a completely different scenario, White et al. [1] propose a probabilistic performance metric for group detection algorithms for biological data. In this case, explicit assignment of labeled individuals (proteins) to groups is evaluated against a ground truth, avoiding whatever issue caused by the visual sensor and the physical layout of the scene. The authors show that several traditional performance metrics are deficient if the size of a group is very small compared to the size of the population of entities (individuals) being considered. The authors propose a new information-theoretic metric, termed *proficiency*, that is, the ratio between the entropy of prior distribution, i.e., the entropy of the prior probability of an entity to belong to some group, and the mutual information between the detection and the ground truth. This metric is theoretically well funded and experimentally validated; nevertheless, its power is limited to the particular case they want to face (the ratio between group size and the population close to 0) which is very common in biological data but not in groups of people. Moreover, the score completely ignores singletons.

16.3 THE GRODE METRICS

The GRODE metrics are focused on (i) capturing the effectiveness of an approach in detecting groups and (ii) understanding *how* the group detections are carried out, by focusing on the cardinalities of the groups present in the scene. These two aims are explained in detail in the next subsections.

16.3.1 Detection Accuracy Measures

A group is not an atomic entity since it is formed by single individuals. Therefore, when a detected group is evaluated against the ground truth, the possibility of having lost some of its individuals should be considered with a certain emphasis. The general hypothesis of the GRODE metrics is that the evaluation of a group detection holds in the space of the individual labels, and not in the image space as happens, for example, in [17]. In fact, the weaknesses of the analysis at the image label have been adequately discussed in the previous section.

The GRODE metrics assume that, in a given scene with N individuals, ID_1, ID_2, ..., ID_N, and at a given time, a detection of groups is defined as a *partition of the whole set of the N individuals into subsets*. An individual is considered as a *singleton*, which is a spurious group (because intuitively a proper group is formed by two or more people), but still it is contained in the partition. Additionally, but not less important, it is assumed that the identities of the individuals are kept without errors by the person detection module (that is, the ID_1 contained in a group detection partition is actually the same ID_1 referred in the ground truth). We decide to assume this hypothesis to avoid too complicated situations. In addition, switches of identities are already modeled by other multiobject tracking metrics [32], so we let the reader to refer to them in the case of interest.

The detection of a group entails that a shared definition of what is a group is given and universally accepted. Unfortunately, as we saw in the previous section, this is not the case, even if attempts in this direction are growing in number [4]. In particular, there is no consensus in the literature of what happens in the case a group loses one of its member: is it the same group diminished or enlarged by an individual, or it is a completely different entity? The sociology literature has a bias toward this second view [25], but for a mere computer vision perspective the misdetection of a minor percentage of a group or the addition of a few individuals is usually accepted, especially when the cardinality of a group is large. For this reason, the adoption of a *tolerant* measure of group detection appears particularly well-suited.

In the GRODE metrics, we consider a group as correctly estimated if at least $\lceil (T \cdot |G|) \rceil$ of their members are found by the grouping method and correctly detected by the tracker, and if no more than $\lceil (1 - T) \cdot |G| \rceil$ false subjects (of the detected tracks) are identified, where $|G|$ is the cardinality of the labeled group G, and $T \in\ ]0.5, 1]$ is an arbitrary threshold, called *tolerance threshold*. In practice, the tolerance threshold individuates how much we are allowed in judging as correct a detection where a person is erroneously missed or added, ranging from 1 (accept only groups where you capture all the members) to 0.5 (capture at least half of the correct people and allow the rest be wrong individuals).

Once the T threshold has been set, we can determine the correctly detected groups (true positives groups – $TP_{(T)}$), the misdetected groups (false negatives groups – $FN_{(T)}$), and the hallucinated groups (false positives groups – $FP_{(T)}$), as well as the consequent precision and recall metrics, here dubbed as $Gprec_{(T)}$ and $Grec_{(T)}$, to highlight the dependence to the tolerant threshold T:

$$Gprec_{(T)} = \frac{TP_{(T)}}{TP_{(T)} + FP_{(T)}},$$

(16.2)

$$Grec_{(T)} = \frac{TP_{(T)}}{TP_{(T)} + FN_{(T)}}.$$

(16.3)

From these, we can calculate the $F_{1(T)}$ score as the harmonic mean of group precision and recall,

$$F_{1(T)} = 2 \cdot \frac{Gprec_{(T)} \cdot Grec_{(T)}}{Gprec_{(T)} + Grec_{(T)}}.$$

(16.4)

All these metrics are computed on a single frame, capturing the features of a method in a particular situation. When it comes to understanding the behavior of an approach along a temporal span, we need to aggregate these figures. Two alternatives are available:

1. *Being single scene independent.* Compute all the false and true detections over the entire time span, and then compute the precision and recall measures by applying the related formulas.
2. *Being single scene dependent.* Compute precision, recall, and F_1 measures at each frame, and then average these scores over all the frames.

In our opinion, the second alternative is the most correct, at least for those methods that operate on a per-frame basis. In the case when a group detection approach works with a memory over the past frames or in an offline batch mode, the first alternative could be taken in account. In such cases, it is mandatory to specify how the final performances are calculated since, obviously, they yield different numbers (think, for example, when the number of people in the scene changes over time).

In addition to these metrics, we consider a metric dubbed *Global Tolerant Matching* (GTM) which is independent of the tolerance threshold T. We compute this new score as the area under the curve (AUC) in the GF_1 vs. T graph with T varying from 1/2 to 1.[1] This value is then multiplied by 2 in order to normalize it at 1.

16.3.2 Cardinality Driven Measures

Due to their intrinsic composite nature, groups should also be evaluated by explicitly considering their cardinalities. Larger groups are more difficult to capture in their entirety, while smaller groups could be easily missed in a very crowded scenario. To focus on these aspects, the first way is to customize the precision, recall, and F_1 metrics for a single cardinality. This could be simply achieved by considering positive and negative detections of ground truth groups of a given cardinality, obtaining the *cardinality specific* group precision, recall, and F_1 measures.

Such an analysis will explicitly communicate the behavior of a given method in capturing particular cardinalities. Possibly, these measures could be aggregated by considering intervals of cardinalities; for example, we could be interested in capturing *small groups* (2–6 people) or *medium groups* (7–12 people) [4]. This could be achieved by simply computing the above metrics, including all of those detections that refer to ground truth groups whose cardinality is in a particular range.

To have a complete control of the behavior of a group detection approach, there is still a gap to fill. In fact, apart from the tolerance with which a ground truth group is matched, it is of interest to capture the tendency (if any) of a method to individuate groups of a given cardinality (say, j), for example, breaking larger groups of cardinality $k > j$, or merging smaller groups of cardinality $i < j$. More than this, we could also verify if this tendency exhibits a structured bias between cardinalities (that is, there is a tendency of merging a group of a specific cardinality into larger groups of another specific size), or if there is no polarization and, in general, the method simply oversegments or undersegments, or, finally, if there is no preference of over- or undersegmenting groups. These behaviors cannot be captured with the above metrics, since they focus uniquely on a given cardinality, discarding any connection between different group sizes.

In order to do this, the GRODE metrics are equipped with the *Histogram of Individuals over Cardinalities* (HIC) matrix (Fig. 16.2). The HIC matrix is computed by considering (over time and for each individual in the scene) the membership of a person of (a) a given ground-truth group and (b) the detected group in which he/she is currently associated, in relation to the cardinalities of these two groups; in other words, an individual belonging at a given time instant to a ground truth group of cardinality i and estimated in a group of cardinality j adds 1 to $HIC(i, j)$, and the process is repeated for all the persons and all the time instants of analysis. The matrix is then normalized over the rows.

---

[1] Please note that we avoid to consider $0 < T < 1/2$, since in this range we are accepting as good those groups where more than a half of the subjects are missing or are false positives, resulting in useless estimates.

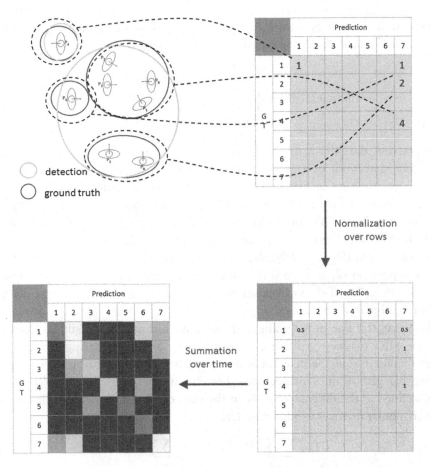

Figure 16.2 Graphical explanation of how the HIC matrix is built. Starting from predicted labels for each individual and ground truths, each individual adds 1 at the position (i, j), where i is the cardinality of the annotated group and j is the cardinality of the predicted group. Then, rows of this matrix are normalized, and the HIC matrix is the result of integration over time of these matrices computed at each frame.

Formally, HIC is defined as

$$HIC(i, j) = \frac{1}{n_j} \sum_{t \in T} \sum_{p \in P} d_{ij}\left(p, G, GT\right) \tag{16.5}$$

where i and j vary over the group cardinalities in the scene, that is, $1 \leq i \leq |g_t|_{max}$ and $1 \leq j \leq |g|_{max}$, with $|g_t|_{max}$ and $|g|_{max}$ being the maximum group cardinality of ground truth and estimated groups, respectively, t is a time instant, i.e., a frame, within the whole set of frames under analysis T, p is a person within the list P of all the detected people at that particular time t, G and GT are the sets of all estimated and ground truth groups at time t, and n_j is a normalization term which sets to 1 the sum over the rows of HIC. Finally, d_{ij} is an accumulation function defined as

$$d_{ij}(p, G, GT) = \begin{cases} 1, & \text{if } \exists g \in G \wedge g_t \in GT: \\ & p \in g \cap g_t \wedge |g| = j \wedge |g_t| = i, \\ 0, & \text{otherwise.} \end{cases} \tag{16.6}$$

In practice, the HIC matrix indicates the tendency of a group detection approach to produce detections of a given cardinality, *highlighting in the off-diagonal entries how the errors are distributed in terms of group cardinalities.* More in detail, having a perfectly diagonal matrix does not necessarily individuate an accurate approach (for that purpose, the *Gprec*, *Grec*, and *GF_1* should be used instead). Think, for example, of a situation in which the ground truth groups are $< ID_1, ID_2 >$, $< ID_3, ID_4 >$, and the detected groups are $< ID_1, ID_3 >$, $< ID_2, ID_4 >$. In this case, the HIC matrix would have the value 4 at position $(2, 2)$, even if the *Grec* with $T > 0.5$ is 0. Conversely, the nonzero entries in the HIC off-diagonal positions necessarily indicate errors in the detections, addressing misplaced individuals.

Following this line of reasoning, a quantitative measure that could be extrapolated from the HIC matrix is a balance between upper and lower triangular submatrices: a high balance means the detector is not biased to merging small groups together or splitting big groups into smaller ones. For this we propose the *upper–lower difference* (*UL*), defined as the difference between the sum of all the elements of the upper and lower triangular matrices extracted by *HIC*,

$$UL = \sum_i \sum_{j>i} HIC(i,j) - \sum_i \sum_{j<i} HIC(i,j), \tag{16.7}$$

and its weighted version (*WUL*) where elements far from the diagonal are weighted more then the closer ones,

$$WUL = \sum_i \sum_{j>i} HIC(i,j) \cdot (j-i) - \sum_i \sum_{j<i} HIC(i,j) \cdot (i-j). \tag{16.8}$$

The rationale of this measure is that we want to penalize more the situations where, for example, an element which is supposed to form a group of 2 people is assigned to a formation consisting of 8 rather than 3 elements. The reason is that drastically differently sized groups are often treated with different models, and thus sharing individuals among them could mean a very low specificity.

With this set of metrics we are now able to evaluate group detectors not only in a global way, but also when analyzing their most important features, i.e., the ability to detect groups of different cardinalities with no bias and the balance between over- and undersegmentation.

Table 16.1 Summary of the features of the datasets used for experiments

Dataset	Data type	Detection	Detection quality
Synthetic Data	synthetic	—	perfect
IDIAP Poster	real	manual	very high
Cocktail Party	real	automatic	high
Coffee Break	real	automatic	low
GDet	real	automatic	very low

16.4 EXPERIMENTS

In this section we present extensive experiments to validate our claims in the previous sections and, in particular, to prove the relationship between the proposed metrics and the behavior we expect from the detector.

16.4.1 Datasets

We used in our experiments five publicly available datasets: two from [15] (Synthetic and Coffee Break), one from [16] (IDIAP PosterData), one from [20] (Cocktail Party), and one from [13] (GDet). A summary of the dataset features is provided in Table 16.1, while a detailed presentation of each dataset follows.

Synthetic Data. A psychologist generated a set of 10 diverse situations, each one repeated 10 times with minor variations, resulting in 100 frames representing different social situations, with the aim to span as many configurations as possible for F-formations. On average 9 individuals and 3 groups are present in the scene, together with some singletons. Proxemic information is noiseless in the sense that there is no clutter in the position and orientation state of each individual.

IDIAP Poster Data. Over 3 hours of aerial videos (resolution 654×439 pixels) have been recorded during a poster session of a scientific meeting. Over 50 people are walking through the scene, forming several groups over time. A total of 82 images were selected with the idea to maximize the crowdedness and variance of the scenes. Images are unrelated to each other in the sense that there are no consecutive frames, and the time lag between them prevents exploiting temporal smoothness. As for the data annotation, a total of 24 annotators were grouped into 3-person subgroups and they were asked to identify F-formations and their associates from static images. Each person's position and body orientation was manually labeled and recorded as pixel values in the image plane – one pixel represented approximately 1.5 cm.

Cocktail Party. This dataset contains about 30 min of video recordings of a cocktail party in a 30 m^2 lab environment involving 7 subjects. The party was recorded using four synchronized angled-view cameras (15 Hz, 1024×768 pixels, jpeg) installed in the corners of the room. Subjects' positions were logged using a particle filter-based body tracker [33] while head pose estimation is computed as in [34]. Every 5 seconds groups in one frame were manually annotated by an expert, resulting in a total of 320 labeled frames for evaluation.

Coffee Break. The dataset focuses on a coffee-break scenario of a social event, with a maximum of 14 individuals organized in groups of 2–3 people each. Images are taken from a single camera with resolution of 1440 × 1080 pixels. People positions have been estimated by exploiting multiobject tracking on the heads, and head detection has been performed afterward with the algorithm of [35], considering solely 4 possible orientations (front, back, left, and right) in the image plane. The tracked positions and head orientations were then projected onto the ground plane. Considering the ground truth data, a psychologist annotated the videos indicating the groups present in the scenes, for a total of 119 frames split in two sequences. The annotations were generated by analyzing each frame in combination with questionnaires that the subjects filled in.

GDet. The dataset is composed of 5 subsequences of images acquired by 2 angled-view low resolution cameras (352 × 328 pixels) with a number of frames spanning from 17 to 132, for a total of 403 annotated frames. The scenario is a vending machines area where people meet and chat while they are having coffee. This is similar to Coffee Break scenario but in this case the scenario is indoor, which makes occlusions many and severe; moreover, people in this scenario know each other in advance. The videos were acquired with two monocular cameras, located at opposite corners of the room. To ensure the natural behavior of people involved, they were not aware of the experiment purposes. For ground truth generation, people tracking had been carried out with the particle filter proposed in [33], while head pose estimation was performed afterward with the method in [35] considering only 4 orientations (front, back, left, and right).

16.4.2 Detection Methods

We compare seven different state-of-the-art methods: one exploiting the concept of *view frustum* (IRPM), two based on dominant sets (IGD and GTCG), three different versions of Hough Voting approaches (linear, entropic, and multiscale HVFF), and one based on graph-cuts technique (GCFF).

Interrelation Pattern Matrix (IRPM). Proposed by Bazzani et al. [13], it uses the head direction to infer the 3D view frustum as an approximation of the focus-of-attention of an individual; this is used together with proximity information to estimate interactions: the idea is that close-by people whose view frustum is intersecting are in some way interacting.

Interacting Group Discovery (IGD). Presented by Tran et al. [21], it is based on dominant sets' extraction from an undirected graph where nodes are individuals and the edges have a weight proportional to how much people are interacting; the attention of an individual is modeled as an ellipse centered at a fixed offset in front of the person, while the interaction between two individuals is proportional to the intersection of their attention ellipses.

Game-Theory for Conversational Groups (GTCG). In [22] the authors develop a game-theoretic framework, supported by a statistical modeling of the uncertainty associated with the position and orientation of people. Specifically, they use a representation of the affinity between candidate pairs by expressing the distance between distributions over the most plausible oriented region of attention. Additionally, they can integrate temporal information over multiple frames by using notions from multipayoff evolutionary game theory.

Hough Voting for F-Formation (HVFF). Under this caption we consider a set of methods based on a Hough Voting strategy to build accumulation spaces and find local maxima of this function to identify F-formations. The general idea is that each individual is associated with a Gaussian probability density function which describes the position of the o-space center he is pointing at. The pdf is approximated by a set of samples, which basically vote for a given o-space center location. The voting space is then quantized and the votes are aggregated on squared cells to form a discrete accumulation space. Local maxima in this space identify o-space centers, and consequently, F-formations. Over the years, three versions of this framework have been presented: in [15] the votes are linearly accumulated by just summing up all the weights of votes belonging to the same cell, in [19] the votes are aggregated by using the weighted Boltzmann entropy function, while in [20] a multiscale approach is used on top of the entropic version.

Graph-Cuts for F-Formation (GCFF). Presented by Setti et al. [4], it proposes an iterative approach that starts by assuming an arbitrarily high number of F-formations: after that, a hill-climbing optimization alternates between assigning individuals to groups using a graph-cut based optimization, and updating the centers of the F-formations, pruning unsupported groups in accordance with a Minimum Description Length prior. The iterations continue until convergence, which is guaranteed.

16.4.3 Detection Accuracy Measures

For each method, the parameters have been set in order to give the best performance on a specific dataset in terms of precision, recall, and GF_1 metrics as defined in Section 16.3.1. A tuning phase has been carried out for each method/dataset combination on half of the sequence by cross-validation, and kept unchanged for the remaining frames.[2]

---

[2] Since the code for Dominant Sets [16] is not publicly available, we used results provided directly from the authors of the method for a subset of data. For this reason, average results over all the datasets are only averaged over 3 datasets, and cannot be taken into account for a fair comparison.

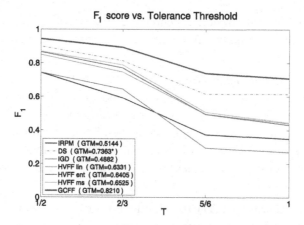

Figure 16.3 Average GF_1 score vs. tolerance threshold T, over all the datasets considered. The Global Tolerant Matching (GTM) score is shown in brackets in the legend. Dominant Sets (DS) results are averaged over 3 datasets only, because of results availability. Best viewed in color. (For interpretation of the references to color in this figure legend, the reader is referred to the web version of this chapter.)

Table 16.2 shows the best results by considering the threshold $T = 2/3$, which corresponds to finding at least $2/3$ of the members of a group with no more than $1/3$ of false subjects; while Table 16.3 presents results with $T = 1$, considering a group as correct if all and only its members are detected.

We can see that GTCG and GCFF methods are the best performing in all the datasets, with GCFF outperforming all the competitors when strong noise in the position and orientation is present (Coffee Break and GDet datasets) or when it is requested to perfectly detect groups ($T = 1$).

From these tables we can also notice that the best GF_1 scores are usually achieved in correspondence with a good balance between *Gprec* and *Grec*, with differences limited to 2%. Nevertheless, IRPM and IGD methods are consistently deficient in recall in every single dataset.

As for the Global Tolerant Matching score, Fig. 16.3 shows the average GF_1 scores for each method computed over all the frames and datasets, where the score value is reported in the legend. From the curves we can appreciate how the GCFF method outperforms the state-of-the-art methods not only in GTM, but consistently for each T-value.

16.4.4 Cardinality Driven Measures

Table 16.4 shows the groups' distribution in terms of number of people for group cardinality in each dataset. While the metrics are suitable for all the datasets, a wide variety of group cardinalities provide more information and allow for deeper discussions; for this reason, in the following of this section we will show detailed results computed

Table 16.2 Average precision, recall and F_1 scores for all the methods and all the datasets ($T = 2/3$). Please note that DS results are averaged over only 3 datasets and thus cannot be taken into account for a fair comparison

	Synthetic			IDIAP Poster			Cocktail Party			Coffee Break			GDet			Total		
	Gprec	Grec	GF$_1$	Gprec	Grec	GF$_1$	Gprec	Grec	GF$_1$	Gprec	Grec	GF$_1$	Gprec	Grec	GF$_1$	Gprec	Grec	GF$_1$
IRPM [13]	0.85	0.80	0.82	0.82	0.74	0.78	0.56	0.43	0.49	0.68	0.50	0.57	0.77	0.47	0.58	0.70	0.49	0.56
DS [16]	0.85	0.97	0.90	0.91	0.92	0.91	—	—	—	0.69	0.65	0.67	—	—	—	0.81	0.83	0.82
IGD [21]	0.95	0.71	0.81	0.80	0.68	0.73	0.81	0.61	0.70	0.81	0.78	0.79	0.83	0.36	0.50	0.83	0.55	0.64
CTCG [22]	1.00	1.00	1.00	0.92	0.96	0.94	0.86	0.82	0.84	0.83	0.89	0.86	0.76	0.76	0.76	0.83	0.83	0.83
HVFF lin [15]	0.75	0.86	0.80	0.90	0.95	0.92	0.59	0.74	0.65	0.73	0.86	0.79	0.66	0.68	0.67	0.68	0.76	0.71
HVFF ent [19]	0.79	0.86	0.82	0.86	0.89	0.87	0.78	0.83	0.80	0.76	0.86	0.81	0.69	0.71	0.70	0.78	0.78	0.77
HVFF ms [20]	0.90	0.94	0.92	0.87	0.91	0.89	0.81	0.81	0.81	0.83	0.76	0.79	0.71	0.73	0.72	0.79	0.79	0.79
GCFF [4]	0.97	0.98	0.97	0.94	0.96	0.95	0.84	0.86	0.85	0.85	0.91	0.88	0.92	0.88	0.90	0.89	0.89	0.89

Table 16.3 Average precision, recall, and F_1 scores for all the methods and all the datasets ($T = 1$). Please note that DS results are averaged over only 3 datasets and thus cannot be taken into account for a fair comparison

	Synthetic			IDIAP Poster			Cocktail Party			Coffee Break			GDet			Total		
	Gprec	Grec	GF$_1$	Gprec	Grec	GF$_1$	Gprec	Grec	GF$_1$	Gprec	Grec	GF$_1$	Gprec	Grec	GF$_1$	Gprec	Grec	GF$_1$
IRPM [13]	0.53	0.47	0.50	0.71	0.64	0.67	0.28	0.17	0.21	0.27	0.23	0.25	0.59	0.29	0.39	0.46	0.29	0.35
DS [16]	0.68	0.80	0.74	0.79	0.82	0.81	—	—	—	0.40	0.38	0.39	—	—	—	0.60	0.63	0.62
IGD [21]	0.30	0.22	0.25	0.31	0.27	0.29	0.23	0.10	0.13	0.50	0.50	0.50	0.67	0.20	0.31	0.45	0.21	0.27
CTCG [22]	0.78	0.78	0.78	0.83	0.86	0.85	0.31	0.28	0.30	0.46	0.47	0.47	0.51	0.60	0.55	0.49	0.52	0.51
HVFF lin [15]	0.64	0.73	0.68	0.80	0.86	0.83	0.26	0.27	0.27	0.41	0.47	0.44	0.43	0.45	0.44	0.43	0.46	0.44
HVFF ent [19]	0.47	0.52	0.49	0.72	0.74	0.73	0.28	0.30	0.29	0.47	0.52	0.49	0.44	0.45	0.45	0.42	0.44	0.43
HVFF ms [20]	0.72	0.73	0.73	0.73	0.76	0.74	0.30	0.30	0.30	0.40	0.38	0.39	0.44	0.45	0.45	0.44	0.45	0.45
GCFF [4]	0.91	0.91	0.91	0.85	0.87	0.86	0.63	0.65	0.64	0.61	0.64	0.63	0.73	0.68	0.71	0.71	0.70	0.71

Table 16.4 Number of individuals for each group's cardinality in each dataset used in the experiments

Cardinality	1	2	3	4	5	6
Synthetic Data	300	360	240	—	—	—
IDIAP Poster	429	910	339	12	5	—
Cocktail Party	174	162	246	176	275	882
Coffee Break	376	464	459	—	—	—
GDet	367	394	372	88	175	78

Table 16.5 GF_1 scores for each cardinality on GDet dataset ($T = 1$)

	$k = 2$	$k = 3$	$k = 4$	$k = 5$	$k = 6$	Avg
# groups	197	124	22	35	13	—
IRPM [13]	0.40	0.59	0.45	0.42	0.35	0.44
IGD [21]	0.15	0.52	0.33	0.54	0.83	0.47
HVFF lin [15]	0.51	0.76	0.03	0.16	0.13	0.32
HVFF ent [19]	0.57	0.73	0.24	0.23	0.13	0.38
HVFF ms [20]	0.56	0.78	0.17	0.41	0.67	0.52
GCFF [20]	0.74	0.87	0.53	0.77	0.88	**0.76**

on the GDet dataset, while in the last part we summarize results obtained on all the datasets.

Results in terms of GF_1 for each group cardinality for the GDet dataset are reported in Table 16.5. As expected from the previous metrics, GCFF is the best performing method for every single group's cardinality, and as a consequence on average over all of them.

Fig. 16.4 shows the *HIC* matrices related to GDet dataset for all the methods used in this comparison. Simply looking at the matrices, the reader can develop an intuition on the behavior of each method. For instance, IRPM and IGD are expected to perform very well when detecting singletons, while HVFF lin is prone to detect small groups of 2 elements. GCFF looks to be the most balanced within the seven methods, while in general all the other six methods show a tendency to oversegment. Moreover, groups of 6 elements are only detected by multiscale HVFF and GCFF. This intuition will be confirmed by the analysis of numerical scores that follows.

Fig. 16.5 reports the upper–lower difference in its absolute and weighted versions. As expected, GCFF is the best performer in both the scores, with UL difference of 0.14, dropping to 0.06 in the weighted version. As already foreseen from the *HIC* matrices of Fig. 16.4, all the methods except for GCFF have a tendency to oversegment; in particular, Hough voting methods are performing very poorly from this point of view. Note that GCFF and GTCG have very similar UL in absolute value, but the way the two methods generate these values is completely different; indeed, looking at the matrices,

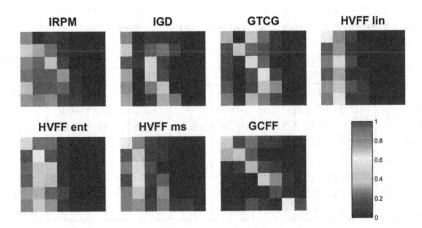

Figure 16.4 HIC matrices for seven state-of-the-art methods on GDet dataset. Best viewed in color. (For interpretation of the references to color in this figure legend, the reader is referred to the web version of this chapter.)

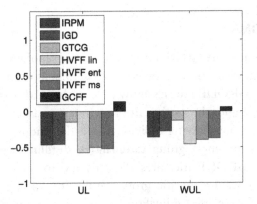

Figure 16.5 Upper–lower difference (*UL*) and weighted upper–lower difference (*WUL*) for all the methods on GDet dataset. Best viewed in color. (For interpretation of the references to color in this figure legend, the reader is referred to the web version of this chapter.)

one can see that GCFF has very few elements off-diagonal, while GTCG has a good balance between over- and undersegmentation (Table 16.6).

Moreover, the weighted version (*WUL*) is more informative than the absolute one, since it takes into account the ability of a detector to approximately detect correct groups. This effect can be seen from the comparison of HVFF ent and HVFF ms. Looking at the absolute difference (*UL*), the multiscale approach seems to be more prone to oversegmentation than the entropic version, but the entropic method tends to split big groups into small ones, while the multiscale only loses some elements in mid-

Table 16.6 Upper–lower difference (*UL*) and weighted upper–lower difference (*WUL*) for all the methods on all the datasets

	Synthetic		IDIAP Poster		Cocktail Party		Coffee Break		GDet	
	UL	WUL	UL	WUL	UL	WUL	UL	WUL	UL	WUL
IRPM [13]	−0.24	−0.16	0.23	0.17	0.02	0.04	0.07	0.08	−0.46	−0.36
DS [16]	−0.09	−0.07	−0.33	−0.25	—	—	−0.13	−0.03	—	—
IGD [21]	0.20	0.11	0.12	0.08	−0.20	−0.16	0.31	0.24	−0.46	−0.27
CTCG [22]	−0.15	−0.10	−0.11	−0.14	−0.10	−0.06	0.02	0.01	−0.15	−0.13
HVFF lin [15]	−0.03	−0.01	0.12	0.07	−0.26	−0.17	−0.14	−0.09	−0.58	−0.46
HVFF ent [19]	0.26	0.18	0.17	0.10	−0.15	−0.10	−0.04	−0.02	−0.50	−0.40
HVFF ms [20]	−0.17	−0.11	−0.21	−0.16	−0.23	−0.12	−0.28	−0.18	−0.52	−0.37
GCFF [4]	−0.01	−0.01	−0.15	−0.14	0.39	0.28	0.12	0.08	0.14	0.06

size groups, leading to smaller values of *WUL*. A similar effect can be seen comparing GTCG and GCFF.

16.5 CONCLUSIONS

In this work we presented the GRODE metrics, a novel set of group detection accuracy measures, based on the evaluation of the correct assignment of single individuals to differently sized groups. GRODE metrics allow us to analyze *exhaustively* the performance of a group detection approach; starting from accuracy measures (precision, recall, F_1) which are *tolerant* (possibly allowing individual false detections), one can focus on analyzing performances for single group cardinalities, distilling specific tendencies of a detection technique. GRODE measures allow us also to understand if errors in the detection are biased toward a specific group cardinality (an approach tends to always individuate groups of a specific cardinality), or in general if an approach over- or undersegments groups. The obtained performance report given by GRODE is useful for many reasons. First of all, it is useful for a deep comparative analysis with other techniques. Secondly, it may allow driving the parametrization of a detection system in a cross-validation loop, or the redesign of the technique itself. Future extensions of the GRODE metrics will consider the problem of including errors coming from person detection modules: the idea is that a group in which not all the components have been detected should be accounted less (because the group detection approach should account for missing data).

REFERENCES

[1] White JV, Steingold S, Fournelle CG. Performance metrics for group-detection algorithms. In: Interface. 2004.

[2] Lee C-H, Hoehn-Weiss MN, Karim S. Grouping interdependent tasks: using spectral graph partitioning to study complex systems. Strateg Manag J 2015. http://dx.doi.org/10.1002/smj.2455.

[3] Garwin T, Crozat MP, Merrell BL, Bebee B. Metrics-based test and evaluation of group detection software for counter-terrorism. In: IEEE aerospace conference. 2006.

[4] Setti F, Russell C, Bassetti C, Cristani M. F-formation detection: individuating free-standing conversational groups in images. Available from arXiv:1409.2702, 2014.

[5] Chang MC, Krahnstoever N, Ge W. Probabilistic group-level motion analysis and scenario recognition. In: International conference on computer vision (ICCV). 2011.

[6] Ge W, Collins RT, Ruback RB. Vision-based analysis of small groups in pedestrian crowds. IEEE Trans Pattern Anal Mach Intell 2012;34:1003–16.

[7] Mazzon R, Poiesi F, Cavallaro A. Detection and tracking of groups in crowd. In: International conference on advanced video and signal based surveillance (AVSS). 2013.

[8] Yamaguchi K, Berg AC, Ortiz LE, Berg TL. Who are you with and where are you going? In: IEEE conference on computer vision and patter recognition (CVPR). 2011.

[9] Bazzani L, Zanotto M, Cristani M, Murino V. Joint individual-group modeling for tracking. IEEE Trans Pattern Anal Mach Intell 2015;37(4):746–59.

[10] Choi W, Savarese S. A unified framework for multi-target tracking and collective activity recognition. In: European conference on computer vision (ECCV). 2012.

[11] Khamis S, Morariu VI, Davis LS. A flow model for joint action recognition and identity maintenance. In: IEEE international conference on computer vision and pattern recognition (CVPR). 2012. p. 1218–25.

[12] Ricci E, Varadarajan J, Subramanian R, Rota Buló S, Ahuja N, Lanz O. Uncovering interactions and interactors: joint estimation of head, body orientation and F-formations from surveillance videos. In: International conference on computer vision (ICCV). 2015.

[13] Bazzani L, Cristani M, Tosato D, Farenzena M, Paggetti G, Menegaz G, Murino V. Social interactions by visual focus of attention in a three-dimensional environment. Expert Syst 2013;30(2):115–27.

[14] Choi W, Chao Y-W, Pantofaru C, Savarese S. Discovering groups of people in images. In: European conference on computer vision (ECCV). 2014.

[15] Cristani M, Bazzani L, Paggetti G, Fossati A, Tosato D, Del Bue A, Menegaz G, Murino V. Social interaction discovery by statistical analysis of F-formations. In: British machine vision conference (BMVC). 2011.

[16] Hung H, Kröse B. Detecting F-formations as dominant sets. In: International conference on multimodal interaction (ICMI). 2011.

[17] Manfredi M, Vezzani R, Calderara S, Cucchiara R. Detection of static groups and crowds gathered in open spaces by texture classification. Pattern Recognit Lett 2014;44:39–48.

[18] Rota P, Conci N, Sebe N. Real time detection of social interactions in surveillance video. In: European conference on computer vision (ECCV). 2012.

[19] Setti F, Hung H, Cristani M. Group detection in still images by F-formation modeling: a comparative study. In: International workshop on image and audio analysis for multimedia interactive services (WIAMIS). 2013.

[20] Setti F, Lanz O, Ferrario R, Murino V, Cristani M. Multi-scale F-formation discovery for group detection. In: International conference on image processing (ICIP). 2013.

[21] Tran KN, Bedagkar-Gala A, Kakadiaris IA, Shah SK. Social cues in group formation and local interactions for collective activity analysis. In: International conference on computer vision theory and applications (VISAPP). 2013.

[22] Vascon S, Eyasu Z, Cristani M, Hung H, Pelillo M, Murino V. A game-theoretic probabilistic approach for detecting conversational groups. In: Asian conference on computer vision (ACCV). 2014.

[23] McPhail C. The myth of the madding crowd. Transaction Publishers; 1991.

[24] Garfinkel H. Studies in ethnomethodology. Prentice-Hall; 1967.

[25] Goffman E. Behavior in public places: notes on the social organization of gatherings. Free Press; 1966.

[26] Kendon A. Conducting interaction: patterns of behavior in focused encounters. Cambridge University Press; 1990.

[27] Bernardin K, Stiefelhagen R. Evaluating multiple object tracking performance: the CLEAR MOT metrics. Int J Image Video Process 2008:1–10. http://dx.doi.org/10.1155/2008/246309.

[28] Hare AP. Group size. Am Behav Sci 1981;24(5):695–708.

[29] Setti F, Cristani M. F-formation detection: individuating free-standing conversational groups in images. In: IEEE international conference on computer vision and pattern recognition workshop (CVPRW). 2015.

[30] Solera F, Calderara S, Cucchiara R. Structured learning for detection of social groups in crowd. In: IEEE international conference on advanced video and signal based surveillance (AVSS). 2013.

[31] Vilain M, Burger J, Aberdeen J, Connolly D, Hirschman L. A model-theoretic coreference scoring scheme. In: Conference on message understanding. 1995.

[32] Li Y, Huang C, Nevatia R. Learning to associate: HybridBoosted multi-target tracker for crowded scene. In: IEEE international conference on computer vision and pattern recognition (CVPR). 2009.

[33] Lanz O. Approximate Bayesian multibody tracking. IEEE Trans Pattern Anal Mach Intell 2006;28(9):1436–49.

[34] Lanz O, Brunelli R. Joint Bayesian tracking of head location and pose from low-resolution video. In: Multimodal technologies for perception of humans. Springer; 2008. p. 287–96.

[35] Tosato D, Spera M, Cristani M, Murino V. Characterizing humans on Riemannian manifolds. IEEE Trans Pattern Anal Mach Intell 2013;35(8):1972–84.

CHAPTER 17

Realtime Pedestrian Tracking and Prediction in Dense Crowds

Aniket Bera*, David Wolinski[†], Julien Pettré[†], Dinesh Manocha*
*Department of Computer Science, University of North Carolina at Chapel Hill, Chapel Hill, NC, USA
[†]INRIA-Rennes, Campus de Beaulieu, Rennes Cedex, France

Contents

17.1 INTRODUCTION

The sensing and computation (*detection*, *tracking*, and *prediction*) of human crowd motion has received considerable attention in the literature. It is a well-studied problem that has many applications in surveillance, behavior modeling, activity recognition, disaster prevention, and crowded scene analysis. Despite many recent advances, it is still difficult to accurately track and predict pedestrians in real-world scenarios, especially as the crowd density increases.

The problem of tracking pedestrians and objects has been studied in computer vision and image processing for three decades. However, tracking pedestrians in a crowded scene is regarded as a hard problem due to the following reasons: intra–pedestrian occlusion (i.e., one pedestrian blocking others), and changes in lighting and pedestrian appearance. Similarly, predicting the trajectory of a pedestrian in a dense environment can also be challenging. In general, pedestrians have varying behaviors and can change their speed to avoid collisions with the obstacles and other pedestrians in a scene. In high density or crowded scenarios, the pairwise interactions between the pedestrians tend to increase significantly which affects their behavior and movement. As a result, the highly dynamic nature of pedestrian movement makes it hard to estimate their current or future positions. Furthermore, many applications such as surveillance, robot navigation, and autonomous driving need realtime prediction capabilities to estimate the positions of the pedestrians.

We give an overview of recent algorithms that can improve the tracking and prediction using improved motion and local navigation models. There has been considerable work done on developing crowd motion models for pedestrians in the areas of computer graphics, robotics, and pedestrian dynamics. Many approaches have been investigated that suggest different ways to model the movement of pedestrians as part of a large crowd. Furthermore, most of these models are parametric, and use different parameters related to shape or physical properties to compute the movement of each human agent. It is important to accurately estimate these parameters based on the actual movement or behavior of each pedestrian in the crowd video.

Most prior realtime or online crowd-sensing algorithms use a single, homogeneous motion model, i.e., a uniform or constant model for the entire trajectory. Every motion model is unique and generally relies upon one or more assumptions: highly coherent motion in terms of velocity or acceleration, or how pedestrian trajectories will change in response to other agents or obstacles.

The simpler motion models assume that agents will ignore any interactions with other pedestrians, instead assuming that they will follow "constant-speed" or "constant-acceleration" paths to their immediate goals. However, the accuracy of this assumption decreases as crowd density in the environment increases (e.g., to 2–4 pedestrians per square meter). More sophisticated pedestrian motion models take into account interactions between the pedestrians, formulated either in terms of attraction or repulsion forces or collision-avoidance constraints.

Sensing the position of moving humans has been studied in robotics and computer vision, e.g., [1–3]. Most of the methods focus on learning motion priors, from the same scene or similar scenes. Some of the methods focus on very crowded high-density scenarios [2,3], assuming that the movement of each individual will follow the overall crowd flow. These methods often depend upon an a priori motion model fitted to capture the crowd flow. However, these models usually fail to accurately capture or predict the trajectory of each individual pedestrian because the motion priors used are typically generalized motions rather than motions specific to individual movements. For example, we frequently observe uncommon pedestrian motions, such as moving against the flow of other agents in a crowd, or quick velocity changes to avoid collisions. In order to address these issues, many pedestrian tracking algorithms use multiagent or crowd motion models based on local collision avoidance [4,5]. These behaviors, such as some people moving against the flow or quick movements to avoid collisions with other people, are common even in moderately crowded scenarios. These multiagent interaction models effectively capture short-term deviations from goal-directed paths, but in order to do so, they must already know each pedestrian's goal positions; they often use handpicked destination information, or other heuristics that require prior knowledge about the environment. As a result, these techniques have important limitations: they are unable to account for unknown environments with multiple destinations, or times when pedestrians take long detours or make unexpected stops. While the destination is important information, it does not predict the pedestrian's complete trajectory. In general, the assumption that the destination information remains constant can often result in large errors in predicted trajectories. Modeling human behaviors is a difficult research problem. There are some computational models of human behaviors, e.g., goal-directed behaviors and collision avoidance [6,7].

In summary, the trajectory of each pedestrian is governed by its intermediate goal location, intrinsic behaviors, as well as local interactions with other pedestrians and obstacles in the scene. In a dense crowd setting, the behavior of each pedestrian changes in response to the environment, the overall crowd density and flow, and the behavior of other pedestrians. We need new dynamic motion models that can indeed capture these variations in the behavior of each pedestrian as part of a dense crowd.

In terms of prediction, we describe an algorithm that learns from the pedestrian local and global movement patterns from sparse 2D pedestrian trajectory data using Bayesian inference. The approach is general, makes no assumptions about pedestrian movement or density, and performs no precomputation. We use the trajectory data information over a sequence of frames to predict the future pedestrian states using Ensemble Kalman Filters (EnKF) and Expectation Maximization (EM). The state information is used to compute movement–flow information of individual pedestrians and coherent pedestrian clusters using a mixture of motion models. The global movement features are combined with local motion patterns computed using Bayesian reciprocal velocity obstacles to compute the predicted state of each pedestrian. The combination of **G**lobal

Figure 17.1 We highlight different motion models (Boids, Social Forces, or reciprocal velocity obstacles) used for the same pedestrian (marked in red) over different frames/time. We believe that it is not possible to model the trajectory of all pedestrians based on a single, homogeneous motion. Instead, we adaptively choose the best-fit model for every pedestrian in the scene that can vary based on the environment or the crowd conditions. (For interpretation of the references to color in this figure legend, the reader is referred to the web version of this chapter.)

and **L**ocal **M**ovement **P**atterns (i.e., GLMP) corresponds to computing dynamically varying individualized motion model for each pedestrian.

17.2 RELATED WORK

In this section, we give a brief overview of prior work on motion models and pedestrian path prediction.

17.2.1 Motion Models

There is an extensive body of work in robotics, multiagent systems, crowd simulation, and computer vision on modeling pedestrian motion in crowded environments. These models can be broadly classified into the following categories (cf. Fig. 17.1): potential-based models, which model virtual agents in a crowd as particles with potentials and forces [6]; boid-like approaches, which create simple rules to steer agents [8]; geometric optimization models, which compute collision-free velocities [9]; and field-based methods, which generate fields based on continuum theory [10]. Many of these models have been used for offline and online pedestrian tracking and trajectory computation. Among these approaches, velocity-based motion models [11–13,9,14] have been successfully applied to the simulation and analysis of crowd behaviors and to multirobot coordination [15]; velocity-based models have also been shown to have efficient implementations that closely match real human paths [16].

17.2.2 Pedestrian Tracking with Motion Models

Prior work in pedestrian tracking [17,3] attempts to improve tracking accuracy by making simple assumptions about pedestrian movement, such as constant velocity and constant acceleration. More recently, long-term motion models and pairwise interaction rules have been combined with tracking to improve the accuracy. Bruce et al. [18] and Gong et al. [19] first estimate pedestrians' destinations and then predict their motions along the path toward the estimated goal positions. Liao et al. [20] extract a Voronoi graph from the environment and predict people's motion along the edges. Many techniques have been proposed for short-term prediction using motion models. Luber et al. [1] apply Helbing's social force model to track people using a Kalman filter based tracker. Mehran et al. [21] also apply the social force model to detect people's abnormal behaviors from videos. Pellegrini et al. [4] use an energy function to build up a goal-directed short-term collision-avoidance motion model. Bera et al. [22–24] use reciprocal velocity obstacles and hybrid motion models to improve the accuracy of pedestrian tracking.

17.2.3 Path Prediction and Robot Navigation

Robots navigating in complex, noisy, and dynamic environments have prompted the development of other forms of trajectory prediction algorithms. Fulgenzi et al. [25] use a probabilistic velocity–obstacle approach combined with the dynamic occupancy grid; this method assumes constant linear velocity motion of the obstacles. DuToit et al. [26] present a robot planning framework that takes into account pedestrians' anticipated future location information to reduce the uncertainty of the predicted belief states. Other techniques use potential-based approaches for robot path planning in dynamic environments [27]. Some methods learn the trajectories from collected data. Ziebart et al. [28] use pedestrian trajectories collected in the environment for prediction using Hidden Markov Models. Bennewitz et al. [29] apply Expectation Maximization clustering to learn typical motion patterns from pedestrian trajectories before using Hidden Markov Models to predict future pedestrian motion. Henry et al. [30] use reinforced learning from example traces, estimating pedestrian density and flow with a Gaussian process. Kretzschmar et al. [31] consider pedestrian trajectories as a mixture probability distribution of a discrete as well as a continuous distribution, and then use Hamiltonian Markov chain Monte Carlo sampling for prediction. Kuderer et al. [32] use maximum entropy based learning to learn pedestrian trajectories and use a hierarchical optimization scheme to predict future trajectories. Many of these methods involve a priori learning, and may not work in new or unknown environments.

Trautman et al. [33] have developed a probabilistic predictive model of cooperative collision avoidance and goal-oriented behavior for robot navigation in dense crowds. Guzzi et al. [34] present a distributed method for multirobot human-like local navigation. Variations of Bayesian filters for pedestrian path prediction have been studied in [35,36]. Some of these methods are not suitable for realtime applications or may not work well for dense crowds.

17.3 PEDESTRIAN STATE

We use the term *pedestrian* to refer to independent individuals or agents in a crowd. We use the notion of *state* to specify the trajectory characteristics of each pedestrian. We assume that the output of the tracker corresponds to discrete 2D positions. Therefore, our state vector, represented using the symbol $\mathbf{x} \in \mathbb{R}^6$, consists of components that describe the pedestrian's movements on a 2D plane:

$$\mathbf{x} = [\mathbf{p} \ \mathbf{v}^c \ \mathbf{v}^{pref}]^{\mathbf{T}}, \tag{17.1}$$

where $\mathbf{p}$ is the pedestrian's position, $\mathbf{v}^c$ is its current velocity, and $\mathbf{v}^{pref}$ is the preferred velocity on a 2D plane. The preferred velocity corresponds to the predicted velocity that a pedestrian would take to achieve its intermediate goal if there were no other pedestrians or obstacles in the scene. We use the symbol $\mathbf{S}$ to denote the current state of the environment, which corresponds to the state of all other pedestrians and the current position of the obstacles in the scene. The state of the crowd, which consists of individual pedestrians, is a union of the set of each pedestrian's state $\mathbf{X} = \bigcup_i \mathbf{x_i}$, where subscript i denotes the ith pedestrian.

The trajectories extracted from a real-world video tend to be noisy and may have incomplete tracks [37]; thus, we use Bayesian-inference technique to compensate for any errors and to compute the state of each pedestrian [38]. At each time step, the observation of a pedestrian computed by a tracking algorithm corresponds to the position of each pedestrian on the 2D plane, denoted as $\mathbf{z}^t \in \mathbb{R}^2$. The observation function $h()$ provides $\mathbf{z}^t$ of each pedestrian's true state $\hat{\mathbf{x}}^t$ with sensor error $\mathbf{r} \in \mathbb{R}^2$, which is assumed to follow a zero-mean Gaussian distribution with covariance Σ_r:

$$\mathbf{z}^t = h(\hat{\mathbf{x}}^t) + \mathbf{r}, \mathbf{r} \sim N(0, \Sigma_r). \tag{17.2}$$

$h(\cdot)$ is the tracking sensor output.

We use the notion of a state-transition model $f(\cdot)$ which is an approximation of true real-world pedestrian dynamics with prediction error $\mathbf{q} \in \mathbb{R}^6$, which is represented as a zero-mean Gaussian distribution with covariance Σ_q:

$$\mathbf{x}^{t+1} = f(\mathbf{x}^t) + \mathbf{q}, \ \mathbf{q} \sim N(0, \Sigma_q). \tag{17.3}$$

We can use any local navigation algorithm or motion model for function $f(\cdot)$, which computes the local collision-free paths for the pedestrians in the scene.

17.3.1 Realtime Multiperson Tracking

We present different algorithms that are based on the use of particle filters to perform realtime pedestrian tracking in moderately crowded scenes and use a notion of a crowd model as a motion prior. Bera and Manocha [22] demonstrated that using a crowd motion model (in this case, specifically using velocity obstacles) can improve pedestrian tracking in dense scenes, compared to using a constant velocity or constant acceleration model.

There has been follow-up work where the different pedestrian dynamics, crowd cluster size, etc., were considered to improve tracking by using a hybrid approach to accelerate computations [23].

These methods are used for realtime tracking and result in improved accuracy compared to prior motion models, but they may not provide sufficient capabilities in terms of capturing different aspects of pedestrian movement or trajectory changes over the entire sequence. In the rest of this chapter, we describe a different approach where different crowd motion models are used to model each pedestrian's trajectory.

The motion model of the crowd parameter estimation is formulated as an optimization problem, and the resulting algorithm solves the optimization problem in a model-independent manner and can be combined with a multiagent pedestrian movement model. Overall, this formulation computes the best-fit mixture.

In order to characterize the heterogeneous, dynamic behavior of each agent, we use an optimization-based scheme to perform the following steps:

- Choose, every few frames, the new motion model that best describes the local behavior of each pedestrian based on tracked data.
- Compute the optimal set of parameters for that motion model that best fit this tracked data.
- Compute the adaptive number of particles for each pedestrian based on a combination of metrics for optimizing performance.

The resulting mixture model is used to predict the next state of the pedestrian for the next frame. In other words, the next state is used as motion prior input for the tracker; it is also combined with a confidence estimation computation [39] to dynamically compute the number of particles. As a final step, the tracker's definitively estimated next state is fed back into the loop.

17.4 MIXTURE MOTION MODEL

In this section, we introduce the notion of a parametrized motion model. We then describe the different parametrized motion models that form the basis of the mixture motion model. Finally, we describe the mixture motion model itself.

17.4.1 Overview and Notations

Data representation The algorithm keeps track of the *state* (i.e., position and velocity) of each pedestrian for the last k timesteps or frames. These are referred to as the k-states

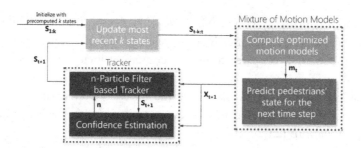

Figure 17.2 Overview of the real time tracking algorithm. The symbols used in this figure are explained in Section 17.4.1. The trajectory computed over prior k frames, expressed as a succession of states, to compute the new motion model; mixture motion model is used to compute the next states using a particle filter.

of each pedestrian. These k-states are initialized by precomputing the states from the first k timesteps. The k-states are updated at each timestep by removing the agents' state from the oldest frame and adding the latest tracker-estimated state.

The mixture motion model is a combination of several independent motion models. This mixture motion model is used to compute the best motion model for the agents during each frame. First, based on an optimization algorithm, the motion models are "configured" to "best" match the recent k-states data and select the best model based on a specific metric. Second, the "best configured" motion model is used to make a prediction on the agents' next state.

The tracker is a particle-filter based tracker that uses the motion prior, obtained from the Mixture of Motion Models, to estimate the agents' next state. This tracker further uses a confidence estimation stage [39] to dynamically compute the number of particles that balance the trade-offs between the computation cost and accuracy.

We use the following notations in the chapter:

- S represents the state (position and velocity) of an arbitrary pedestrian as computed by the tracker;
- X represents the state (position and velocity) of an arbitrary pedestrian inside a crowd motion model;
- m represents the "best configured" motion model from the Mixture of Motion Models $\{f1, f2, \dots\}$;
- Bold fonts are used to represent values for all the pedestrians in the crowd; for example, $\mathbf{S}$ represents the states (positions and velocities) of all pedestrians as computed by the tracker;
- Subscripts are used to indicate time; for example, m_t represents the "best configured" motion model at timestep t, and $\mathbf{S}_{t-k:t}$ represents all states of all agents for all successive timesteps between $t - k$ and t, as computed by the tracker.

The "best configured" motion model can then be used as follows: $X_{t+1} = m_t(X_t)$ or $\mathbf{X}_{t+1} = m_t(\mathbf{X}_t)$ to compute the motion of one arbitrary pedestrian or all pedestrians, respectively.

17.4.2 Particle Filter for Tracking

We use particle filters as the underlying tracker approach. The particle filter is a parametric method that solves non-Gaussian and nonlinear state estimation problems [40]. Particle filters are frequently used in object tracking, since they can recover from lost tracks and occlusions. The particle tracker's tracking uncertainty is represented in a Markovian manner by only considering information from present and past frames.

Here, "best configured" motion model m_t is considered as well as the error Q_t in the prediction that this "best configured" motion model generated. Additionally, the observations of the tracker can be represented by a function $h(\cdot)$ that projects the state X_t to a previously computed state S_t. Moreover, we denote the error between the observed states and the ground truth as R_t. We can now phrase them formally in terms of a standard particle filter as below:

$$S_{t+1} = m_t(X_t) + Q_t, \tag{17.4}$$

$$S_t = h(X_t) + R_t. \tag{17.5}$$

Particle filtering is a Monte Carlo approximation of the optimal Bayesian filter, which monitors the posterior probability of a first-order Markov process,

$$p(X_t|S_{t-k:t}) = \alpha p(S_t|X_t) \int_{X_{t-1}} p(X_t|X_{t-1})p(X_{t-1}|S_{t-k:t-1})dX_{t-1}, \tag{17.6}$$

where X_t is the process state at time t, S_t is the observation, $S_{t-k:t}$ is all of the observations through time t, $p(X_t|X_{t-1})$ is the process dynamical distribution, $p(S_t|X_t)$ is the observation likelihood distribution, and α is the normalization factor. Since the integral does not have a closed form solution in most cases, particle filtering approximates the integration using a set of weighted samples $X_t^{(i)}, \pi_t^{(i)}{}_{i=1,...,n}$, where $X_t^{(i)}$ is an instantiation of the process state, known as a particle, and $\pi_t^{(i)}$'s are the corresponding particle weights. With this representation, the Monte Carlo approximation of the Bayesian filtering equation is given as

$$p(X_t|S_{t-k:t}) \approx \alpha p(S_t|X_t) \sum_{i=1}^{n} \pi_{t-1}^{(i)} p(X_t^{(i)})p(X_{t-1}^{(i)}), \tag{17.7}$$

where n refers to the number of particles.

In the formulation, we use the motion model to infer dynamic transition, $p(X_t|X_{t-1})$, for particle filtering.

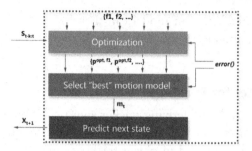

Figure 17.3 The parameter optimization algorithm used in Fig. 17.2. Based on the error metric, we compute optimal parameters for each motion model. The best motion model (from RVO2, Social Forces, Boids, or LIN) is used for trajectory extraction and to predict the next state.

17.4.3 Parametrized Motion Model

A motion model is defined as an algorithm f, which, from a collection of agent states $\mathbf{X}_t$, derives new states $\mathbf{X}_{t+1}$ for these agents, representing their motion over a timestep toward the agents' immediate goals $\mathbf{G}$,

$$\mathbf{X}_{t+1} = f(\mathbf{X}_t, \mathbf{G}). \qquad (17.8)$$

Motion algorithms tend to use several parameters that can be tuned in order to change the agents' behaviors. We assume that each parameter can have a different value for each pedestrian. By changing the value of these parameters, we get some variation in the resulting trajectory prediction algorithm. We use $\mathbf{P}$ to denote all the parameters of all the pedestrians. Typically, for a crowd of 50 pedestrians, the dimension of $\mathbf{P}$ could be anywhere in the range of 150–300 depending on the motion model. In the formulation, we denote the resulting parametrized motion model as

$$\mathbf{X}_{t+1} = f(\mathbf{X}_t, \mathbf{G}, \mathbf{P}). \qquad (17.9)$$

17.4.4 Mixture of Motion Models

We now present the algorithm to compute the mixture motion model, which essentially corresponds to computing the "best" motion model at any given timestep (Fig. 17.3). In this case, the "best" motion model is the one that most accurately matches agents' immediately past states, as per a given error metric. This "best" motion model is determined by an optimization framework, which automatically finds the parameters $\mathbf{P}$ that minimize the error metric. Wolinski et al. [41] designed an optimization framework for evaluating crowd motion models, but it computes the optimal parameters in an offline manner for a single homogeneous simulation model. Instead, the approach is designed for interactive applications, and iteratively computes the heterogeneous motion model every few frames and chooses the most optimized crowd parameters at the given time instance. The computation cost is considerably lower and can be used for realtime applications.

17.4.5 Formalization

Formally, at any timestep t, we define the agents' $(k+1)$-states (as computed by the tracker) $\mathbf{S}_{t-k:t}$ as

$$\mathbf{S}_{t-k:t} = \bigcup_{i=t-k}^{t} \mathbf{S}_i. \tag{17.10}$$

Similarly, a motion model's corresponding computed agents' states $f(\mathbf{S}_{t-k:t}, \mathbf{P})$ can be defined as

$$f(\mathbf{S}_{t-k:t}, \mathbf{P}) = \bigcup_{i=t-k}^{t} f(\mathbf{X}_i, \mathbf{G}, \mathbf{P}), \tag{17.11}$$

initialized with $\mathbf{X}_{t-k} = \mathbf{S}_{t-k}$ and $\mathbf{G} = \mathbf{S}_t$.

At timestep t, considering the agents' k-states $\mathbf{S}_{t-k:t}$, computed states $f(\mathbf{S}_{t-k:t}, \mathbf{P})$, and a user-defined error metric $error()$, the algorithm computes

$$\mathbf{P}_t^{opt,f} = \underset{\mathbf{P}}{\operatorname{argmin}}\, error(f(\mathbf{S}_{t-k:t}, \mathbf{P}), \mathbf{S}_{t-k:t}), \tag{17.12}$$

where $\mathbf{P}_t^{opt,f}$ is the parameter set which, at timestep t, leads to the closest match between the states computed by the motion algorithm f and the agents' k-states.

For several motion algorithms $\{f1, f2, \ldots\}$, we can then compute the algorithm which best matches the agents' k-states $\mathbf{S}_{t-k:t}$ at timestep t,

$$m_t = f_t^{opt} = \underset{f}{\operatorname{argmin}}\, error(f(\mathbf{S}_{t-k:t}, \mathbf{P}_t^{opt,f}), \mathbf{S}_{t-k:t}), \tag{17.13}$$

and consequently, the best (as per the error in the $error(\cdot)$ metric itself) prediction for the agents' next state obtainable from the motion algorithms for timestep $t+1$ is

$$\mathbf{X}_{t+1} = m_t(\mathbf{S}_t). \tag{17.14}$$

For more details regarding different error metrics and detailed experiments, we refer the reader to [39].

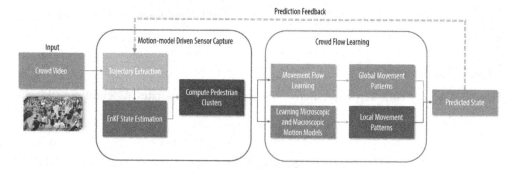

Figure 17.4 Prediction overview.

Figure 17.5 Global vs. local movement patterns. The blue trajectories indicate prior tracked data. The red dots indicate local predicted patterns retrieved from learning macro- and microscopic simulation models. The shaded (green–blue) path represent the global movement patterns learned from the path data in that cluster. (For interpretation of the references to color in this figure legend, the reader is referred to the web version of this chapter.)

17.5 REALTIME PEDESTRIAN PATH PREDICTION

In this section, we present a realtime algorithm that learns movement flows from real-world pedestrian 2D trajectories that are extracted from video. The approach involves no precomputation or learning, and can be combined with realtime pedestrian trackers.

Fig. 17.4 gives an overview of the approach, including computation of movement flows and using them for pedestrian prediction. The input of the method consists of a live or streaming crowd video. We extract the initial set of trajectories using an online particle-filter based pedestrian tracker. These trajectories are time-series observations of the positions of each pedestrian in the crowd. The output is the predicted state of each agent that is based on learning the local and global pedestrian motion patterns (Fig. 17.5).

Pedestrian clusters The approach is targeted toward computing the movement flows of pedestrians in dense settings. It is not uncommon for some nearby pedestrians to have similar flows. As a result, we compute clusters of pedestrians in a crowd based on their positions, velocity, inter-pedestrian distance, orientations, etc. In particular, we use a bottom-up hierarchical clustering approach, as they tend to work better for small clusters. Initially, we assign each pedestrian to a separate cluster that consists of a single pedestrian. Next, we merge these clusters based on computing the distance between various features highlighted above.

The approach is based on group-expand procedure [42] and we include many pedestrian movement-related features to compute the clusters. We compute a connectivity graph among the pedestrians and measure the graph density based on intra-cluster proximity [23]. Eventually, we use a macroscopic model to estimate the movement of each cluster and use this model to predict their global movement.

17.5.1 Global Movement Pattern

A key aspect of the approach is to compute global movement patterns that can be used to predict the state of each pedestrian. These movement patterns describe the trajectory-level motion or behavior at a certain position at time frame t. The patterns include the movement of the pedestrian during the past w frames, which we call *time window*, and the intended direction of the movement (preferred velocity) at this position.

In our formulation, we represent each movement feature vector as a six-dimensional vector,

$$\mathbf{b} = [\mathbf{p}\ \mathbf{v}^{avg}\ \mathbf{v}^{pref}]^T, \tag{17.15}$$

where $\mathbf{p}$, $\mathbf{v}^{avg}$, and $\mathbf{v}^{pref}$ are each two-dimensional vectors representing the current position, average velocity during past w frames, and estimated preferred velocity computed as part of state estimation, respectively. $\mathbf{v}^{avg}$ can be computed from $(\mathbf{p}^t - \mathbf{p}^{t-w})/w * dt$, where dt is the time step.

We use the notion of average velocity over the last w frames as that provides a better estimate of pedestrian movement. In a dense setting, some pedestrians may suddenly stop or change their local directions as they interact with other pedestrians. As a result, the duration of the time window, w, is set based on the characteristics of a scene. If we use small time windows, the movement flows will be able to capture the details in dynamically changing scenes. On the other hand, larger time windows tend to smooth out abrupt changes in pedestrian motion and are more suitable for scenes that have little change in pedestrians' movement.

At every w steps, we compute the new trajectory features for each pedestrian in the scene, using Eq. (17.15). Moreover, we group the similar features and find K most common trajectory patterns, which we call *global movement patterns*. We use recently observed behavior features to learn the time-varying movement flow. This set of K

global movement patterns $B = \{B_1, B_2, \ldots, B_K\}$ is computed as follows:

$$\underset{B}{\mathrm{argmin}} \sum_{k=1}^{K} \sum_{b_i \in B_k} dist(b_i, \mu_k), \tag{17.16}$$

where b_i is a movement feature vector, μ_k is a centroid of each flow movement pattern, and $dist(b_i, \mu_k)$ is a distance measurement between the arguments. In our case, the distance between two feature vectors is computed as

$$\begin{aligned} dist(b_i, b_j) = {} & c_1 \left\| \mathbf{p}_i - \mathbf{p}_j \right\| \\ & + c_2 \left\| (\mathbf{p}_i - \mathbf{v}_i^{avg} w \ dt) - (\mathbf{p}_j - \mathbf{v}_j^{avg} w \ dt) \right\| \\ & + c_3 \left\| (\mathbf{p}_i + \mathbf{v}_i^{pref} w \ dt) - (\mathbf{p}_j - \mathbf{v}_j^{pref} w \ dt) \right\|, \end{aligned} \tag{17.17}$$

which corresponds to the weighted sum of the distance among three points: current positions, previous positions, and estimated future positions that are extrapolated using v^{pref}, c_1, c_2, and c_3 as the weight values. Comparing the distance between the positions rather than mixing the points and the vectors eliminates the need to normalize or standardize the data. We use the movement feature of the cluster to compute the predicted state at time t, $\mathbf{S}_t^g$.

17.5.2 Local Movement Pattern

During each frame, some of the pedestrians are modeled as discrete agents, while the clusters are treated using macroscopic techniques. Based on the observations and state information, we estimate the motion model for these discrete agents and pedestrian clusters. For each individual pedestrian represented as a discrete agent, we compute the motion model that best fits its position as tracked over recent frames, i.e., we compute the features per-agent and predict motion patterns locally. We choose the "best" local motion model from a fixed set of choices. The common choices are based on social forces, reciprocal velocity obstacles or Boids. In our case, the "best" motion model is the one that most accurately matches the immediate past states based on a given error metric. This "best" motion model is computed using a local optimization algorithm [43], which automatically finds the motion model parameters that minimize that error metric (refer to Eqs. (17.10)–(17.14)). For more details please refer to [44].

17.5.3 Prediction Output

For every pedestrian, we compute both the global and local movement patterns separately. In practice, we observed that for lower density scenarios, local movement patterns are more useful than global patterns and vice versa. The final predicted state is a weighted average of the individual predicted states generate from the local and global patterns as

$$\mathbf{S}_t^p = (1 - w) * \mathbf{S}_t^l + w * \mathbf{S}_t^g, \tag{17.18}$$

where $\mathbf{S}_t^p$ is the final predicted state at time t, $\mathbf{S}_t^l$ is the state predicted from the local patterns, and the $\mathbf{S}_t^g$ is the state predicted from global patterns or from the movement flows. As a general rule of thumb, w varies from 0 to 1 and is computed based on the pedestrian density. We use a larger weight for higher density. In order to perform long-term predictions (5–6 seconds or even longer), we tend to increase w as the global movement patterns provide better estimates for pedestrian position.

17.6 IMPLEMENTATION AND RESULTS

17.6.1 Pedestrian Tracking

In this section we present some implementation details and highlight the performance of different crowd video datasets.

17.6.2 Evaluation

We use the **CLEAR MOT** [45] evaluation metrics to analyze the performance analytically. We use the **MOTP** and the **MOTA** metrics. **MOTP** evaluates the alignment of tracks with the ground truth while **MOTA** produces a score based on the amount of false positives, missed detections, and identity switches. The higher the value of MOTA/MOTP, the better. These metrics have become standard for evaluation of detection and tracking algorithms in the computer vision community, and we refer the interested reader to [45] for a more detailed explanation.

We analyze these metrics across the density groups and the different motion models (Table 17.1).

17.6.3 Tracking Results

We highlight the performance of the algorithm based on a Mixture of Motion Models on different benchmarks, comparing the performance of the algorithm with single, homogeneous motion model methods: constant velocity model (LIN), LTA [4], Social Forces [5], Boids [8], and RVO2 [46]. LIN models the velocities of pedestrians as constant, and is the underlying motion model frequently used in the standard particle filter. The other four models compute the pedestrian states based on optimizing functions, which model collision avoidance, destinations of pedestrians, and the desired speed. In the implementation, we replace the state transition process of a standard particle-filtering algorithm with different motion models.

Table 17.1 We compare the MOTA and MOTP values across the density groups and the different motion models

	LIN			Boids			Social Forces			RVO2			MMM		
	LD	MD	HD	LD	MD	HD	LD	MD	HD	LD	MD	HD	LD	MD	HD
MOTP	64.42%	52.82%	40.31%	67.24%	57.10%	43.14%	70.52%	61.33%	49.88%	72.19%	63.17%	51.31%	73.98%	69.23%	54.29%
MOTA	49.42%	35.3%	31.37%	50.59%	26.42%	30.88%	53.28%	44.19%	33.51%	53.95%	48.81%	35.83%	54.18%	50.16%	38.83%

Table 17.2 We compare the percentage of successful tracks (ST) and ID switches (IS) of Mixture Motion Model algorithm (MMM) with homogeneous motion models – LIN, Boids, Social Forces, LTA, RVO2, and a baseline mean-shift tracker. MMM provides higher accuracy compared to homogeneous motion models and lesser ID switches. The benefits of the approach are higher, as the crowd density increases. These datasets are publicly available at http://gamma.cs.unc.edu/RCrowdT/

| | High density | | | | | | | | | | Medium density | | | | | | Low density | | | |
| | NDLS-1 | | IITF-1 | | IITF-3 | | IITF-5 | | NPLC-1 | | NPLC-3 | | IITF-2 | | IITF-4 | | NDLS-2 | | NPLC-2 | |
	ST	IS	ST	IS	ST	IS	ST	IS	ST	IS	ST	IS	ST	IS	ST	IS	ST	IS	ST	IS
LIN	53	17	63	27	51	35	59	18	67	15	60	29	36	22	52	36	68	23	69	21
Boids	58	15	66	23	56	33	65	14	73	13	65	26	40	19	52	35	70	22	72	19
Social Forces	56	16	66	26	52	33	62	15	74	11	68	23	41	19	59	31	75	18	72	14
LTA	54	17	65	22	51	32	60	17	68	11	62	28	42	18	54	32	69	23	70	20
RVO2	57	14	69	20	53	29	64	13	71	10	64	26	42	18	53	32	72	20	74	16
MeanShift	27	32	31	38	23	52	34	29	39	36	41	31	22	33	39	45	31	28	45	28
MMM	63	12	73	19	57	27	67	10	77	7	71	20	44	16	63	28	79	17	78	14

Table 17.3 We compare the percentage of successful tracks (ST) and average tracking frames per second (FPS) of the Mixture of Motion Models algorithm using adaptive particle filtering (MMM)

	High density												Medium density	
	NDLS-1		IITF-1		IITF-3		IITF-5		NPLC-1		NPLC-3		IITF-2	
	ST	FPS	ST	FPS	ST	FPS	ST	FPS	ST	FPS	ST	FPS	ST	FPS
MMM	63	27	73	28	57	26	67	26	77	28	71	26	44	26

	Medium density						Low density							
	IITF-4		NDLS-2		NPLC-2		seq_hotel		seq_eth		zara01		zara02	
	ST	FPS	ST	FPS	ST	FPS	ST	FPS	ST	FPS	ST	FPS	ST	FPS
MMM	63	27	79	28	78	26	252	28	267	29	63	27	68	28

We evaluate some challenging datasets [22] which are available publicly and also some standard datasets from the pedestrian tracking community. These videos were recorded at 24–30 fps. We manually annotated these videos and corrected the perspective effect by camera calibration. We also compare the performance to a baseline mean-shift tracker (Table 17.2).

We show the number of correctly tracked pedestrians and the number of ID switches. A track is counted as "successful" when the estimated mean error between the tracking result and the ground-truth value is less than 0.8 meters in ground space. The average human stride length is about 0.8 meters and we consider the tracking to be incorrect if the mean error is more than this value. The method provides 9–18% higher accuracy over LIN for medium density crowds (Table 17.3).

17.6.4 Pedestrian Prediction

We highlight the prediction results using GLMP algorithm and compare its performance with prior methods. We have applied it to the 2D trajectories generated from different crowd videos and compared the prediction accuracy with the ground truth data that was also generated using a pedestrian tracker. The underlying crowd videos have different pedestrian density corresponding to low (i.e., less than 1 pedestrian per squared meter), medium (1–2 pedestrians per squared meter), and high (more than 2 pedestrians per squared meter). We highlight the datasets, their crowd characteristics, and the prediction accuracy of different realtime algorithms for short-term and long-term prediction in Table 17.4. We also analyze the accuracy of the approach based on varying the pedestrian density (Fig. 17.7). The performance of the method with noisy data (i.e., sensor noise) is also analyzed. Finally, we perform a qualitative and quantitative comparison to other realtime pedestrian path prediction algorithms.

We include comparisons to constant velocity (ConstVelocity) and constant acceleration (ConstAccel) motion models, which are widely used for pedestrian tracking and prediction in robotics and computer vision [47]. We also compare the accuracy with recent methods that use more sophisticated motion models (LTA and ATTR) to compute

Table 17.4 Crowd scene benchmarks. We highlight many attributes of these crowd videos, including density and the number of tracked pedestrians. We use the following abbreviations about some characteristics of the underlying scene: Background Variations (BV), Partial Occlusion (PO), Complete Occlusion (CO), and Illumination Changes (IC). We highlight the results for short-term prediction (1 s) and long term prediction (5 s). We notice that our GLMP algorithm results in higher accuracy for long-term prediction and dense scenarios

Dataset	Challenges	Density	# Tracked	ConstVelocity		Kalman Filter		BRVO		GLMP	
				1 s	5 s	1 s	5 s	1 s	5 s	1 s	5 s
NDLS-1	BV, PO, IC	High	131	55.3%	32.0%	53.1%	37.9%	56.5%	42.0%	60.2%	51.2%
IITF-1	BV, PO, IC, CO	High	167	63.5%	33.4%	63.9%	39.1%	65.3%	41.8%	71.2%	50.5%
IITF-3	BV, PO, IC, CO	High	189	61.1%	29.1%	63.6%	31.0%	67.6%	37.5%	68.4%	45.7%
IITF-5	BV, PO, IC, CO	High	71	59.2%	28.8%	61.7%	29.1%	62.9%	30.1%	64.6%	40.0%
NPLC-1	BV, PO, IC	Medium	79	76.1%	63.9%	78.2%	65.8%	79.9%	69.0%	82.3%	72.5%
NPLC-3	BV, PO, IC, CO	Medium	144	77.9%	70.1%	79.1%	71.9%	80.8%	74.4%	84.3%	78.1%
Students	BV, IC, PO	Medium	65	65.0%	58.2%	66.9%	61.0%	69.1%	63.6%	72.2%	66.8%
Campus	BV, IC, PO	Medium	78	62.4%	57.1%	63.5%	59.0%	66.4%	59.1%	69.6%	59.5%
seq_hotel	IC, PO	Low	390	74.7%	67.8%	76.7%	68.3%	76.9%	69.2%	79.5%	70.1%
Street	IC, PO	Low	34	78.1%	70.9%	78.9%	71.0%	81.4%	71.2%	83.8%	72.7%

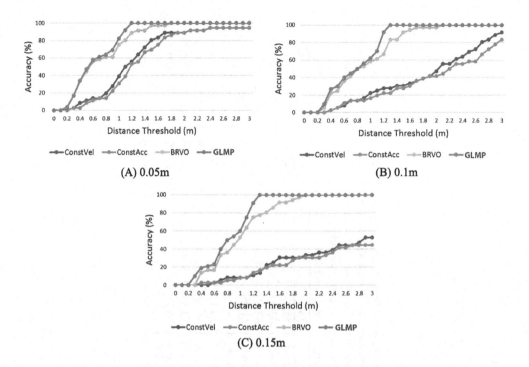

Figure 17.6 Prediction accuracy vs. sensor error (higher is better). We increase the sensor noise (Gaussian) from (A) to (C) and highlight the prediction accuracy across various distance thresholds. The X-axis represents the percentage of correctly predicted paths within varying accuracy thresholds. In this, GLMP results in more accurate predictions, as compared to BRVO, Constant Velocity, Constant Acceleration. As the sensor noise increases (C), we observe more significant benefits.

local movement patterns [4,5]. Finally, we also compare the accuracy with the Bayesian reciprocal velocity obstacle (BRVO) algorithm [38] that computes a more individualized motion model for estimating local movement patterns.

17.6.5 Noisy Data

Sensor noise is an important concern in pedestrian prediction algorithms. In order to evaluate the impact of noise, we add synthetic noise to the datasets and compare the performance of GLMP vs. other algorithms on these benchmarks: IITF [22], ETH, and Campus [4] datasets.

Fig. 17.6 compares the prediction accuracy of GLMP, constant velocity, constant acceleration and BRVO, by comparing the predicted positions to the actual ground truth data extracted using pedestrian trackers. We use these noise levels, 0.05, 0.1, and 0.15 m, to simulate different sensor variations. During the prediction step, we assume that no further information is given when we are predicting the future state, and our

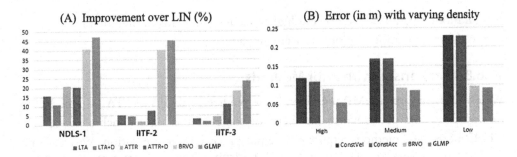

Figure 17.7 (A) We compare the improvements of GLMP, LTA, ATTR, and BRVO over LIN (linear velocity) model. GLMP outperforms LTA, ATTR with 24–47% error reduction rate in all three different scenarios. (B) Errors for varying pedestrian densities (lower is better). In low-density scenarios, local movement patterns (e.g., BRVO) are able to predict the positions well, but are more accurate than constant velocity and constant acceleration. We observe improved accuracy with GLMP, as the pedestrian density increases.

best guess is that the pedestrians move according to their preferred velocity computed using the movement patterns. For GLMP, the pedestrian's movement direction changes when there is any interaction with obstacles or other pedestrians as observed based on local and global movement patterns. Fig. 17.6 shows the fraction of correctly predicted paths within varying accuracy thresholds. At an accuracy threshold of 0.5 m, GLMP has higher accuracy than BRVO and offers considerable benefits over constant velocity/constant acceleration models even with little noise. As the noise increases, the benefit in prediction accuracy using GLMP also increases.

17.6.6 Long-Term Prediction Accuracy

Being able to predict a trajectory over a longer time-horizon is important for service robots and autonomous vehicles. GLMP is able to perform long-term prediction (5–6 seconds) with much higher accuracy than prior methods (see Table 17.4). We define prediction accuracy as the ratio of the number of "successful" predictions and total number of tracked pedestrians in the scene. We use the algorithm for long and short term prediction across a large number of datasets, highlighted in Table 17.4.

17.6.7 Varying the Pedestrian Density

We use a variation of crowd videos with different densities (Low, Medium, and High) and compare GLMP's error to that of BRVO, constant velocity and constant acceleration models (see Fig. 17.7). Both the constant velocity and constant acceleration models have large variations in error for different regions of the scenario with varying densities. In contrast, the GLMP approach performs well across all densities because it can dynamically adapt the parameters for each agent for each frame and learn global as well as local motion patterns. We observe higher accuracy benefits in high density scenarios due to the computation of global movement patterns.

17.6.8 Comparison with Prior Methods

We directly compare GLMP results with the prediction results of LTA [4] and ATTR [5], which report performance numbers for some of the same benchmarks. Unlike GLMP, both these methods require offline pre-processing or annotation. We also compare GLMP with BRVO along with LTA and ATTR on Street, NDLS, IITF, seq_hotel, and seq_eth datasets, all sampled every 1.5 seconds, and measure mean prediction error for every agent in the scene during the entire video sequence.

The metric used was error reduction comparison, and is measured as improvement in percentage of error reduction over the LIN model for different algorithms. The results are shown in Fig. 17.7. GLMP outperforms LTA and ATTR with 24–47% error reduction rate across the three different scenarios. LTA and ATTR use the ground truth destinations for prediction; LTA+D and ATTR+D use destinations learned offline, as explained in [5]; ATTR+D uses grouping information learned offline. Even though GLMP is an online and realtime method, it shows significant improvement in prediction accuracy on all the datasets, producing less error than other approaches.

17.7 CONCLUSION

We present an overview of a collection of interactive approaches for computing trajectory level behavior features from crowd videos and demonstrate its application to surveillance and training applications. The approaches are general, can handle moderately dense crowd videos, and can compute and predict the trajectory (past, present, and future) for each agent during each time step. A key benefit of these approaches is that it can capture dynamically changing movement behaviors of pedestrians and therefore can be used for dynamic or local behavior analysis.

Limitations The performance and accuracy of this algorithm are governed by the tracking algorithm, which can be noisy, sparse, or may lose tracks. Furthermore, current realtime methods may not work well in very dense crowd videos, e.g., those with thousands of agents in a single frame.

ACKNOWLEDGMENTS

This work was supported by National Science Foundation award 1305286, ARO contract W911NF-16-1-0085, and a grant from the Boeing company.

REFERENCES

[1] Luber M, Stork J, Tipaldi G, Arras K. People tracking with human motion predictions from social forces. In: Proc. of the IEEE international conference on robotics and automation (ICRA). 2010. p. 464–9.

[2] Rodriguez M, Ali S, Kanade T. Tracking in unstructured crowded scenes. In: Proc. of the IEEE 12th international conference on computer vision. 2009. p. 1389–96.

[3] Kratz L, Nishino K. Tracking pedestrians using local spatio-temporal motion patterns in extremely crowded scenes. IEEE Trans Pattern Anal Mach Intell 2011;99:1.

[4] Pellegrini S, Ess A, Schindler K, Van Gool L. You'll never walk alone: modeling social behavior for multi-target tracking. In: Proc. of the IEEE 12th international conference on computer vision. 2009. p. 261–8.

[5] Yamaguchi K, Berg A, Ortiz L, Berg T. Who are you with and where are you going? In: Proc. of the 2011 IEEE conference on computer vision and pattern recognition (CVPR). 2011. p. 1345–52.

[6] Helbing D, Molnar P. Social force model for pedestrian dynamics. Phys Rev E 1995;51(5):4282.

[7] Van den Berg J, Lin M, Manocha D. Reciprocal velocity obstacles for real-time multi-agent navigation. In: ICRA. 2008. p. 1928–35.

[8] Reynolds CW. Steering behaviors for autonomous characters. In: Game developers conference. 1999. Available from http://www.red3d.com/cwr/steer/gdc99.

[9] Van den Berg J, Guy S, Lin M, Manocha D. Reciprocal n-body collision avoidance. Int J Robot Res 2011:3–19.

[10] Treuille A, Cooper S, Popović Z. Continuum crowds. In: ACM SIGGRAPH 2006. ISBN 1595933646, 2006. p. 1160–8.

[11] Karamouzas I, Heil P, van Beek P, Overmars M. A predictive collision avoidance model for pedestrian simulation. Motion Games 2009:41–52.

[12] Karamouzas I, Overmars M. A velocity-based approach for simulating human collision avoidance. In: Intelligent virtual agents. Springer; 2010. p. 180–6.

[13] Van den Berg J, Patil S, Sewall J, Manocha D, Lin M. Interactive navigation of individual agents in crowded environments. In: Symp. on interactive 3D graphics and games (I3D 2008). 2008.

[14] Pettré J, Ondřej J, Olivier AH, Cretual A, Donikian S. Experiment-based modeling, simulation and validation of interactions between virtual walkers. In: Symp. on computer animation. 2009. p. 189–98.

[15] Snape J, van den Berg J, Guy S, Manocha D. The hybrid reciprocal velocity obstacle. IEEE Trans Robot 2011;27(4):696–706.

[16] Guy SJ, Chhugani J, Curtis S, Dubey P, Lin M, Manocha D. PLEdestrians: a least-effort approach to crowd simulation. In: Symp. on computer animation. 2010. p. 119–28.

[17] Cui J, Zha H, Zhao H, Shibasaki R. Tracking multiple people using laser and vision. In: Proc. of the IEEE/RSJ international conference on intelligent robots and systems (IROS). IEEE; 2005. p. 2116–21.

[18] Bruce A, Gordon G. Better motion prediction for people-tracking. In: Proc. of the international conference on robotics and automation (ICRA), New Orleans, USA. 2004.

[19] Gong H, Sim J, Likhachev M, Shi J. Multi-hypothesis motion planning for visual object tracking. In: 2011 IEEE international conference on computer vision (ICCV). IEEE; 2011. p. 619–26.

[20] Liao L, Fox D, Hightower J, Kautz H, Schulz D. Voronoi tracking: location estimation using sparse and noisy sensor data. In: Proc. of the 2003 IEEE/RSJ international conference on intelligent robots and systems (IROS), vol. 1. IEEE; 2003. p. 723–8.

[21] Mehran R, Oyama A, Shah M. Abnormal crowd behavior detection using social force model. In: Proc. of the IEEE conference on computer vision and pattern recognition (CVPR). 2009. p. 935–42.

[22] Bera A, Manocha D. Realtime multilevel crowd tracking using reciprocal velocity obstacles. In: ICPR. 2014.

[23] Bera A, Manocha D. REACH: realtime crowd tracking using a hybrid motion model. In: ICRA. 2015.

[24] Bera A, Galoppo N, Sharlet D, Lake A, Manocha D. Adapt: real-time adaptive pedestrian tracking for crowded scenes. In: ICRA. 2014.

[25] Fulgenzi C, Spalanzani A, Laugier C. Dynamic obstacle avoidance in uncertain environment combining PVOS and occupancy grid. In: 2007 IEEE international conference on robotics and automation. 2007. p. 1610–6.

[26] Du Toit N, Burdick J. Robotic motion planning in dynamic, cluttered, uncertain environments. In: 2010 IEEE international conference on robotics and automation (ICRA). 2010. p. 966–73.

[27] Pradhan N, Burg T, Birchfield S. Robot crowd navigation using predictive position fields in the potential function framework. In: American control conference (ACC). 2011. p. 4628–33.

[28] Ziebart BD, Ratliff N, Gallagher G, Mertz C, Peterson K, Bagnell JA, et al. Planning-based prediction for pedestrians. In: Proc. of the 2009 IEEE/RSJ international conference on intelligent robots and systems (IROS). Piscataway (NJ): IEEE Press; ISBN 978-1-4244-3803-7, 2009. p. 3931–6.

[29] Bennewitz M, Burgard W, Cielniak G, Thrun S. Learning motion patterns of people for compliant robot motion. Int J Robot Res 2005;24(1):31–48.

[30] Henry P, Vollmer C, Ferris B, Fox D. Learning to navigate through crowded environments. In: 2010 IEEE international conference on robotics and automation (ICRA). 2010. p. 981–6.

[31] Kretzschmar H, Kuderer M, Burgard W. Learning to predict trajectories of cooperatively navigating agents. In: 2014 IEEE international conference on robotics and automation (ICRA). IEEE; 2014. p. 4015–20.

[32] Kuderer M, Kretzschmar H, Sprunk C, Burgard W. Feature-based prediction of trajectories for socially compliant navigation. In: Robotics: science and systems. 2012.

[33] Trautman P, Ma J, Murray R, Krause A. Robot navigation in dense human crowds: the case for cooperation. In: 2013 IEEE international conference on robotics and automation (ICRA). 2013. p. 2153–60.

[34] Guzzi J, Giusti A, Gambardella L, Theraulaz G, Di Caro G. Human-friendly robot navigation in dynamic environments. In: 2013 IEEE international conference on robotics and automation (ICRA). 2013. p. 423–30.

[35] Schneider N, Gavrila DM. Pedestrian path prediction with recursive Bayesian filters: a comparative study. In: Pattern recognition – 35th German conference, proceedings. 2013. p. 174–83.

[36] Mogelmose A, Trivedi MM, Moeslund TB. Trajectory analysis and prediction for improved pedestrian safety: integrated framework and evaluations. In: Intelligent vehicles symposium (IV). IEEE; 2015. p. 330–5.

[37] Enzweiler M, Gavrila DM. Monocular pedestrian detection: survey and experiments. IEEE Trans Pattern Anal Mach Intell 2009:2179–95.

[38] Kim S, Guy SJ, Liu W, Wilkie D, Lau RWH, Lin MC, Manocha D. BRVO: predicting pedestrian trajectories using velocity-space reasoning. Int J Robot Res 2015;34(2):201–17.

[39] Bera A, Wolinski D, Pettré J, Manocha D. Real-time crowd tracking using parameter optimized mixture of motion models. Available from arXiv:1409.4481, 2014.

[40] Arulampalam MS, Maskell S, Gordon N, Clapp T. A tutorial on particle filters for online nonlinear/non-Gaussian Bayesian tracking. IEEE Trans Signal Process 2002:174–88.

[41] Wolinski D, Guy SJ, Olivier AH, Lin MC, Manocha D, Pettré J. Parameter estimation and comparative evaluation of crowd simulations. In: Eurographics. 2014.

[42] McPhail C, Wohlstein RT. Using film to analyze pedestrian behavior. Sociol Methods Res 1982;10(3):347–75.

[43] Bera A, Kim S, Manocha D. Efficient trajectory extraction and parameter learning for data-driven crowd simulation. In: Graphics interface. 2015.

[44] Bera A, Kim S, Randhavane T, Pratapa S, Manocha D. GLMP – realtime pedestrian path prediction using global and local movement patterns. In: ICRA. 2016.

[45] Bernardin K, Stiefelhagen R. Evaluating multiple object tracking performance: the clear mot metrics. EURASIP J Image Video Process 2008;2008(1):1–10.

[46] Van den Berg J, Guy SJ, Lin M, Manocha D. Reciprocal n-body collision avoidance. In: Robotics research. 2011.

[47] Del Bimbo A, Dini F. Particle filter-based visual tracking with a first order dynamic model and uncertainty adaptation. Comput Vis Image Underst 2011;115(6):771–86.

SUBJECT INDEX

Printed in the United States
By Bookmasters